Almost Eternal

# Art and Material Culture in Medieval and Renaissance Europe

*Edited by*

Sarah Blick
Laura D. Gelfand

VOLUME 10

The titles published in this series are listed at *brill.com/amce*

# Almost Eternal

*Painting on Stone and Material Innovation in Early Modern Europe*

*Edited by*

Piers Baker-Bates
Elena Calvillo

BRILL

LEIDEN | BOSTON

Cover illustration: Cavaliere d'Arpino, *Perseus Liberating Andromeda*, oil on lapis lazuli, c. 1593-1594, St Louis, St Louis Art Museum (photo: ©St Louis Art Museum).

Library of Congress Cataloging-in-Publication Data

Names: Baker-Bates, Piers, editor. | Calvillo, Elena M., editor.
Title: Almost eternal : painting on stone and material innovation in early modern Europe / edited by Piers Baker-Bates, Elena Calvillo.
Description: Boston : Brill, 2018. | Series: Art and material culture in Medieval and Renaissance Europe, ISSN 2212-4187 ; Volume 10 | Includes bibliographical references and index.
Identifiers: LCCN 2017054706 (print) | LCCN 2017057132 (ebook) | ISBN 9789004361492 (E-book) | ISBN 9789004315051 (hardback : alk. paper)
Subjects: LCSH: Stone painting--Europe. | Painting, European--16th century. | Painting, European--17th century. | Diffusion of innovations--Europe.
Classification: LCC ND1580 (ebook) | LCC ND1580 .A46 2018 (print) | DDC 759.94--dc23
LC record available at https://lccn.loc.gov/2017054706

Typeface for the Latin, Greek, and Cyrillic scripts: "Brill". See and download: brill.com/brill-typeface.

ISSN 2212-4187
ISBN 978-90-04-31505-1 (hardback)
ISBN 978-90-04-36149-2 (e-book)



This book is printed on acid-free paper and produced in a sustainable manner.

*To our families and in memory of Brenda Nott*

# Contents

# Preface: The Study of Painting on Stone Comes of Age

It is a genuine pleasure to see the realization of this fine volume that has been in progress for a number of years. It comes at a time when the study of painting on stone surfaces is emerging from its initial developmental period. Having attracted growing interest among collectors, dealers, museum curators, and academic scholars, the subject of painting on stone has stimulated sophisticated inquiry and posed complex questions that will deepen our understanding of the late Renaissance and early baroque periods.

So much has happened in the field since I first turned my attention to the practice in 2000. In January of that year, the Saint Louis Art Museum bought its Cavaliere d'Arpino on lapis lazuli that graces the cover of this volume. The painting was bought at auction in January 2000 at Christie's. When I traveled to New York to preview the sale, the questions that were being asked about the picture were pretty rudimentary. No one questioned its attribution to d'Arpino since its style was so obviously consonant with the Cavaliere's other work, and Herwarth Röttgen, the great authority on the artist, had accepted it and assigned it a date of 1594. Among the curators and dealers with whom I discussed the painting, there were two main issues of concern. The first, as at any auction, was price. So few stone paintings of such quality had come up for sale that everyone agreed it was a difficult assessment, resulting in a fairly low estimate ($30,000 to $60,000). The other, more telling question was whether or not it was something worthy of a museum collection. I encountered several colleagues who responded incredulously when told this small painted lapis lazuli oval was the piece that I sought to acquire. There was little discussion about the rarity of such paintings or how such an object would inform our understanding of late sixteenth-century Roman art.

Luckily for me, the picture had been hung in an easily missed corner at Christie's. That placement certainly confirmed the impression that it wasn't really regarded as a true "museum piece." In the end, I decided that it was a wonderful picture and an intriguing object. After returning from New York, I conferred with our director and with our head curator who both agreed that we should pursue its purchase. We were fortunate to cast the winning bid and secure it for the museum.

In buying this small gem, I thought I knew what I was getting. After all, I had written my doctoral dissertation on Guido Reni's fresco cycle for the Annunciation Chapel in the Quirinal Palace that he painted in 1609-10. My investigation of that space had led me to examine contemporary Roman chapel interiors that were decorated during the pontificates of Clement V and Paul V. In studying the painted narrative cycles undertaken during those years (1585 to 1610), I had noted the rich marble revetments and elaborate gilded appointments of the spaces that they adorned. I therefore assumed our small lapis panel reflected that same taste for costly decoration and sumptuous stone.

Once the acquisition was secured, I began to explore the practice of painting on stone more thoroughly since, like many art historians who focus on the sixteenth-century in Italy, I knew little about it. I soon discovered what a number of other scholars already understood. The Cavaliere d'Arpino's choice of lapis lazuli as the support for his picture did not grow out of a taste for lavish surfaces. Rather, it represented an advanced stage of a different phenomenon, one that emerged earlier in the sixteenth century, owing in large part to the inventive powers of Sebastiano del Piombo, who figured out how to make paint adhere to stone.

I was fortunate to be able to teach myself about painted stone through the efforts of earlier investigations undertaken by talented and insightful scholars who illuminated several aspects of this most fascinating subject. The earliest study was undertaken in Florence and focused on the Medici, who were enthusiastic collectors and connoisseurs of both *pietre dure* (cut and pieced stone) as well as painted stone. Their collections are housed in the Opificio delle Pietre Dure and in the Galleria Palatina, Palazzo Pitti. As early as 1971, examples of stone paintings from the Florentine collections were exhibited and published. The resulting catalogue made photographs and information available to scholars, collectors, and dealers, proving increasingly valuable as more of these works were discovered and analyzed.

The Saint Louis Art Museum's purchase of the Cavaliere d'Arpino came at an opportune time, since it was around that year that interest in painted stone began to accelerate. An important exhibition was mounted in Milan devoted to a private local collection of stone paintings. Its catalogue was the first publication to offer a wide sampling of the types of stone surfaces and varied approaches to stone painting exhibited by sixteenth- and seventeenth-century painters. The collection focused on Italian paintings, but included French, Flemish, and German examples as well. I learned of the project only after the exhibition closed, but was able to benefit from the rich trove of material included in the catalogue. The Milanese collector who assembled those works has proven to be a wonderful colleague and resource, providing access to his extensive holdings that has allowed many students of stone paintings to develop their knowledge and increase their understanding of this material.

During the following years, several articles and essays presented individual artists who worked with stone or specific examples of the craft. In 2006, the first comprehensive study of painting on stone in Italy appeared. Anne-Laure Collomb completed her doctoral dissertation at the Université Lumière-Lyon, an ambitious and at times revelatory study that she published in 2012. In 2007, Hana Seifertová published the first English-language overview of the practice, based on her study that undoubtedly began with the fine collection housed in the National Gallery, Prague. This year, Mario Casaburo added his published dissertation to the mix, offering an expanded discussion in terms of geographical breadth as well as providing important investigations of sources for stone in the early modern period.

In looking back over the last two decades, perhaps the most important work of discovery, attribution, and stone analysis has been done by the dealers who have brought so many paintings on stone into the discussion. The market for paintings often plays a surprising role in decisions about exhibitions and publications. Through the years, works have appeared on the market that have completely reoriented our understanding of the late 16th and early 17th century in terms of the use of stone. My own work in preparing an exhibition of paintings on stone has benefitted enormously from dealers' investigations and willingness to share images and information, and I owe these dealers a huge debt of gratitude as I have worked to assemble my own database of some 1350 examples. Art historians traditionally have a love/hate relationship with the market. For those of us interested in paintings on stone surfaces, however, we must acknowledge the love. For me, that means thanking Alexis and Nicolas Kugel, Alessandra di Castro and her father Alberto di Castro, Paul Smeets, Luca Baroni, Adam Willliams, Carlo Orsi, Filippo Benappi, Alan Salz, and Bob Haboldt. They have brought to the market compelling and beautiful examples of painted stones, and generously shared insights and expertise, helping to further a proper understanding of how artists utilized the stone surfaces they painted as well as an appreciation for the results they achieved.

In light of such an assessment of the state of study devoted to stone paintings, I can fully acknowledge the promised impact of the papers that have been brought together in this important and groundbreaking volume. Several of the essays focus on Rome and artists working in Rome. The importance of that city in the development of painting on stone cannot be overstated. This fact has been acknowledged, but now we are afforded nuanced investigations into the situation and implication of its Roman origins. Another essay expands the discussion to Siena. This volume brings a richer investigation of Sebastiano del Piombo and the meaning of stone for him. It also provides a valuable look at Titian's technique which complements the small amount of technical studies that have appeared so far. Another essay looks at other materials that artists utilized for their supports, an important and necessary context if we are to properly assess the significance and intention of artists who chose stone. This volume also helps to flush out one of the real geniuses of stone painting, Antonio Tempesta, who has been celebrated for his prints but less often for his innovative approach to painted stone. Finally, several papers offer rich inquiries into the meaning of stone paintings, in my mind the most interesting and potentially edifying aspect of the subject.

Together, these studies will contribute to our knowledge so that paintings on stone may start to play a more prominent role in art historical assessments of Renaissance art. In too many museums, stone paintings have languished unappreciated in storage, sometimes due to confusion about assigning curatorial responsibility. I once tried to examine firsthand one of the most beautiful examples of painted lapis lazuli, only to discover that the curator who was responsible for the panel in question did not have access to the space in which the painting was stored. Until recently, some stone paintings were not even considered fine art, while others were in the care of decorative arts curators not trained to understand them. There are still many instances where stone paintings need conservation yet little has been published on important issues surrounding their care such as the use of varnish, how surfaces were prepared to receive paint, or consensus as to the desired aesthetic effect of the finished work. The essays comprising this volume mark an important moment in the study of stone painting by reflecting a new level of sophisticated analysis. The result will not only enrich the way we understand artistic production and intellectual discourse in the sixteenth and seventeenth centuries. It may also reap tangible rewards in the more widespread display and increased attention to the care of these incredible objects.

*Judith Mann*

# Acknowledgments

This volume has been a long time in gestation, and, like all projects that develop over several years, it has benefitted from the involvement of many people whose counsel and assistance have proven invaluable. It seems invidious to single out anyone, but there are certain people, besides the members of our families to whom this book is dedicated, who must be acknowledged by name. We are especially grateful to Judith Mann, who has been working on the topic of painting on stone supports for many years and who has shared her knowledge and experience with us unstintingly. Her generosity has extended to writing the preface to this volume.

The first essays in this collection developed from a session at the Renaissance Society of America's annual conference in San Diego, 2013; five additional essays were added between 2013 and 2016, when most of the authors were able to present their papers at a public symposium held at the British School at Rome on May 17. This second convening and workshop was supported generously both by the British School and by the University of Richmond and would not have come to fruition without their encouragement and resources. The Director of the British School, Professor Christopher Smith, provided a warm welcome to the public and scholars alike, and Tom True, the Vice Director, gave the symposium a thoughtful and eloquent introduction, in addition to coordinating our arrangements at the British School. We would like to extend particular thanks to him for his generosity and enthusiasm. The help of Stefania Peterlini, the Director's Assistant and Permissions Officer at the School, was also invaluable over a number of years for obtaining access to the many paintings on stone still *in situ* in Rome. Former Dean Kathleen Skerrett of the University of Richmond School of Arts & Sciences saw fit to provide funding for the symposium with a mind to support the development of this publication.

We the editors would also like to thank their fellow contributors, whose thoughtful essays have made this volume what it is, and also Professor Laura Gelfand, who gave a paper at the RSA panel in San Diego but was not able to contribute an essay for the volume. Without question our thanks are due to the anonymous reader of the original manuscript, whose knowledgeable criticism substantially strengthened and improved the final work. Any faults that remain are entirely ours. All of the authors have benefitted from conversations with a wider network of colleagues, whose names are too numerous to include here but are noted in the individual essays. Each of the authors also profitted from the diligence and fine-grained attention of David Routt, whose editing and good judgment undoubtedly improved every aspect of the manuscript.

We owe a debt of gratitude to the staff of a number of museums and galleries who have given generously of their time and expertise in discussing and showing their paintings on stone. We are especially grateful to the museums that worked with all of our authors and provided the images that so amply demonstrate the beauty and variety of these works on stone. The Churches of San Silvestro in Quirinale and Santa Pudenziana in Rome were also especially welcoming and informative, and we will fondly remember the conversations in which their staff engaged us. We would also like to thank Simonetta Antellini (Direttrice Ufficio Esportazione Oggetti d'Antichità e d'Arte a Roma), who kindly gave a sneak preview of the still to be published results of the restoration of Sebastiano's *Visitation* in Santa Maria del Popolo.

Several institutions supported the research and writing that resulted in this volume. The Villa I Tatti, where Elena Calvillo was a Hanna Kiel Fellow during 2013-2014, provided an invaluable haven for the completion of her essay and research for the introduction, while the University of Richmond awarded her a summer research fellowship in 2016 and further funding for the editing and illustrations of the volume. The Renaissance Society of America was also essential to the completion of this volume; its award to Piers Baker-Bates of the Samuel H. Kress Mid-Career Research and Publication Fellowship generously funded most of the illustrations for the final publication. Finally, last but in no way least, we would like to thank Brill for the assistance and encouragement we have received from Marcella Mulder and the editors of this series, Laura Gelfand and Sarah Blick, who happily saw the potential of this volume to make a contribution to this growing field of study.

*Piers Baker-Bates*
*Elena Calvillo*

# List of Illustrations

## Figures

## Plates

# Notes on Contributors

*Nadia Baadj*
(PhD Michigan) is Assistant Professor of Art History and Rosalind Franklin Fellow at the University of Groningen. She is the author of *Jan van Kessel I (1626-1679): Crafting a Natural History of Art in Early Modern Antwerp* (2016). She has held fellowships at the Metropolitan Museum of Art, the Victoria & Albert Museum, and the Max Planck Institute for the History of Science, Berlin. She is a member of the editorial board of the Brill series *Studies in Art & Materiality*.

*Piers Baker-Bates*
(PhD Cantab 2006) is a Visiting Research Associate in Art History at the Open University. He is the co-author of the *Spanish Presence in Sixteenth-Century Italy* (Ashgate, 2015) and author of *Sebastiano del Piombo and the World of Spanish Rome* (Routledge, 2016), as well as a number of scholarly articles on Sebastiano and the manifold sixteenth-century cultural relationships between Italy and Spain. He was also involved with the 2017 Sebastiano and Michelangelo exhibition at the National Gallery, London. He has held Fellowships, among others, at the British School at Rome, the Dutch Institute in Florence and from the Society for Renaissance Studies and the Renaissance Society of America.

*Elena Calvillo*
(PhD Johns Hopkins University) is Associate Professor of Art History at the University of Richmond, Virginia. Her essays on Giulio Clovio, Francisco de Holanda, and Sebastiano del Piombo have appeared in *The Art Bulletin*, *Renaissance Quarterly* and most recently in *The Spanish Presence in Sixteenth-Century Italy* (Ashgate, 2015) and *Trust and Proof: Translators in Early Modern Print Culture* (Brill, 2017). She is completing a manuscript on Clovio, *At the Cardinal's Court: Artistic Service in Sixteenth-century Rome*. She has been a fellow of the Kress Foundation, the American Philosophical Society and the Villa I Tatti.

*Ana González Mozo*
is a Doctor of Fine Arts and a research specialist both in the technical aspects of art and in the contributions of infrared reflectography systems and digital image processing to the study of painting. From 1990 to 1996, her research was carried out in the Faculty of Fine Arts in Madrid. Since 1996 she has worked at the Prado Museum, where she has participated in numerous research and

exhibition projects. As of 1999 she has been working for the Office of Technical Documentation, as part of the research group that analyzes the technical and creative process of works of art as well as the scientific methods used to study paintings, specially focused on Italian Painting.

*Anna Marazuela Kim*
(PhD Virginia) is a Visiting Research Fellow at the Institute of Advanced Studies, University College London. Prior to that she was a Fellow of the research group BildEvidenz: Geschichte und Ästhetik at the Free University, Berlin and Andrew Mellon Postdoctoral Fellow at the Courtauld Institute of Art. Her research engages the deep structures of our complex relation to images, from Plato to the present, focusing on the icon, idol and iconoclasm as historical and conceptual phenomena. Currently she is working on a book project titled: *Reforming the Image: Idols and Iconoclasm in Early Modern Rome.*

*Helen Langdon*
(PhD London), formerly Assistant Director of the British School at Rome, is an art historian with a particular interest in the Italian Seicento. Her books include *Claude Lorrain* (1989), *Caravaggio: a Life* (1998) and *Caravaggio's Cardsharps; Trickery and Illusion* (2012). In 2010 she curated the exhibition, Salvator Rosa, at Dulwich Picture Gallery. She has contributed many articles to learned journals and has been a frequent reviewer for the Burlington Magazine.

*Johanna Beate Lohff*
studied Art History, Italian Philology and Philosophy at the University Cologne and at the Università degli Studi di Bologna / Sede à Forlì, graduating with a master thesis on Sebastiano del Piombo's paintings on stone in 2003. She gained scholarships of the German Academic Exchange Service (DAAD) and the Bibliotheca Hertziana Max Planck Institut in Rome between 2007-2010. She finished her dissertation with a focus on Antonio Tempesta's paintings on stone at the University of Bonn in 2011, awarded with the Lempertz-Award in History of Art, University of Bonn in 2012, publication as a book at Hirmer Editor, Munich in 2015. Lectureships at the Ruhr-University, Bochum. Johanna Beate Lohff works as a freelancer for several Museums in the Ruhr Area.

*Christopher J. Nygren*
(PhD Johns Hopkins University) is Assistant Professor of Renaissance & Baroque Art at the University of Pittsburgh. He was a Kress Pre-Doctoral Fellow at the Kunsthistorisches Institut in Florence, and held Mellon Post-Doctoral

Fellowships at the University of Pennsylvania and Emory University. His work focuses on the interaction between art and devotion in Renaissance Italy, and his essays have appeared in *Renaissance Quarterly*, *Word & Image*, and *The Art Bulletin*, among other journals. He is completing a book manuscript on Titian's Icons.

*Susanne Wegmann*
studied Art History, Archeology and religious studies in Regensburg. She earned her doctoral degree in 2001 with a thesis on the iconography of Purgatory. Subsequently she was a research fellow at the Zentralinstitut für Kunstgeschichte in Munich and at the VW-Research Group KultBild in Münster. From 2007 until 2013 she worked at the Martin-Luther-University Halle-Wittenberg, where she obtained her Habilitation. In 2013-2014, she was guest professor at the University in Leipzig, and Assistant Professor at the University in Freiburg/Brsg in 2015. Since 2017, she has been Professor of Art and Cultural History at the Cologne Institute of Conservation Sciences (Techische Hochschule Köln). Her research focuses on the visual arts and art theory in the middle ages and early modern art particularly with regard to art and theology and the use of images in Lutheran confessionalization.

*Giulia Martina Weston*
(PhD The Courtauld Institute of Art) is Associate Lecturer at the Courtauld Institute. She is also an independent art consultant and contributing editor of several online journals. She co-organized the Courtauld Fifth Early Modern Symposium and co-edited a volume on mid 17th-century Italian art (*I Pittori del Dissenso*). She has published on the display of art in Baroque Rome, art and technology in post-Galilean Italy, and the British afterlife of the Neapolitan painter Salvator Rosa. She recently published the monograph *Niccolò Tornioli (1606-1651). Art and Patronage in Baroque Rome* (Artemide, 2016).

# Introduction

*Piers Baker-Bates and Elena Calvillo*

## Eternal Paintings?

On the night of 21 September 1936, almost exactly four hundred years after arriving in Úbeda (Andalusia), Sebastiano del Piombo's *Pietà* [**Plate 1**] in the funerary chapel of Francisco de los Cobos in San Salvador was bayoneted by an anti-clerical mob. Two other historically significant works suffered that evening: a statue of the young St. John the Baptist, whose attribution to Michelangelo has recently been widely accepted after an extensive restoration, was smashed into pieces, and a *retablo* by Alonso Berruguete perished in flames. That Sebastiano's painting on slate survived testified to one of the first claims made about the technique of oil painting on stone supports, which its earliest commentators said would withstand the destructive forces of time as "pittura poco meno che eterna," or painting little less than eternal.[1] Although this judgment was to be more than proved four centuries later, the question of durability and eternity has in many ways condemned the innovation and what motivated its development to simplistic explanations.

One context invoked to historicize Sebastiano's technique, one that lends particular irony to the event in 1936 given the sculptural casualty, has been the *paragone* between sculpture and painting, which commonly praised sculpture's durability over that of painting.[2] Another explanation for Sebastiano's

---

1 The author of this description, Vittore Soranzo, continued by noting that the artist had tested his technique and found it durable; see his report to Pietro Bembo in *Delle Lettere da diversi Re et Principi et Cardinali et altri huomini dotti a Mons. Pietro Bembo scritte* (Venice, 1560), p. 110 "dovete sapere che Sebastianello nostro Venetiano ha trovato un segreto di pingere in marmo a olio bellissimo il quale farà la pittura poco meno che eterna. I colori ... si uniscono col marmo di maniera che quasi impetriscono ... ha fatto ogni prova & è durevole. Ne ha fatto una imagine di Christo & halla mostrato a N. Sig." Michael Hirst was the first to emphasize the significance of Soranzo's letter, *Sebastiano del Piombo* (Oxford, 1981), p. 124.

2 The *paragone* has been the subject of several recent studies, since Leatrice Mendelsohn's *Paragoni: Benedetto Varchi's* Due Lezzioni *and Cinquecento Art Theory* (Ann Arbor, 1982) See Ben Thomas, "'The lantern of painting': Michelangelo, Daniele da Volterra and the *paragone*," *Apollo*, 154, no. 474 (2001), 46-53; Rudolph Preimesberger, *Paragons and Paragone: Van Eyck, Raphael, Michelangelo, Caravaggio, Bernini*, trans. Sabine Eiche and Fiona Elliott (Los Angeles, 2011); Sefy Hendler, *La Guerre des arts Le Paragone Peinture-Sculpture en Italie* XV-XVII *siècle* (Rome, 2013); *Daniele da Volterra e la prima pietra del Paragone*, ed. Andrea G. de Marchi,

 | DOI 10.1163/9789004361492_002

use of stone supports relates to formal appraisals of his late Roman practice; several art historians have explained the use of slate as a means to achieve the dark tonality and accomplished modeling, or *rilievo*, of these works, which have often been interpreted as a kind of austere proto-Counter-Reformatory response to the Sack of Rome in 1527.[3] Although Sebastiano's formally spare works do seem to have been understood by his contemporaries as evocative of the personal and intense devotional responses that would later be associated with the Counter Reformation, especially in Spain, the use of stone as a support was more likely employed for reasons that extended beyond formal choices and the emotive effects of a somber palette.[4] Many of the essays in this volume offer alternatives to these first explanations of the early modern phenomenon of oil painting on stone.

The 'Material Turn' in the research, theory and methodology of art history, represented by this and another forthcoming Brill series *Art and Materiality*, has initiated a discourse from which Sebastiano's technical and intellectual innovation and its legacy throughout Europe in the following two centuries might be better understood, beyond the fact of its innovation, for its signifying potential.[5] The essays that follow present close considerations of stone's particular ability to convey metaphorical meaning and ontological relationships through its materiality.[6] While the first chapters in Part One consider Sebastiano's invention, its initial reception, and historical context, the middle chapters in Part Two present various forms of painting on stone and its appeal to collectors throughout Europe at a moment when novelty in materials and

(Rome, 2014); and Christiane Hessler, *Zum Paragone Malerei, Skulptur und Dichtung in der Rangstreitkultur des Quattrocento* (Berlin, 2014). Also see Christopher J. Nygren, "Titian's *Ecce Homo* on Slate: Stone, Oil and the Transubstantiation of Paint," *Art Bulletin* 99, 1 (2017), 36-66.

3 See for example the works of Marcia Hall, *Color and Meaning* (Cambridge, 1992), pp. 137-42, 168; *The Sacred Image in the Age of Art, Titian, Tintoretto, Barocci, El Greco, Caravaggio*, (New Haven and London, 2011), p. 147-51.

4 Elena Calvillo, "Authoritative Copies and Divine Originals: Lucretian Metaphor, Painting on Stone, and the Problem of Originality in Michelangelo's Rome," *Renaissance Quarterly*, 66, no. 2 (2013), 453-508, especially pp. 483-85; Piers Baker-Bates, *Sebastiano del Piombo and the World of Spanish Rome* (New York and London, 2016), pp. 172-73, 180-81; Nygren, "Titian's *Ecce Homo*," p. 49.

5 See also the essays in *The Matter of Art: Materials, Practices, Cultural Logics, c. 1250-1750*, eds. Christy Anderson, Anne Dunlop, and Pamela H. Smith (Manchester, U.K., 2015).

6 For the most recent discussions of the technique in terms of early modern and contemporary theories of representations, see Anne-Laure Collomb, "Les traits artistiques et la peinture sur pierre, XVIe et XVIIe siècle," *Histoire de l'art* 52 (2003), 111-20; Calvillo, "Authoritative Copies and Divine Originals"; Nygren, "Titian's *Ecce Homo* on Slate," which draws from material from his dissertation, "Vibrant Icons: Titian's Art and the Tradition of Christian Image-making," (PhD diss., Johns Hopkins University, 2011).

techniques and the conjunction of *ars* and *natura* were increasingly valued. The final chapters in Part Three address other technical developments in media and painting supports that seem related to oil painting on stone and to have been motivated by similar metaphors and concepts.

Previous investigations of this subject have appeared as broad surveys, whether in the form of exhibitions, such as that organized by the late Marco Chiarini at Palazzo Pitti in 1970 or the more recent exhibition in Palazzo Reale in Milan in 2000-01, or monographs such as those by Hana Seifertova (2007), Anne-Laure Collomb (2012), and most recently Mario Casaburo (2017).[7] The essays here focus on the ways in which stone supports might function as markers of artistic inventiveness and innovation and as agents of meaning; whether the lithic material signifies metaphorically or metonymically, suggests transubstantiation or consubstantiality, represents nature or the supernatural, the following chapters provide a variety of historical and theoretical emphases through specific case studies. Before turning to these, it would be best to begin with a brief history of the technical invention by Sebastiano del Piombo in Rome, which almost all of the authors reference before moving on to their specific examples of its legacy and efflorescence.

## The Invention of Oil Painting on Stone and its Early Reception

Almost twenty years after arriving in Rome from his native Venice in 1511, the painter Sebastiano Luciano, or Sebastiano del Piombo as he became known for his position as papal *piombatore* (keeper of the lead seals), began to paint in oil on stone surfaces. Despite speculation that the Sack of Rome and the consequent loss of works of art led Sebastiano to create 'almost eternal' paintings, no explicit evidence remains to indicate how and why Sebastiano turned

7 For the first major exhibition of oil paintings on stone, see Marco Chiarini's pioneering *Pittura su Pietra*, exhibition catalogue, Galleria Palatina (Florence, 1970). This precedent was followed thirty years later by *Pietra Dipinta: Tesori nascosti del '500 and '600 da una collezione*, ed. Marco Bona Castellotti (Milan, 2000). A number of recent mongraphs have also been devoted to the subject; see Hana Seifertova, *Malby na kameni: Umělý experiment v 16. Na začátku 17 století/ Painting on Stone: an Artistic Experiment in the 16th and Early 17th centuries* (Prague, 2007); Anne-Laure Collomb, *Splendours de l'Italie: la Peinture sur Pierre à la Renaissance* (Rennes, 2012), whose research began as her doctoral dissertation; as did that of Johanna Beate Lohff's recent book, *Malerei auf Stein Antonio Tempestas Bilder auf Stein im Kontext der Kunst- und Naturtheorie seiner Zeit* (Munich, 2015), and Mario Casaburo's most recent monograph, *Pittura su Pietra: Diffusione, Studie Materiali, ed Tecniche Artistiche* (Florence, 2017). Also see, Angela Cerasuolo, *Diligenza e Prestezza: La tecnica nella pittura e nella letteratura artistica del Cinquecento*, (Florence, 2014).

his ambitions to this new material support.[8] Nonetheless, on 8 June 1530, the papal *cameriere*, Vittore Soranzo, wrote from Rome to his mentor Pietro Bembo:

> You should know that our little Sebastiano the Venetian has found the secret with which to paint in oils on marble in the most beautiful fashion, which will make his paintings little less than eternal. As soon as the colors are dry these unite with the marble as if they were turned to stone; every test has been tried, and it has proved durable. Using this method Sebastiano has made an image of Christ and shown it to our Lord.[9]

This is the first recorded notice of Sebastiano's invention, although it is probable, as Baker-Bates's essay discusses, that his experiments had already begun a number of years earlier. Sebastiano himself first mentions the practice in a letter of just over a year later, 22 July 1531, to Michelangelo:

> Pardon me that I have not sent you the head of the Pope: I painted it on a canvas from the life, and the Pope wanted me to make another of these, on a stone. And as soon as it has been copied, I shall send it to you.[10]

Within the space of a year, then, we have evidence for both of Sebastiano's characteristic uses of stone as a medium, for sacred subjects and for portraiture.

Between 1530 and 1539, Sebastiano seems to have produced at least ten paintings on stone supports: the presumably lost image of Christ referred to in Soranzo's letter, a portrait of Clement VII referred to in the letter to Michelangelo, which could have been a work other than the unfinished portrait in the Museo di Capodimonte, the portraits of Baccio Valori (Palazzo Pitti) and Giulia Gonzaga, whose original on stone is probably lost, the *Pietà* for Úbeda [**Plate 1**], the *Nativity of the Virgin* in the Chigi Chapel of Sta. Maria del Popolo [**Plate 2**], the *Madonna del Velo* [**Plate 3**]—also at the Capodimonte, the three surviving

8 For brief summaries of Sebastiano's invention, see Hirst, *Sebastiano del Piombo*, pp. 124-26; Seifertova, *Painting on Stone*, pp. 17-20, in which she also cites André Chastel's interpretation of these works as indicative of Sebastiano's trauma and somber response to the Sack of Rome in 1527.

9 *Delle Lettere*, p. 110, as in note 1.

10 *Il Carteggio di Michelangelo*, ed. Paola Barocchi and Renzo Ristori, 5 vols. (Florence,1965-83), 3:318.

*portacroce*'s in Madrid, St. Petersburg, and Budapest [**Plates 10, 11,** and **12,** respectively], and the portrait of Paul III and Ottavio Farnese [**Plate 13**] in the Gallerie Nazionali of Parma. Given that the inventory of Sebastiano's house includes several pieces of stone for painting, as Baker-Bates notes in his essay, this is probably an incomplete list for a period in his career that has usually been characterized as relatively unproductive.[11] Certainly, by his death in 1547, Sebastiano had left a series of both partially painted and unpainted "quadri de preta," as described in his inventory. Among these were copies of his portraits of "papa Clemente, uno in preta" and "la Signora Giulia Gonzaga in preda."[12] Sebastiano's invention, as these numbers and the immediate critical reception documented by Soranzo and Vasari suggest, was quickly embraced by patrons in Rome, Florence, and further abroad in Spain. Or, perhaps, one should rather say Sebastiano's reinvention or rediscovery of painting on stone, as a similar practice might have been in existence already in Imperial Rome, and, as Ana González Mozo's discussion of the technical and theoretical precedents for Sebastiano's innovation suggests, his experimentation was firmly grounded in long-standing solutions to the environmental challenges to polychrome sculpture and mural painting in Venice.[13]

In terms of the ancient precedents and Sebastiano's reinvention of an antique form, we might begin by looking at the beginning of Book 35 in Pliny the Elder's *Natural History*, which briefly mentions painting on marble. His language, however, is frustratingly vague and open to interpretation:

---

11 For the inventory, see Michael Hirst, *Sebastiano del Piombo*, pp. 154-57; also see Baker-Bates in this volume, p. 56.

12 Hirst, *Sebastiano del Piombo*, pp. 155-56. That 'preda' means stone is indicated by certain modifiers in the language, such as "un quadro de preda de marmore gialo" and "una preda de porfiro," or "quadro ritrato de la Signora Giulia Gonzaga in preda"; we know from other primary sources that Sebastiano's portrait of Giulia Gonzago was on stone.

13 In addition to González Mozo's essay in this volume, see Angela Cerasuolo, *Diligenza e Prestezza: La tecnica nella pittura e nella letteratura artistica del Cinquecento* (Florence, 2014), pp. 133-39. An English translation is now available *Literature and Artistic Practice in the Sixteenth Century: the Processes of Painting in the Treatises and the Works* (Leiden, 2017). Collomb, Casaburo and Calvillo have also followed this notion of a revival of an antique technique; see Collomb "Les traits artistiques," pp. 111-12; and Casaburo, "La Pittura su Pietra in Italia. Primi Scenari di Ricerca: Appunti sulle Tecniche, i Materiali e le Opere (1530-1630)," in *Per la Conoscenza dei Beni Culturali*, III (2010), 239-57, esp. p. 41; and Calvillo in chapter three of this volume (pp. 113-14) and in "Authoritative Copies," p. 487, in a slightly different context relating to Pliny's description of a touchstone. Most recently, Fabio Barry has published a study that substantially addresses the antique origins of painted stone, "Painting in Stone": Early Modern Experiments in a Metamedium,' The Art Bulletin 99, no. 3 (2017), 30-61. Unfortunately, this study was published after these essays were submitted and thus is not engaged in a way that it merits.

> We are no longer content with panels, nor with surfaces displaying broadly a range of mountains in a bedchamber; we have begun even to paint on the masonry (lapidem) This was invented in the principate of Claudius, while in the time of Nero a plan was discovered to give variety to uniformity by inserting markings that were not present in the embossed marble surface, so that Numidian stone might show oval lines and Synnadic marble be picked with purple, just as fastidious luxury would have liked them to be by nature.[14]

Other textual evidence enumerating or describing types of painting supports suggests that ancient artists did use stone surfaces.[15] Given the austerity of Sebastiano's late *oeuvre*, noted both in his time and ours, it is ironic that Pliny's passage is one part of a wider attack on contemporary luxury in Imperial Rome. It is easy to imagine how later Roman and Byzantine Church interiors comprised of walls and floors of cut, ornamental stones, or glittering mosaics might evoke the sense of luxury decried by Pliny, but the initial panel paintings by Sebastiano and the first generation of artists to use his technique seem concerned with something quite different from material opulence. Still, painting on stone seems to have been connected, at least in theory, to the ongoing practice and resurgence of mosaic in Renaissance Venice and Rome; as Vasari states, mosaic too was a medium considered 'quasi eterno.'[16] The first essay in the Part Three of this volume by Anna Kim addresses this important context

14 We thank Peter Fane-Saunders for reminding us that Pliny does not refer explicitly to figurative representations, but rather notes the use of paint to amplify the curious markings of exotic marbles beyond the powers of nature to produce. In other places, Pliny notes the amazing figures to be found made by nature, as in noted in Nygren's essay in this volume, p. 145; see *Natural History*, 37.iii.5 and 36.4.14.For the interpretation of the passage in question at the beginning of book 35, see Gaio Plinio Secondo, *Storia Naturale v Mineralogia e Storia dell'Arte Libri 33-37*, ed. and trans. Antonio Corso, Rossana Mugellesi, and Gianpiero Rosati (Turin, 1988), 294-95; the English translation used here is from Loeb Classical Library, Pliny the Elder, *Natural History*, ed. G.P Goold, vol. 9, Books XXXIII-XXXV (Cambridge, MA and London, 1952; reprint 1999), pp. 260-63.

15 See Nathaniel Jones, "Ancient Painted Panels: Terminology and Appearance," *Mnemosyne* 67 (2014), 295-304, esp. p. 304 for the appearance of stone as a support in the Delian Inventories. Sorcha Carey also clearly interprets the Plinian passage as describing paintings on marble surfaces; see her *Pliny's Catalogue of Culture: Art and Empire in the Natural History* (Oxford, 2003), pp. 107-8.

16 Giorgio Vasari, *Le Vite de'più eccellenti pittori scultori e architettori nelle redazioni del 1550 e 1568*, ed. Rosanna Bettarini and Paola Barocchi, 9 vols. (Florence, 1966-87), 1:148, "per la sua lunghissima vita si puo quasi chiamare eterno." Also see Alexander Nagel and Christopher Wood's discussion in *Anachronic Renaissance* (New York, 2010), p. 130.

for understanding what might have been expected of oil painting on stone at the time of its development around 1530 and its implications for the ontology of art over the next century, while the following essay by Giulia Weston returns us to the art of mosaic through a consideration of Niccolò Tornioli's technical experiments and failures in the mid Seicento and suggests that the connection between mosaic and painting on stone persisted into the next century.

The first three essays concerning Sebastiano's technique address, in different ways, how his invention would have been understood as an accomplishment in terms of both theory and practice. Leaving aside its metaphorical and ontological significance, it was something both difficult to do and difficult to conceive, and thus it should not be surprising that it was popular with the intellectuals and poets of his time, many of whom were his friends and peers in the ambit of the papal court. Piers Baker-Bates traces Sebastiano's life-long experimentation with materials and the way in which his choices aligned with the interests of his patrons in Rome and Spain. Ana González Mozo provides important new technical analysis, especially of the surface preparation of Sebastiano's stone supports, and demonstrates the extent to which his invention was both solidly based in ancient theory and traditional Venetian practices and well-suited to produce effects appropriate to his subject matter. Elena Calvillo considers the *concetti* and metaphoric potential of Sebastiano's use of stone for the construction and communication of identity, for his first patron of the technique, Clement VII, his papal successors, and for Sebastiano himself as the keeper of the papal seals and official portraitist.

Sebastiano's invention appears to form one part of a larger contemporary engagement with different materials as artistic media. Soranzo's letter to Bembo indicates that it was a noteworthy development at the papal court, while Vasari explicitly states that it was popular, both for its novelty and the perception that the paintings might become eternal.[17] Given Vasari's generally negative presentation of Sebastiano's career, the extent to which he acknowledges the importance and ingenuity of this new technique is noteworthy.[18]

---

17 Vasari, *Le Vite*, 5:97; In the 1550 Torrentino edition, Vasari notes the novelty of the technique as a factor in its popularity writing, "aveva cominciato un nuovo modo di colorire in pietra; *la qual novità* piaceva molto a' popoli, considerando che tali pitture diventassero eterne" (emphasis ours); in the Giuntina edition, almost 40 years after the technique was first described—and thus no longer new, Vasari emphasized only the perception of durability, "avendo poi cominciato questo pittore un nuovo modo di colorire in pietra, ciò piaceva molto a' popoli, parendo che in quell modo le pitture diventassero eterne."

18 Vasari also discusses the technique extensively in his introductory section that describes the different practices of painting, sculpture and architecture; Vasari, *Le Vite*, 1:135ff.

Another source for the intellectual reception of the technique and point of intersection for Sebastiano's practice and the literary circles of his days, as a number of scholars including Christopher Nygren in this volume have noted, are the poems written by Francesco Maria Molza and Gandolfo Porrino that include praise of the naturalism of Sebastiano's portrayal on slate of Giulia Gonzaga.[19] A little more than a decade later, Benedetto Varchi would cite this work and the poems dedicated to it in his lectures on the *paragone*, published in 1549—the year before Vasari praises Sebastiano's invention and its popularity, as noted above.[20] Painting on stone had rapidly become a part of the discourse of art, as evidenced by Daniele da Volterra's double-sided *David and Goliath* [**Plate 6**] painted for Giovanni della Casa between 1550 and 1555 and directly linked to a theoretical treatise that della Casa was writing on the *paragone*.[21]

## Dissemination and Efflorescence of the Technique after 1530

The essays in Part Two of this volume trace the intellectual and aesthetic appeal of these works and the way in which Sebastiano and his immediate successors such as Titian and Daniele da Volterra found equally ambitious counterparts in the next generation of artists painting in oil on stone supports, figures such as Cavaliere d'Arpino, Antonio Tempesta, and Hans von Aachen. Christopher Nygren's essay bridges Sebastiano's and Titian's work on the dark stone supports and the more spectacular use of figurated stone in the work of d'Arpino and Jacques Stella. Nygren's emphasis on and development of the notion

19 *Delle Poesie Volgari e Latine di F.M. Molza*, ed. Pierantonio Serassi, 3 vols, (Bergamo, 1747-54) 1:135-61. See Mendelsohn, *Paragoni*, pp. 129, 276 n. 136; Costanza Barbieri, "I ritratti di Sebastiano a *paragone*: pitture scolpite, elogi per immagini," in *Sebastiano del Piombo, 1485-1547*, pp. 53-7; and "'Tu, che lo stile con mirabil cura pareggi col Martello," Fortunae e sfortune di Sebastiano," in *La* Pietà *di Sebastiano a Viterbo. Storia e tecniche a confronto*, edited by Costanza Barbieri, Ernrico Parlato, and Simona Rinaldi (Rome, 2009), pp. 51-65, especially pp. 55-56 and p. 61. Most recently, Nygren here and more extensively in his essay "Titian's *Ecce Homo* on Slate" puts greater pressure on the way in which these paintings might actually have been understood within the context of the *paragone* between painting and sculpture.

20 Benedetto Varchi, "Lezzione nella quale si disputa della maggioranza delle arti e qual sia più nobile, la scultura o la pittura," in *Trattati d'Arte del Cinquecento*, ed. Paola Barocchi, 3 vols, (Bari, 1960-62), 1:3-82, esp. pp. 40-41.

21 See Thomas, "Daniele da Volterra and the *paragone*", pp. 47-49; and De Marchi, *Daniele da Volterra e la prima pietra del Paragone*, p. 14.

similitude provides a coherent intellectual and material history for works that ostensibly seem very different but whose conceits are deeply connected. Johanna Beate Lohff's study of the avid collecting of Antonio Tempesta's works in early 17th-century Italy and Vienna and the theoretical context of Tempesta's practice foregrounds the intersection of art and nature in his paintings on various stones support. Her essay thus continues Nygren's discussion of ancient notion of *ars et natura*, while Susanna Wegmann's contribution considers the way in which the material support of Hans von Aachen's paintings bears significance for his subject matters in ways very much in line with his Italian predecessors. Both his sacred and profane subjects saw their meaning enhanced or effectively emphasized through the properties and figuration of the stones that he used. Nadia Baadj's chapter continues Lohff's consideration of collecting and Wegmann's attention to the Northern scientific inquiry and the *kunstkammer* but first takes as its focus Antwerp, where the dissemination of paintings on stone may be seen in the spectacular Jesuit Church of St. Charles Borromeo. Baadj and Wegmann also importantly relate the practice of painting on stone supports with other material innovations in the North and Italy, such as paintings on copper, which Baadj observes is the quintessential support for the Northern 'period eye' and rightly notes the important connection made between paintings on stone and their counterparts on metal.[22] Her discussion of the Northern interest in paintings on metal supports and the cultural values that practice represents introduce the third section of this volume, which concerns techniques related to oil painting on stone supports.

The final three essays in this volume return the focus to Italy. Before turning to these, it might be useful to make a few general observations about the geography of the dissemination and evolution of the technique, beginning with its immediate and long-lasting impact in Rome.

### *In Rome*

Although most altarpieces painted on stone supports date from the second half of the 16th century, Sebastiano's invention was quickly adopted by a number of artists in his and the succeeding generation in Rome, beginning with Titian's *Ecce Homo* [**Plate 5**] painted in 1547 for Charles V,[23] Francesco Salviati's work in the Chigi Chapel in Sta. Maria del Popolo in the 1540s and the Cappella

22 As Baadj observes (p. 250 of this volume), this connection was made by Vasari in his life of Sebastiano; see the *Le Vite*, 5:98-99.

23 Piers Baker-Bates, "The 'Cloister Life' of the Emperor Charles V: Art and Ideology at Yuste," *Hispanic Research Journal* 14, no. 5 (2013), 427-45, esp. 435-37; Nygren, "Titian's *Ecce Homo* on Slate," pp. 36-41 and 43.

del Pallio in the Palazzo della Cancelleria, *c.* 1549 [**Plate 15**].[24] Another example of Sebastiano's early influence is the *Baptism of Christ* in San Pietro in Montorio, which forms the altarpiece of the Ricci chapel, planned from 1556 onwards. This painting has often been attributed to Daniele da Volterra but was instead executed by Michele Alberti, who worked for Daniele using his designs and under his immediate direction, in a collaborative manner perhaps inspired by that of Sebastiano del Piombo and Michelangelo for the Borgherini Chapel *Flagellation* in the same church.[25] The Ricci *Baptism* is painted on a series of marble blocks and is an obvious attempt to emulate Sebastiano's earlier *Flagellation* in oils on the wall,[26] the central design of which was provided by Michelangelo.[27] That Daniele da Volterra had just completed the slate *David*

24 Salviati worked more closely with Sebastiano in the early 1530s than is acknowledged and succeeded Sebastiano in the Chigi Chapel. For Salviati's work in there, see the forthcoming publication on the restoration of the chapel; and Antonio Pinelli, "La Cappella delle tombe scambiate Novità sulla Cappella Chigi in Santa Maria del Popolo," in *Francesco Salviati et la Bella Maniera. Actes des colloques de Rome et de Paris (1998)*, ed. Catherine Monbeig Goguel, Phillipe Costamagna, and Michel Hochmann (Rome, 2001), pp. 253-85. There is very little written about the interesting but formally problematic altarpiece for the Cappella del Pallio apart from Patricia Rubin, "The Private Chapel of Cardinal Alessandro Farnese in the Cancelleria, Rome," *Journal of the Courtauld and Warburg Institutes* 54 (1987), 82-112, esp. pp. 83-87; and Clare Robertson, *Il Gran Cardinale Alessandro Farnese Patron of Arts* (New Haven and London, 1992), pp. 151-56.

25 See Teresa Pugliatti, *Giulio Mazzoni e la decorazione a Roma nella cerchia di Daniele da Volterra* (Rome, 1984), pp. 169-70 and 177-78; Letizia Treves, "Daniele da Volterra and Michelangelo: a collaborative relationship," *Apollo*, 154, no. 474 (2001), 36-45, esp. 37; Thomas, "Daniele da Volterra and the *paragone*," pp. 51-52; and Morten Steen Hansen, *In Michelangelo's Mirror Perino del Vaga, Daniele da Volterra, Pellegrino Tibaldi* (University Park, 2013), pp. 80 and 84-95.

26 A distinction should also be made, though it is often not, between the relatively more common altarpieces painted in oil on the wall, such as the *Flagellation* in San Pietro in Montorio, which also used a method developed originally by Sebastiano, and those for which the wall was specially prepared with a slate or other stone surface. As Baker-Bates discusses in this volume, Leonardo and Raphael also worked in oils for murals, but Sebastiano seems to have perfected the technique; also see Baker-Bates, "Sebastiano del Piombo. The Technique of Genius," in *Michelangelo/Sebastiano A Meeting of Minds*, (London, 2017), pp. 74-85.

27 For the division between the designer of the altarpiece and its executant, which represented "a radical separation between *disegno* and execution"; see Steen Hansen, *In Michelangelo's Mirror*, p. 91. Despite his collaboration with Michelangelo, some of the most extensive praise that Vasari gives Sebastiano centers on the Borgherini Chapel, specifically emphasizing the painter's masterful execution and technique. See Vasari, *Le Vite*, 5:89-90.

*and Goliath* [**Plate 6a** and **b**] for Giovanni della Casa's treatise on the *paragone* between painting and sculpture, and had furthermore taken the step that Sebastiano never made, moving from painting to sculpture, seems significant. The marble surface for the Ricci Baptism was most likely a deliberate choice reflecting that decision, from a drive to rival Sebastiano and to present his work in the context of the *paragone* after his recent experience work for della Casa.

After these projects, which were all completed within a decade of Sebastiano's death, a second generation of painters and of altarpieces on stone emerged. In fact, almost all the most significant artists active at Rome in that period engaged Sebastiano's invention at least once—and on a large scale and through the use of several blocks of stone (whose seams can now be seen quite easily).[28] It should be emphasized that altarpieces of such great size in this technique are not seen outside of Rome; developments elsewhere were in a smaller, more intimate (and collectible) scale, as the works discussed in all the essays of Part Two indicate. One such work from the second half of the 16th century is the high altar of the Basilica of San Lorenzo in Damaso painted *in situ* by the brothers Zuccaro begun in 1564.[29] Despite its commission by Cardinal Alessandro Farnese (1520-89) and the prestige of its artists, the painting has received little attention as an extraordinary example of the legacy of Sebastiano's technique. The subject is *The Virgin and Child appearing to Saint Peter, Saint Damasus, Saint Lawrence, and Saint Paul; the Martyrdom of Saint Lawrence in the Background.* This altarpiece should be singled out, as even Sebastiano had not attempted to paint quite such a complex altarpiece of this scale on stone; it consists of stone blocks and measures 8.93 by 3.97 metres. The years after the closing of session of the Council of Trent saw the prolifera-

28 Many of these Roman stone altarpieces still await substantial study, are in a poor state of repair, and lack good photography. Nonetheless, great strides have been made recently. See Marcia Hall, *Color and Meaning* (Cambridge and New York, 1992), for an incomplete list, see pp. 168-69 and 254, n. 16; Hana Seifertová, *Painting on Stone*; Fabio Fernetti, "Verso un Censimento dei Dipinti Cinquecenteschi su Pietra o Lavagna," in *Sebastiano del Piombo e la Capella Borgherini nel contesto della pittura Rinascimentale*, ed. Santiago Arroyo Esteban, Bruno Marrocchini, and Claudio Seccaroni (Rome, 2010), pp. 54-57; Collomb, *La Peinture sur Pierre*, chapter two "Les Peintures d'Autel," pp. 63-96; and Casaburo, "Le Grande Pale su Ardesia Romane," which illustrates the use of the blocks very well on p. 108, fig. 2, and notes the payments for the blocks and their installation for the Zuccaro *Coronation of the Virgin* in San Lorenzo in Damaso, n. 8, pp. 108-9.

29 Clare Robertson, *'Il Gran Cardinale'*, pp. 162-69; Cristina Accidini Luchinat, *Taddeo e Federico Zuccari fratelli pittori del Cinquecento*, 2 vols (Milan, 1998), vol. 1:272-74.

tion of such altarpieces in Roman churches, many of which were tied to the Farnese and the artists working for them.[30]

This volume does not attempt to survey these altarpieces, as they appear in Seifertova's, Collomb's and now Casaburo's volumes, but it might prove useful to note the most significant examples to give a simple indication of their number and of the range of artists involved in their production.[31] These include Taddeo Zuccaro's, *Conversion of Saul* (1565) in San Marcello al Corso, Rome, and his *Raising of the Son of the Widow of Naim* and *Christ healing the Blind Man* (1570) in Orvieto Cathedral; Girolamo Muziano's *Deposition* (1568 onwards) with surrounding painted decoration in copper and Marcello Venusti's, *Stories of Saint John the Baptist* (1571-73), both in Santa Caterina dei Funari. Venusti in particular became known for his adoption of this technique from his *Saint Catherine* (1557-60) in Sant'Agostino to his *Madonna between Saints Peter and Paul* (1577-78) in the Palazzo dei Conservatori and his *San Giacomo Maggiore* in Santa Maria sopra Minerva. There are also examples by Giovanni de' Vecchi, the *Saint Jerome* in Santa Maria in Aracoeli and by Girolamo Siciolante da Sermoneta, the *Martyrdom of Saint Catherine* (1570) in the Cesi chapel in Santa Maria Maggiore; and, probably best known, that of Rubens, the *Madonna della Vallicella* (1608) in the Chiesa Nuova. Finally, one of the last and most virtuosic examples in late Cinquecento Rome, which will be discussed at the conclusion of Baker-Bates' essay, is Scipione Pulzone's *Assumption of the Virgin*, in the Bandini Chapel in San Silvestro al Quirinale.

Recent conservation of works by Sebastiano, Daniele da Volterra, and Scipione Pulzone have partially corrected the paucity of knowledge concerning the practice of oil painting on stone, but more technical analysis to enable comparisons between various artists working in and outside of Rome would be instructive.[32] Furthermore, González Mozo's essay suggests the way in which

---

30 Although some commissions are easily tied to the first generation of artists and patrons connected to Sebastiano's innovation, a precise genesis of other works is hard to trace.

31 This volume was completed before editors had access to Mario Casaburo's recent monograph *Pittura su Pietra* (Florence, 2017) became available to them, but this author's dissertation was partly available through two articles, which are cited; see here "Le Grandi Pale su Ardesia Romane tra Cinquecento e Seicento: Aspetti Storici di Esecuzione, Conservazione e Restauro," in *Per La Conoscenza dei Beni Culturali*, 4 (2012), pp. 105-19. Also see Collomb, *La Peinture sur Pierre*, pp. 63-86.

32 See for example, *Sebastiano del Piombo y España*, ed. Manuela Mena Marqués (Madrid, 1995); *Sebastiano del Piombo e la Cappella Borgherini nel contesto della pittura rinascimentale*, ed. Santiago Arroyo Esteban, Bruno Marocchini, Claudio Seccaroni (Florence, 2010); *Sebastiano del Piombo e la pittura su pietra: il ritratto di Baccio Valori*, ed. Alessandro Cecchi, Marco Ciatti, and Oriana Sartiani (Florence, 2014); Cristiana Bigari, "Note sul restauro

such information might further our knowledge not just of the practice of these artists but also the way in which that practice acted both to produce and convey meaning. To what extent later artists had Sebastiano's original conception of the technique in mind is very difficult to establish, though it seems that many understood the potential of stone, or its affordances—as Nygren following Gibson's theory proposes, in a similarly ambitious way.[33] Some, like Daniele da Volterra and Francesco Salviati, were undoubtedly inspired by Sebastiano's direct example. With others the links were more indirect, possibly through the interests of patrons and other artists who had known Sebastiano, as is suggested by the works produced for the Farnese, who would have been attracted to both the practical/physical and poetic possibilities of stone to embody particular ideas, as Calvillo argues in the final essay of Part One. Certainly, by the end of the 16th century, as all the essays in Part Two demonstrate, small paintings on stone, especially figurated stones, had become highly sought after collector's pieces and were to be found in the most prestigious collections in Italy and Imperial centers.[34]

### *Beyond Rome*

Although Sebastiano's innovation is first described in Rome, several of his works on stone leave that city either as diplomatic gifts or as direct commissions, so that by the mid-16th century, his works are in both Spanish and French collections. In this case, the artists themselves may not have traveled, but the works of art and the technique did. The geographic dissemination of the technique depended on both artistic cross-fertilisation and the interest of learned patrons and collectors. The monographs by Seifertova, Collomb, and Casaburo provide more detailed histories of these movements than this volume of case studies, though in their variety they reflect the different locations where this

dell'Assunzione della Vergine in San Silvestro al Quirinale," in *Scipione Pulzone e il suo tempo*, ed. Alessandro Zuccari (Rome, 2015), pp. 233-47.

33 Gibson's theory of affordances and its potential for art historical methodology is discussed very clearly by Ann Sophie Lehmann's in her essay, "The matter of medium: some tools for an art-theoretical interpretation of materials," pp. 21-40, especially 31-34, in *The Matter of Art*, as in note 5.

34 The practice thus also fits within the recent disciplinary consideration of the geographies of art history. See Thomas DaCosta Kaufmann, *Towards a Geography of Art* (Chicago and London, 2004); David Young Kim, *The Traveling Artist in the Italian Renaissance Geography, Mobility and Style* (New Haven and London, 2014).

practice flourished.[35] As noted in the Preface, Judith Mann at the St. Louis Museum of Art has planned a much anticipated exhibition of paintings on stone, which will constitute a significant addition to the geographic and historical survey provided by these recent monographs.

Beyond Rome, there were other areas on the Italian peninsula that seemed particularly interested in these paintings; of note are Tuscany and the Veneto. In the former, we see parallel developments in *pietre dure* in Florence and *commessi* in Siena, as Giulia Weston's essay observes. In the Veneto, our first encounter with the technique is through Titian, whose work Nygren introduces briefly here and discusses extensively in a recent essay on the Prado *Ecce* Homo.[36] Titian was undoubtedly influenced by his first-hand experience of Sebastiano's invention during his visit to Rome between 1545 and 1546, when the two artists met for the first time in many years, and, it should be noted, the *Ecce Homo* was made to go to Emperor Charles V in Spain, where it has remained.[37] A tradition of painting on stone also developed separately at Verona, reaching its apex at the end of the century with a group of artists including Marcantonio Bassetti, Pasquale Ottino, and Alessandro Turchi, called l'Orbetto, whose use of stone supports seems to have increased after their respective moves to Rome.[38] All, nonetheless, had learned the technique originally at Verona from their master, Felice Brusasorci.[39]

## Related Material Innovations

The essays in Part Three of this volume introduce artists like Turchi, who painted on a variety of surfaces in addition to slate, such as *pietra di paragone*

---

35 Seifertová, *Painting on Stone*, traces the development by area in Venice, Verona, Florence, and also by artist; see her chapters between pp. 71 and 146. Collomb, *La Peinture sur Pierre*, pp. 97-162; and Casaburo, *Pittura su Pietra*, also address different geographic centers.

36 See Nygren, "Titian's *Ecce Homo* on Slate."

37 Baker-Bates, "The 'Cloister Life," esp. pp. 427-45 (435-37); and Calvillo in this volume.

38 For a general overview of this Northern center, see Marco Bona Castellotti, "Il collezionismo dei dipinti su pietra a Verona," in *Pietra Dipinta* ed. Marco Bona Castellotti (Milan, 2000); and Collomb, *Peinture sur la pierre*, pp. 118-30. For specific studies, see among others *Alessandro Turchi, detto l'Orbetto 1578-1649*, ed. Daniela Scaglietti Kelescian (Milan, 1999); Davide Dossi, "Marcantonio Bassetti, Venezia e la pietra di paragone," ARTE Documento, 30 (2014), 96-99.

39 For Brusasorci, see Seifertova, *Painting on Stone*, pp. 78-91.

and copper.[40] As Helen Langdon's essay on Salvator Rosa demonstrates, such a wide range of materials used for painting supports was diffuse among his contemporaries. Turchi was encouraged in this by a sophisticated patron, in his case Cardinal Scipione Borghese who, like the Farnese before him, was fascinated by such technical experimentation.[41] Niccolò Tornioli's drive to innovate was also supported by the patronage of a cardinal, and the materials that he chose were closely tied, as Weston shows, to traditions in both Siena and Rome. Tornioli's spectacular failure in a mosaic commission at St. Peter's brings us back to Rome, whose mosaics, like those of Venice, probably inspired Sebastiano's consideration of the stone supports. Earlier interest in mosaic thus begins Part Three with an essay by Anna Kim, whose insights concerning a different kind of "almost eternal" painting might just as easily have begun this volume. Kim's argument about the micromosaic of Sta. Croce in Gerusalemme and the way in which that medium provides a structural, material parallel for the reliquary it ornaments provides a key to the potential meaning of Sebastiano's invention and the sacred images produced by him and his contemporaries on stone.[42] Weston's essay also reinforces the extent to which these types of material innovations signaled artistic ambition and intellectual discovery, a point that is also beautifully illustrated in this volume's final essay by Helen Langdon on Salvator Rosa's experimentation with different kinds of support. Langdon observes the way in which Rosa's use of stone reflected his interaction with artists painting on stone in both Rome and Florence and postulates that Rosa's unconventional use of a very conventional support, wood, reflects his personal inventiveness and intellectual prowess.

## Early Modern Innovation and Recent Scholarship on Materials

As noted above, the first record of Sebastiano's innovation, the letter written by Soranzo to Bembo in June of 1530, emphasizes the invention and discovery of

40 On *pietra di paragone* see Andrea G. de Marchi, "Tracce per la storia su lavagna e su altre pietre nerastre," in *Daniele da Volterra e la prima pietra del Paragone*, ed. Andrea G. de Marchi (Rome, 2014), pp. 45-49.

41 These works are only now beginning to gain the attention that they probably enjoyed in the Seicento. The Fitzwilliam Museum in Cambridge has recently acquired a work representative of this group of artists who developed the practice in Verona: *The dead Christ supported by the Virgin Mary Magdalene*, which was painted by Bassetti when the artist was in Rome around 1616.

42 Nygren's painstaking analysis in his *Art Bulletin* essay on Titian's *Ecce Homo* (p. 49) also presents the micromosaic as fundamental influence.

painter's new method.[43] Given Sebastiano's reputation during his lifetime as a masterful colorist and portraitist, one might assume that his works would also have been praised for their naturalism; however, both this source and Vasari's *Vita* specifically praise Sebastiano for his discovery of something that would contribute to the physical preservation of the art object, arguably as much for the intellectual, technical achievement as for the purported 'almost eternal' state of the works. Even before Sebastiano's death, contemporary sources including Sebastiano himself indicate that the works on slate were in fact fragile and difficult to move, so the language that centers on notions of the eternal was probably more rhetorical or symbolic than literal.[44] While this idea should certainly be understood in the context of the *paragone* between sculpture and painting and its argument concerning the longevity of the respective arts, the limited way in which the *paragone* has been treated until recently has not allowed for adequate discussion of the practical and theoretical richness and ambition of Sebastiano's innovation.

Focus on technical process and material, as conceived in the work of Alfred Gell, aims to free the work of art from the burden of beauty and meaning,[45] but these burdens were almost certainly tied in Sebastiano's mind to the invention, which signaled mastery on several levels.[46] Anna Kim's essay ties the cultic and technical value of the micromosaic to Gell's theory and the possibility of enchantment and connection to the supernatural and divine. Certain materials' ability to invoke magical transformation, as Nygren's and Landgon's essays suggest, was not limited to religious subject matter. Still, given the frequency with which these works were associated with *acheiropoieta*, miraculous works not made by human hands, such as the Veronica, this way of thinking about material seems particularly rich for the devotional works produced in stone.

---

43 The framing of this technical endeavor as an intellectual achievement is discussed by Calvillo, "Authoritative Copies," pp. 486, 500.

44 See for example the correspondence between Niccolò Sernini and Ferrante Gonzaga's secretary concerning the Úbeda Pietà, published and analysed by Michael Hirst, "Sebastiano's Pietà for the *Comendador Mayor*," 114, no. 834 (1974), 585-93, esp. p. 595; see p. 589 for the letter that discusses the fragility of the slate; and Calvillo's discussion of it, "Authoritative copies," pp. 484-86.

45 See Alfred Gell, "The technology of enchantment and the enchantment of technology," in *The Art of Anthropology: Essays and Diagrams*, ed. Eric Hirsch (London, 1999), pp. 159-86. See Lehmann's useful synopsis of Gell's theory for the art historical theoretical toolbox in "Matter of the Medium," pp. 30-31.

46 Gell's theory has undoubtedly provided a rich foundation for new art scholarship treating materiality, the point here is only to insist on the 16th-century emphasis on artistic agency and intention.

How the stones were used, the extent to which they were visible as supports, and to what degree they were materially valuable are also significant.

The earliest textual sources for Sebastiano's practice record that he used marble and other stones for his paintings' supports and ornamental frames.[47] Extant physical evidence, as noted by Baker-Bates, testifies to Sebastiano's use of peperino and slate, to Titian's use of slate and marble, and several contemporary and early followers' use of both stones.[48] In these first cases, if the stone is not hidden, it is also not foregrounded as a contributor to the paintings' composition, in contrast to the later use of certain stones' figuration. González Mozo's study demonstrates the extent to which the material body of the stone, as opposed to only the surface appearance, contributed to the both the theory and practice of the earliest practitioners of this technique. She suggests the way in which Sebastiano's figures, and to a certain degree Titian's figure of *Ecce Homo,* seem to emerge from the stone, *seem* to be built from and depend upon the material of the stone.[49] She contrasts this to later figures on figurated stones that appear almost to float above the stone surface or exist as overlays. Nygren's discussion of the similitude between the material and the subject represented on the material support, Anna Kim's observation of a similar convergence of material and subject in the micromosaic of Santa Croce in Gerusalemme, and Calvillo's examination of the interaction of the support, metaphor and portrait all propose the stone's ability to produce meaning through its material identity, through something akin to consubstantiality and transubstantiation.

Such a use of material both asserts artistic ambition on the part of the original early modern practitioners and inventors, or re-inventors, and encourages a contemporary approach to and interpretation of materials that has been explained by the editors and contributors of the recent volume *The Matter*

---

47 See Soranzo's comments to Bembo cited above and in note 2, which describe the stone as marble. Vasari's *Vita* is vague, but his section on painting techniques cites a number of stones, Sebastiano's inventory is another important source, as is the inventory of the Farnese familiare Fulvio Orsini, noted in Baker-Bates'and Calvillo's essays.

48 We might think of the works by, or designed by, Daniele da Volterra cited above in San Pietro in Montorio (marble) and the double-sided panel (slate) for Giovanni della Casa, or use of slate for the *Ecce Homo* and marble for the *Mater Dolorosa* for Charles v and now in the Prado.

49 This relationship between the body of stone and the development of figures seemingly from it is also treated by Nygren's discussion of Titian's painting in his "Titian's *Ecce Homo* on Slate," p. 49.

*of Art*, both in the introduction to the volume and in its individual essays.[50] These scholars, and Christopher Nygren in this volume, provide rich examples of the way in which Bruno Latour's theory of materiality and Actor-Network Theory might help us to discern the agency of materials, as makers of meaning in works of art, mediators as opposed to intermediaries. How painting's medium or support, whether stone, copper or wood, embodies or signals artistic innovation and meaning particular to the subject represented is the central concern of several essays in this volume. Other essays, such as those by Baadj, Lohff, and Wegmann, examine the extent to which these innovations reflect and shed light on patterns in patronage, material value, and collecting; as curiosities and novelties that engaged early modern discourse concerning the interaction between *ars* and *natura*, the natural sciences and the immense influx of new materials arriving in Europe from global exploration. The following chapters also address innovation beyond the use of stone supports including Sebastiano's interest in perfecting oil murals, as Baker-Bates discusses, the status of mosaics and the emergence of metal supports such as copper and, in the case of Langdon's study, a new way and inventive approach to wood as a support, one in which the artist's personal experience, knowledge, and relationships informed his participation in the larger field of material experimentation.

## The Ancient and Early Modern Reception and Theory

If Sebastiano del Piombo is the primary figure for any history of early modern painting on stone, Pliny the Elder is a close second, as an important source for the theory and reception of stone's use both in antiquity and in the period discussed in this volume. Beyond his criticism of marble as a material embodying ancient *luxuria*, Pliny also addresses the dynamic between *ars* and *natura* that is essential to any discussion of the late 16th-century use of semi-precious figurated stones such as lapis lazuli, alabaster, and agate.[51] Semi-precious stones

50 See the introduction by Anderson, Dunlop and Smith to *The Matter of Art*, pp. 1-17, in particular pp. 3-4; especially useful is the theoretical 'toolbox' proposed by Ann-Sophie Lehmann in "The matter of medium: some tools for an art-theoretical interpretation of materials," pp. 21-40, esp. pp. 26-30 and 33-34 for Latour's theories.

51 Pliny invoked the theme of luxuria both when the author indicts the practice of overlaying marble with color at the beginning of Book 35, as noted in the beginning of this introduction, and following in the introduction to Book 36 the quarrying of mountains on a whim, mountains that at one time it was remarkable "even to have succeeded in crossing them." See Pliny, *Natural History* (35.1.2-3), 1952, pp. 260-63; 1962 (36.1.1), pp. 2-3: "caedimus

and gemstones also represent for Pliny "nature's grandeur ... gathered together within the narrowest limits,"[52] offering a kind of beauty and variety that made artistic invention unnecessary, or excessive.[53] What the material meant for the work of art according to Pliny ranged from signaling excess to the parallel methods and works of artists and nature.

The 16th- and 17th-century artist using semi-precious and figurated stones would probably have been aware of Pliny's disdain for luxury materials, since the ancient author begins Book 36 with a very general criticism of the use of stones, which he considered the "the prime folly in our behavior," even though this book did not even address "gems, amber and vessels of rock-crystal and fluorspar."[54] His criticism of material excess and artistic rivalry with nature, however, seems to have provided more inspiration than discouragement for early modern artists, though they might have felt the need to address his comments. Calvillo observes that Francisco de Holanda's citation of Pliny distinguishes carefully between the ancient practice and critique of painting on stone and the reception of his own day, which he implies is far more positive.[55] Nygren and Lohff note, in different ways, the extent to which artists understood and exploited the collaborative and rich potential of stones' natural figures and patterns.

Certainly the conjunction between artistic invention and natural wonder to be found in the works of artists like Antonio Tempesta, Hans van Aachen, and Jacques Stella appealed to the late 16th-century patrons collecting for their *Kunst-* and *Wunderkammern*. The intersection of material and aesthetic value, however, was present from time of Sebastiano's re-invention, which would

hos trahimusque nulla alia quam deliciarum causa, quos transcendisse quoque mirum fuit." For the theme of *luxuria* and collecting in Pliny, also see Chapter Four, "The Problem of Totality: Collecting Greek Art, Wonders, and Luxury," of Sorcha Carey's *Pliny's Catalogue of Culture*, pp. 75-101. As noted above, Carey also addresses Pliny's criticism of painting on marble in Book 35 in her next chapter, "The Artifice of Nature," pp. 102-37; see pp. 107-8.

52 Pliny, 1962, 37.1.1, pp. 164-65.

53 Ibid. In this passage, Pliny writes that according to some the natural beauty of certain stones is such that "they think it a crime to tamper with (them)...by engraving them as signets ... Hence very many people find that single gemstone alone is enough to provide them with a supreme and perfect aesthetic experience of the wonders of Nature."

54 Ibid. We thank Peter Fane-Saunders for noting that earlier artists might not have been as awareness of Pliny's critique of luxury materials because of the use of epitomes that contained passages focused on wonders.

55 See Calvillo in this volume, pp. 113-14; and Francisco de Hollanda, *On Antique Painting*, trans. Alice Sedgwick Wohl (University Park, 2013), p. 229.

have been understood as providing the naturalism praised by Pliny in modest materials; that is to say that the worth of a work was determined by the artist's skill rather than by the value of its material.[56] That naturalistic representation—in addition to the conceptual invention that signaled such notions of eternity, antiquity, and material transubstantiation—would have been at least part of the appeal of Sebastiano's works is suggested by a text written a few years before Soranzo's letter to Bembo. Written by Paolo Giovio after he had escaped Rome and taken refuge on the island of Ischia at the invitation of Vittoria Colonna, the dialogue *Notable Men and Women of our Time* contains a passage that sheds light on the reception of mosaic in the early Cinquecento and how it might have been compared to contemporary painting:

> And by God, it's indisputably true that what our friend Ghiberti used to say about Vittoria is not only clever but also true. Since he honors her most reverently and esteems her greatly, he would say this about her in comparison to other women: that it was the lot of all besides her alone to be like the images of a mosaic composed of small pieces of stone. To those looking from a distance, the figures and human faces of this work of art seem admirable and beautiful; but if you approach the work more closely and examine it more minutely and with focused attention, immediately seams appear to view, and both the enormous gaps and the crowded awkwardness, the almost haphazard workmanship of the crudely inserted pebbles—all discerned clearly by the eyes—shatter one's entire opinion of the work.[57]

Sebastiano's portraits on stone combined the durability (at least rhetorically if not practically) and the antiquity of mosaic with the powerful illusionism of the *maniera moderna* and the naturalism valued in Pliny's discussion of portraits.

56 Pliny's ambivalence extends to man's inventiveness in general, which he sees as inflating the value of objects, 1952, 33.2.4, pp. 4-5: "Alas for the prodigality of our inventiveness! In how many ways have we raised the prices of objects! The art of painting has come in addition, and we have made gold and silver dearer by means of engraving! Man has learnt to challenge nature in competition!" Even painting, seemingly preferred by Pliny, is not excused from this drive.

57 Paolo Giovio, *Notable Men and Women of our Time*, ed. and trans. Kenneth Gouwens, The I Tatti Library, vol. 56 (Cambridge, 2013), p. 505. For an example of contemporary portraiture in mosaic, see that of Cardinal Pietro Bembo by Valerio Zuccato from 1542 in the Museo Nazionale del Bargello, inv. 873.

In this sense, the early modern painter's invention offers a corrective to different antique media rather than a simple reinvention of a method noted by the ancient author. The visual integrity of his works, like the figure of Vittoria Colonna, held up to close inspection, inviting the kind of examination and intimate engagement that develop for devotional works, portraits, and, later, collectors' pieces. Where one might admire the extent to which Sebastiano's and Titian's figure seem to emerge from their dark grounds, others might marvel at the way in which Tempesta's stones provided the rough waves of the red sea or the Cavaliere's lapis produced the seascape of beyond Andromeda. In either case, close inspection and contemplation of the subject's relationship to the material supporting its figures rewarded the viewer, in the both chapel and the *Kunstkammer*.

## In Conclusion

Ancient and early modern theory provides much in way of explaining the interest that stone held for Sebastiano del Piombo and the artists and collectors who developed and prized his initial invention. While neither has been neglected in past studies of painting on stone, more recent theory related to the 'Material Turn' is now expanding and enriching the scholarship on this subject, yielding—in conjunction with the close historical analysis—rich analyses of the potential of materials to act as agents of meaning. The essays in this volume touch on some the key works and moments in the history of early modern material innovation, but there is ample room for further investigation, both in terms of other works on stone and other materials that appeared in both secular and religious spaces in the 16th and 17th centuries. What is written here intends to provide a brief introduction to the history, theory and historiography of the topic; in what follows the ten authors will explore different aspects of the practice in order to share the fruit of new approaches and insights. In this way, the enduring quality of paintings on stone once again extends beyond the physical to the conceptual.

## Bibliography

Baker-Bates, Piers. "Sebastiano del Piombo. The Technique of Genius." In *Michelangelo/ Sebastiano A Meeting of Minds*, edited Matthias Wivel, pp. 74-85. London, 2017.

Baker-Bates, Piers. "The 'Cloister Life' of the Emperor Charles V: Art and Ideology at Yuste." *Hispanic Research Journal* 14, no. 5 (2013), 427-45.

Baker-Bates, Piers. *Sebastiano del Piombo and the World of Spanish Rome*. New York and London, 2016.

Barbieri, Costanza. "Sebastiano's Portraits in *Paragone*: Sculpted Paintings, Praise through Images." In *Sebastiano del Piombo, 1485-1547*, edited by Carlo Strinati, pp. 53-57. Rome, 2008.

Barbieri, Costanza. "'Tu, che lo stile con mirabil cura pareggi col martello.' Fortune e sfortune di Sebastiano." In *La* Pietà *di Sebastiano a Viterbo. Storia e tecniche a confronti*, edited by Costanza Barbieri, Enrico Parlato, and Simona Rinaldi, pp. 50-65. Rome, 2009.

Barry, Fabio. "'Painting in Stone": Early Modern Experimentation in a Metamedium,' The Art Bulletin, 99, no. 3 (2017): 30-61.

Bigari, Cristiana. "Note sul restauro dell'Assunzione della Vergine in San Silvestro al Quirinale." In *Scipione Pulzone e il suo tempo*, edited by Alessandro Zuccari, pp. 233-47. Rome, 2015.

*Il Carteggio di Michelangelo*, edited by Paola Barocchi and Renzo Ristori, 5 vols. Florence,1965-83.

Calvillo, Elena. "Authoritative Copies and Divine Originals: Lucretian Metaphor, Painting on Stone, and the Problem of Originality in Michelangelo's Rome," *Renaissance Quarterly*, 66, no. 2 (2013), 453-508.

Carey, Sorcha. *Pliny's Catalogue of Culture: Art and Empire in the* Natural History. Oxford, 2003.

Casaburo, Mario. *Pittura su Pietra: Diffusione, Studie Materiali, ed Tecniche Artistiche.* Florence, 2017.

Casaburo, Mario. "Le Grandi Pale su Ardesia Romane tra Cinquecento e Seicento: Aspetti Storici di Esecuzione, Conservazione e Restauro," *Per La Conoscenza dei Beni Culturali*, IV (2012), 105-19.

Casaburo, Mario. "La Pittura su Pietra in Italia. Primi Scenari di Ricerca: Appunti sulle Tecniche, i Materiali e le Opere (1530-1630)," *Per la Conoscenza dei Beni Culturali*, III (2010), 239-57.

Cerasuolo, Angela. *Diligenza e Prestezza: La tecnica nella pittura e nella letteratura artistica del Cinquecento*. Florence, 2014.

Collomb, Anne-Laure. "Les traits artistiques et la peinture sur pierre, XVIe et XVIIe siècle," *Histoire de l'art* 52 (2003), 111-20.

Collomb, Anne-Laure. *Splendours de l'Italie: la Peinture sur Pierre à la Renaissance.* Rennes, 2012.

*Daniele da Volterra e la prima pietra del Paragone*, ed. Andrea G. de Marchi. Rome, 2014.

Delle Lettere da diversi Re et Principi et Cardinali et altri huomini dotti a Mons. *Pietro Bembo scritte* (Venice, 1560).

Dossi, Davide. "Marcantonio Bassetti, Venezia e la pietra di paragone." ARTE Documento, 30 (2014), 96-99.

Fernetti, Fabio. "Verso un Censimento dei Dipinti Cinquecenteschi su Pietra o Lavagna." In *Sebastiano del Piombo e la Capella Borgherini nel contesto Della pittura Rinascimentale*, edited Santiago Arroyo Esteban, Bruno Marrocchini, and Claudio Seccaroni, pp. 54-57. Rome, 2010.

Gell, Alfred. "The technology of enchantment and the enchantment of technology." In *The Art of Anthropology: Essays and Diagrams*, edited by Eric Hirsch, pp. 159-86. London, 1999.

Giovio, Paolo. *Notable Men and Women of our Time*, edited and translated by Kenneth Gouwens. Cambridge, 2013.

Hall, Marcia. *Color and Meaning*. Cambridge, 1992.

Hall, Marcia. *The Sacred Image in the Age of Art, Titian, Tintoretto, Barocci, El Greco, Caravaggio*. New Haven and London, 2011.

Hansen, Morten Steen. *In Michelangelo's Mirror Perino del Vaga, Daniele da Volterra, Pellegrino Tibaldi*. University Park, 2013.

Hendler, Sefy. *La Guerre des arts Le Paragone Peinture-Sculpture en Italie* XV-XVII *siècle*. Rome, 2013.

Hessler, Christiane. *Zum Paragone Malerei, Skulptur und Dichtung in der Rangstreitkultur des Quattrocento*. Berlin, 2014.

Hirst, Michael. *Sebastiano del Piombo*. Oxford, 1981.

Hirst, Michael. "Sebastiano's Pietà for the *Comendador Mayor*," 114, no. 834 (1974), 585-93.

Hollanda, Francisco de. *On Antique Painting*, translated by Alice Sedgwick Wohl. University Park, 2013.

Jones, Nathaniel. "Ancient Painted Panels: Terminology and Appearance." *Mnemosyne* 67 (2014), 295-304.

Kaufmann, Thomas DaCosta. *Towards a Geography of Art*. Chicago and London, 2004.

Kim, David Young. *The Traveling Artist in the Italian Renaissance Geography, Mobility and Style*. New Haven and London, 2014.

Lehmann, Ann-Sophie. "The matter of medium: some tools for an art-theoretical interpretation of materials." In *The Matter of* Art, pp. 21-40.

Lohff, Johanna Beate. *Malerei auf Stein Antonio Tempestas Bilder auf Stein im Kontext der Kunst- und Naturtheorie seiner Zeit*. Munich, 2015.

Marchi, Andrea G. de "Tracce per la storia su lavagna e su altre pietre nerastre." In *Daniele da Volterra e la prima pietra del Paragone*, edited by Andrea G. de Marchi, pp. 45-49. Rome, 2014.

*The Matter of Art: Materials, Practices, Cultural Logics, c. 1250-1750*, edited by Christy Anderson, Anne Dunlop, and Pamela H. Smith. Manchester, 2015.

Mendelsohn, Leatrice. *Paragoni: Benedetto Varchi's* Due Lezzioni *and Cinquecento Art* Theory. Ann Arbor, 1982.

*Delle Poesie Volgari e Latine di F.M. Molza*, ed. Pierantonio Serassi, 3 vols. Bergamo, 1747-54.

Nagel, Alexander and Wood, Christopher. *Anachronic Renaissance*. New York, 2010.

Nygren, Christopher J. "Titian's *Ecce Homo* on Slate: Stone, Oil and the Transubstantiation of Paint," *Art Bulletin* 99, 1 (2017), 36-66.

Nygren, Christopher J. "Vibrant Icons: Titian's Art and the Tradition of Christian Image-making." PhD diss., Johns Hopkins University, 2011.

*Pietra Dipinta: Tesori nascosti del '500 and '600 da una collezione*, Palazzo Reale, Milan, exhibition catalogue, edited by Marco Bona Castellotti. Milan, 2000.

Pinelli, Antonio. "La Cappella delle tombe scambiate Novità sulla Cappella Chigi in Santa Maria del Popolo." In *Francesco Salviati et la Bella Maniera. Actes des colloques de Rome et de Paris (1998)*, edited by Catherine Monbeig Goguel, Phillipe Costamagna, and Michel Hochmann, pp. 253-85. Rome, 2001.

*Pittura su Pietra*, Palazzo Pitti, exhibition catalogue, edited by Marco Chiarini. Florence, 1970.

Plinio Secondo, Gaio. *Storia Naturale v Mineralogia e Storia dell'Arte Libri 33-37*, edited and translated by Antonio Corso, Rossana Mugellesi, and Gianpiero Rosati. Turin, 1988.

Pliny the Elder. *Natural History*. Vol. 9, *Books 33-35*. Edited by G.P. Goold and translated by H. Rackham. London and Cambridge, MA, 1952; reprint edition, Cambridge, MA, 1999.

Preimesberger, Rudolph. *Paragons and Paragone: Van Eyck, Raphael, Michelangelo, Caravaggio, Bernini*, translated by Sabine Eiche and Fiona Elliott. Los Angeles, 2011.

Pugliatti, Teresa. *Giulio Mazzoni e la decorazione a Roma nella cerchia di Daniele da Volterra*. Rome, 1984.

Robertson, Clare. *Il Gran Cardinale Alessandro Farnese Patron of Arts*. New Haven and London, 1992.

Rubin, Patricia. "The Private Chapel of Cardinal Alessandro Farnese in the Cancelleria, Rome," *Journal of the Courtauld and Warburg Institutes* 54 (1987), 82-112.

*Sebastiano del Piombo e la Cappella Borgherini nel contesto della pittura rinascimentale*, edited by Santiago Arroyo Esteban, Bruno Marocchini, Claudio Seccaroni. Florence, 2010.

*Sebastiano del Piombo e la pittura su pietra: il ritratto di Baccio Valori*, ed. Alessandro Cecchi, Marco Ciatti, and Oriana Sartiani. Florence, 2014.

*Sebastiano del Piombo y España*, ed. Manuela Mena Marqués. Madrid, 1995.

Seifertova, Hana. *Malby na kameni: Uměcký experiment v 16. Na začátku 17 století/Painting on Stone: an Artistic Experiment in the 16th and Early 17th centuries*. Prague, 2007.

Thomas, Ben. "'The lantern of painting': Michelangelo, Daniele da Volterra and the *paragone*," *Apollo*, 154, no. 474 (2001), 46-53.

Treves, Letizia. "Daniele da Volterra and Michelangelo: a collaborative relationship," *Apollo*, 154, no. 474 (2001), 36-45.

*Alessandro Turchi, detto l'Orbetto 1578-1649*, ed. Daniela Scaglietti Kelescian. Milan, 1999.

Varchi, Benedetto. "Lezzione nella quale si disputa della maggioranza delle arti e qual sia più nobile, la scultura o la pittura." In *Trattati d'Arte del Cinquecento*, edited by Paola Barocchi, 3 vols. Bari, 1960-62.

Vasari, Giorgio. *Le Vite de'più eccellenti pittori scultori e architettori nelle redazioni del 1550 e 1568*, edited by Rosanna Bettarini and Paola Barocchi, 9 vols. Florence, 1966-87.

# PLATES

*1-15*

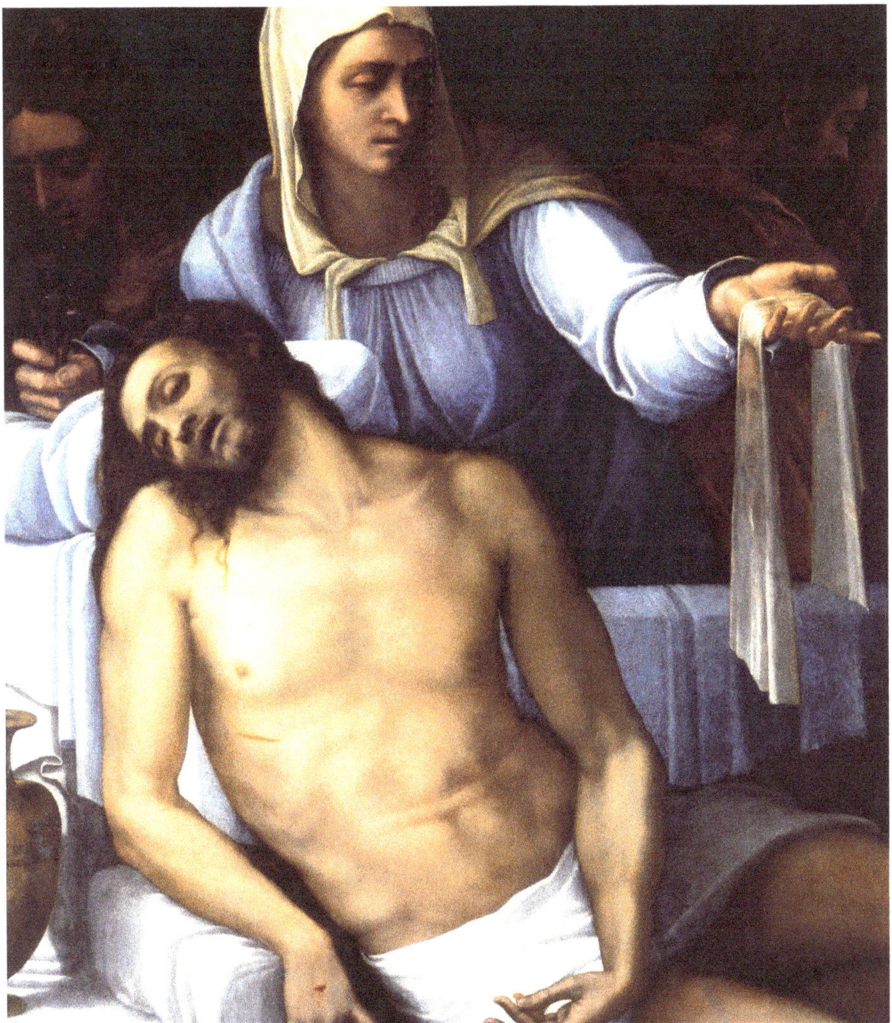

PLATE 1 *Sebastiano del Piombo,* Pietà, *oil on slate, 1533-1539, Museo del Prado, Madrid, on loan to the Prado from the Fundacion de Casa Ducal de Medinaceli* (PHOTO: © MUSEO DEL PRADO, MADRID).

 | DOI 10.1163/9789004361492_003

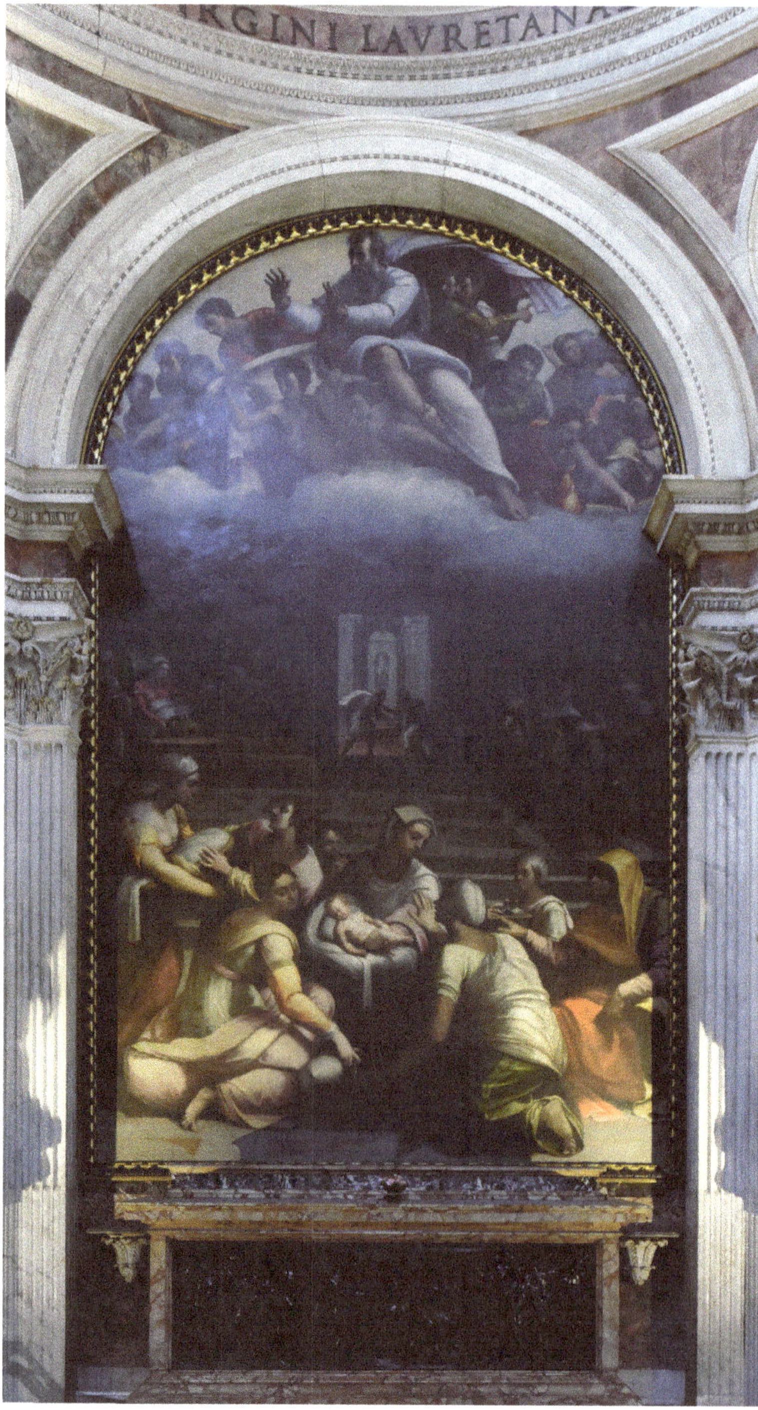


PLATE 2 *Sebastiano del Piombo,* Nativity of the Virgin, *oil on peperino, 1530-, Rome, Santa Maria del Popolo* (PHOTO: DAVID MARSHALL).

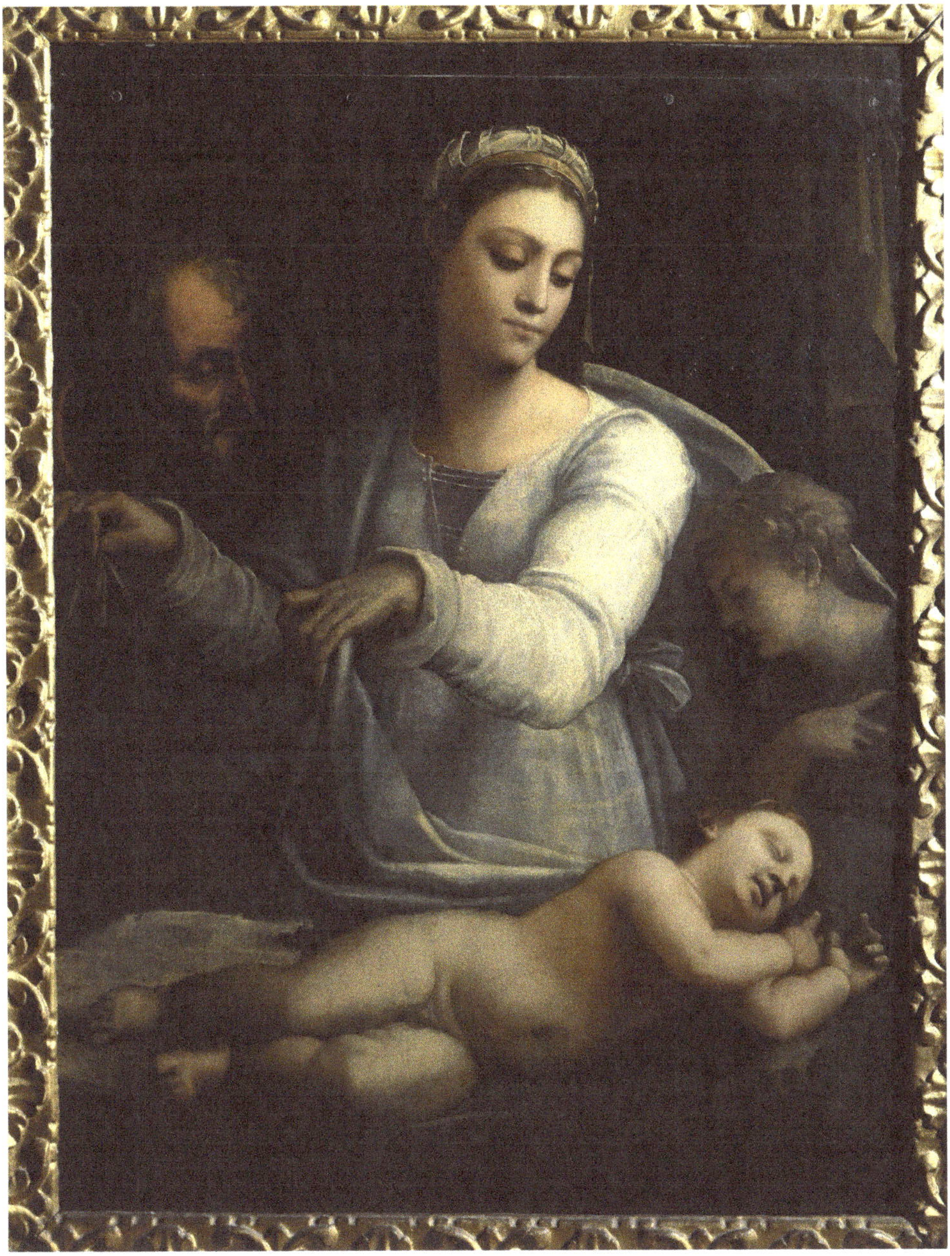

PLATE 3 *Sebastiano del Piombo,* Madonna del Velo, *oil on slate, ca. 1535, Naples, Museo di Capodimonte* (PHOTO: ©SCALA, FLORENCE, 2015).

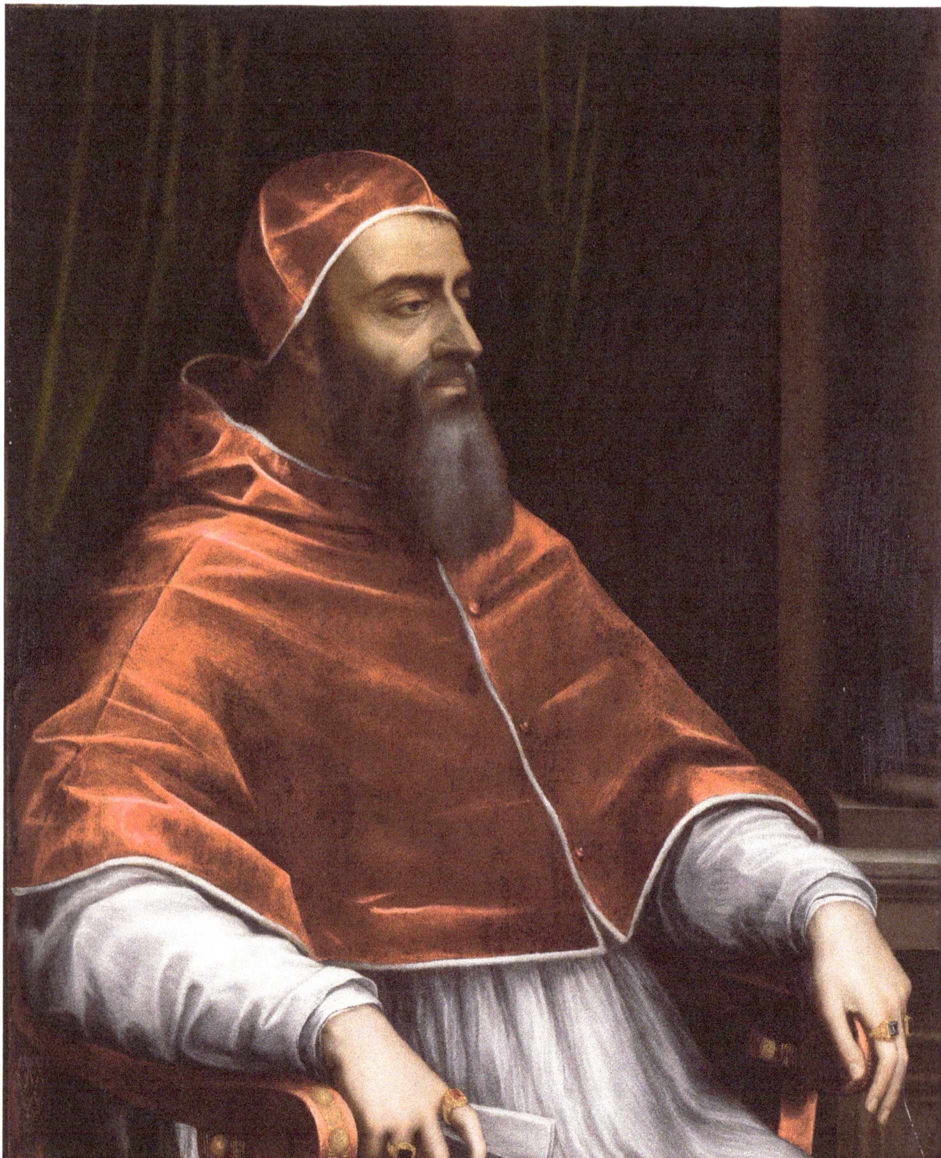

PLATE 4 *Sebastiano del Piombo,* Clement VII, *oil on slate, ca. 1531, Los Angeles, Getty Museum* (PHOTO: DIGITAL IMAGE COURTESY OF THE GETTY'S OPEN CONTENT PROGRAMME).

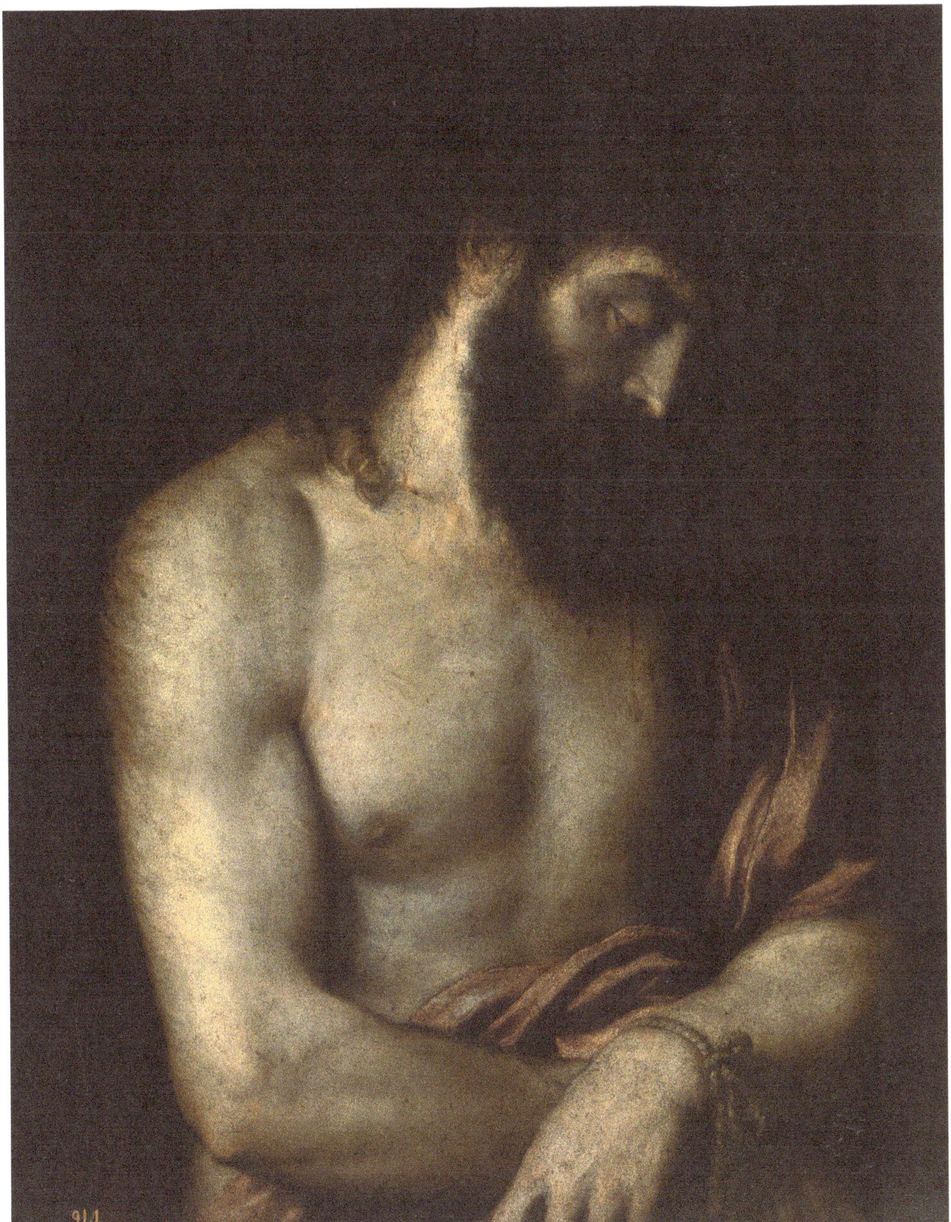

PLATE 5 *Titian,* Ecce Homo, *oil on slate, 1547, Madrid. Museo Nacional del Prado* (PHOTO: © MUSEO DEL PRADO, MADRID).

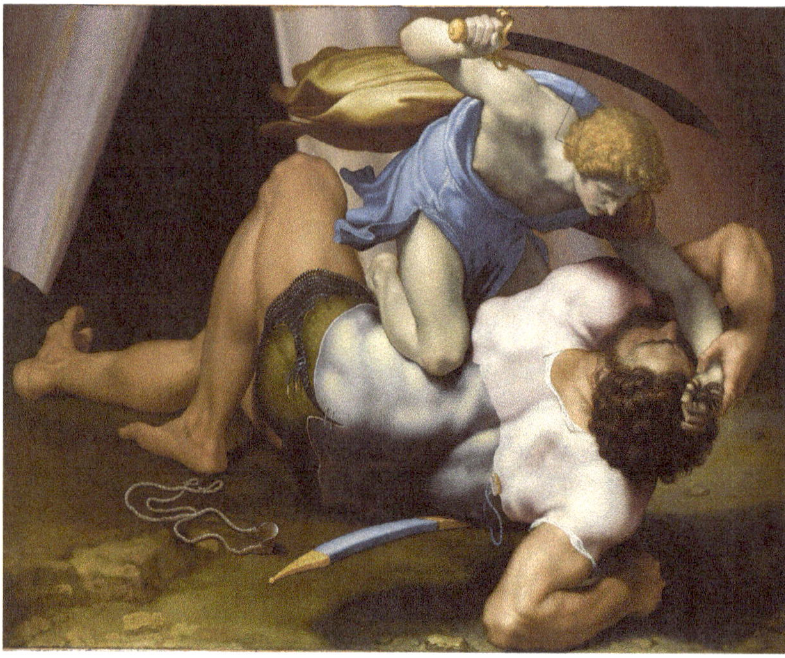

6a

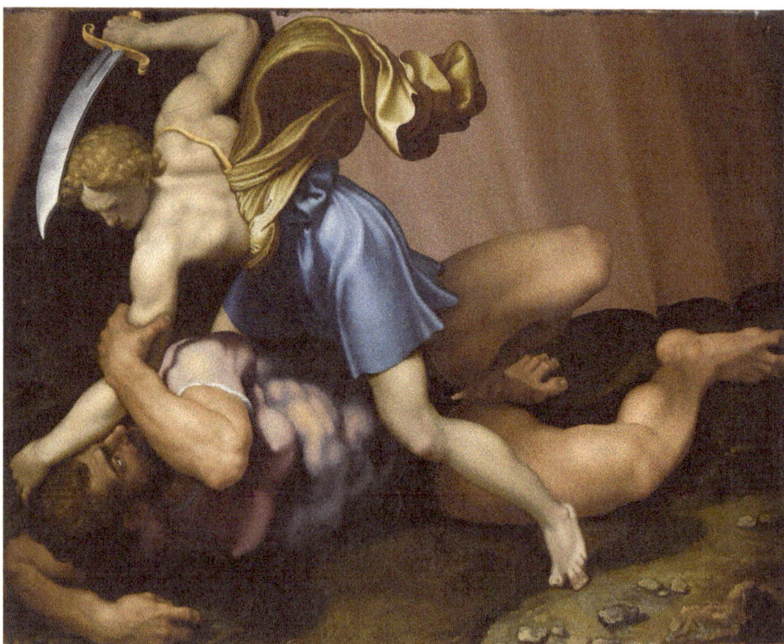

6b

PLATE 6 *Daniele da Volterra,* David and Goliath, *oil on slate, 1550-1555, Paris, Musée du Louvre* (PHOTO: © PARIS, MUSÉE DU LOUVRE)

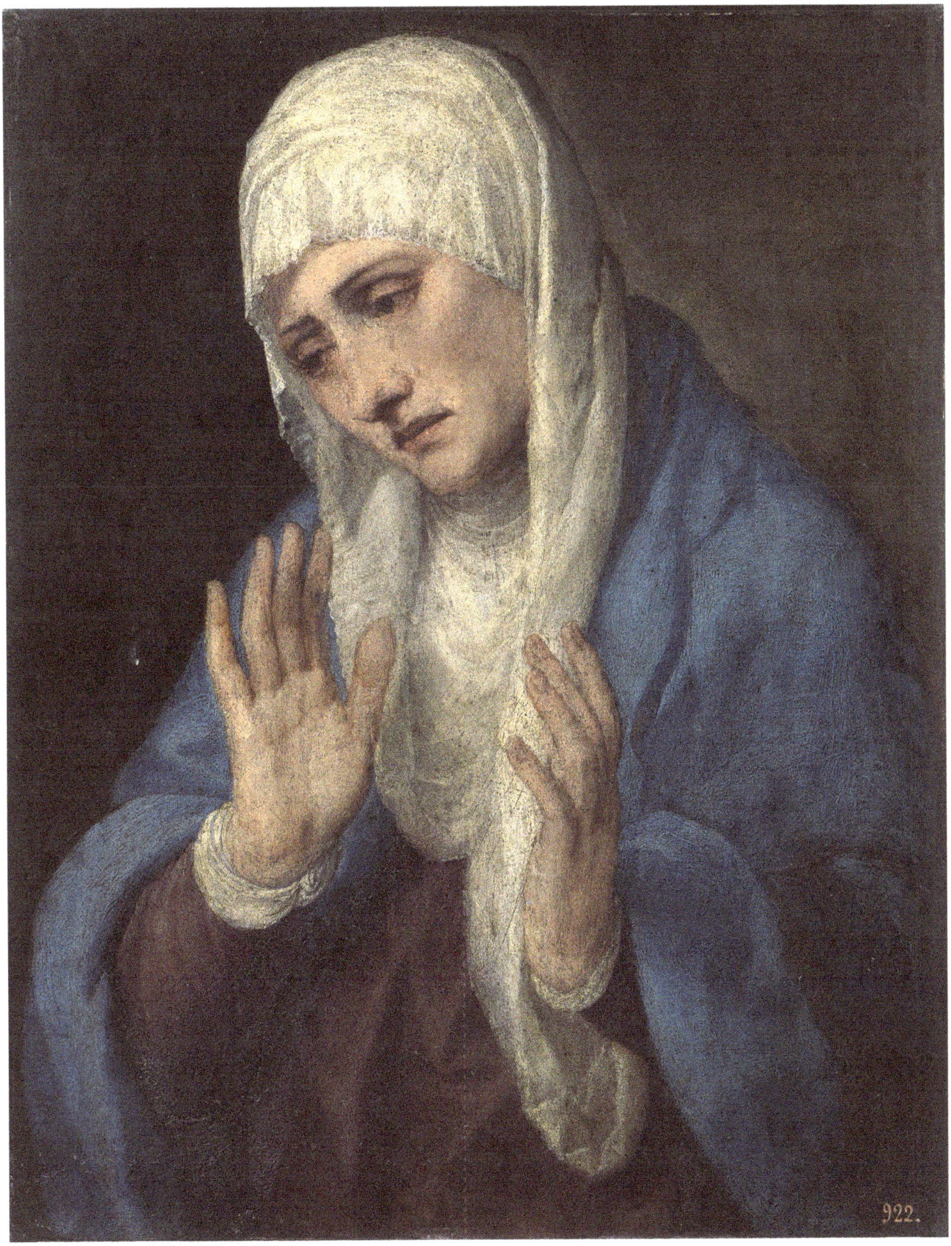


PLATE 7 *Titian,* Our Lady of Sorrows, *oil on marble, 1555, Madrid, Museo Nacional del Prado* (PHOTO: © MUSEO DEL PRADO, MADRID).

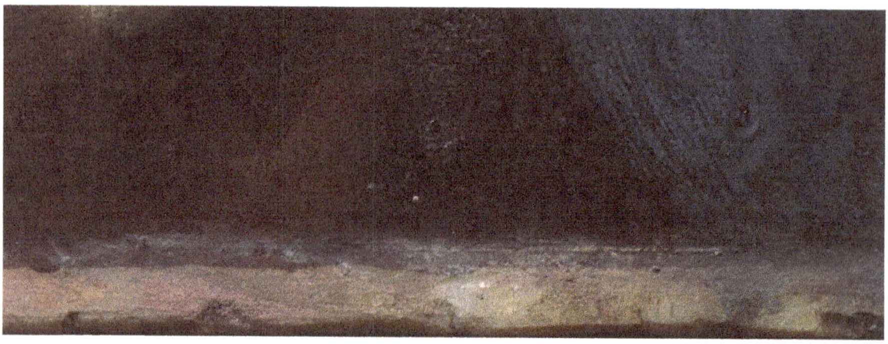

PLATE 8 *Titian,* Our lady of Sorrows, *oil on marble, 1555, Madrid, Museo Nacional del Prado. Detail of the border* (PHOTO: © MUSEO DEL PRADO, MADRID).

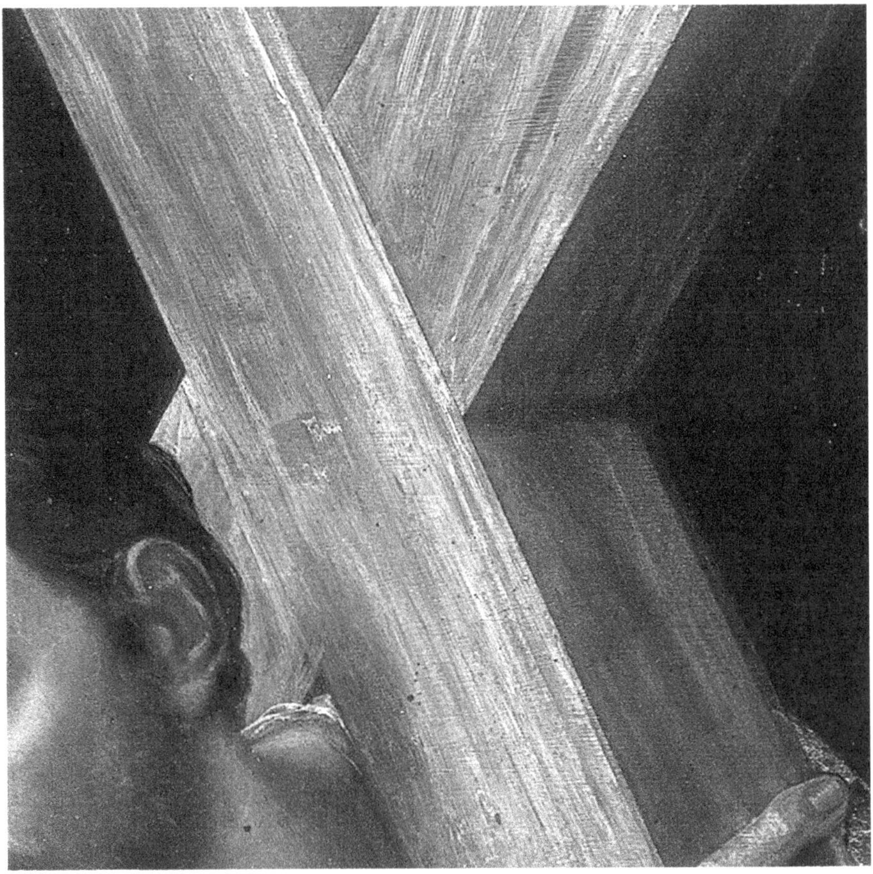

PLATE 9 *Sebastiano del Piombo,* Christ carrying the Cross, *oil on slate, 1532-35, Madrid, Museo Nacional del Prado. Infrared Reflectogram. Detail* (PHOTO: © MUSEO DEL PRADO, MADRID).

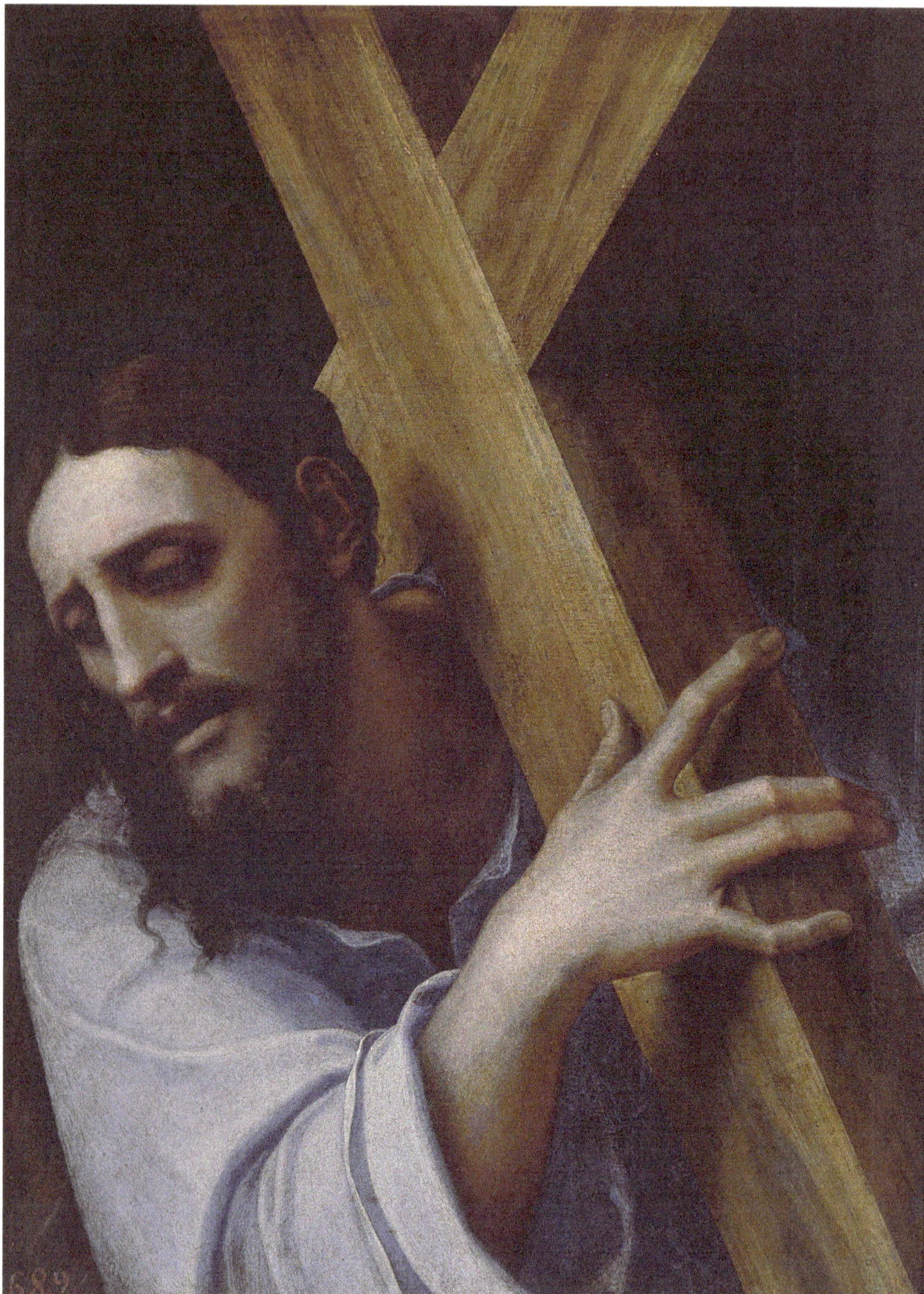

PLATE 10 *Sebastiano del Piombo,* Christ carrying the Cross, *oil on slate, 1532-35 Madrid, Museo Nacional del Prado. Detail* (PHOTO: © MUSEO DEL PRADO, MADRID).

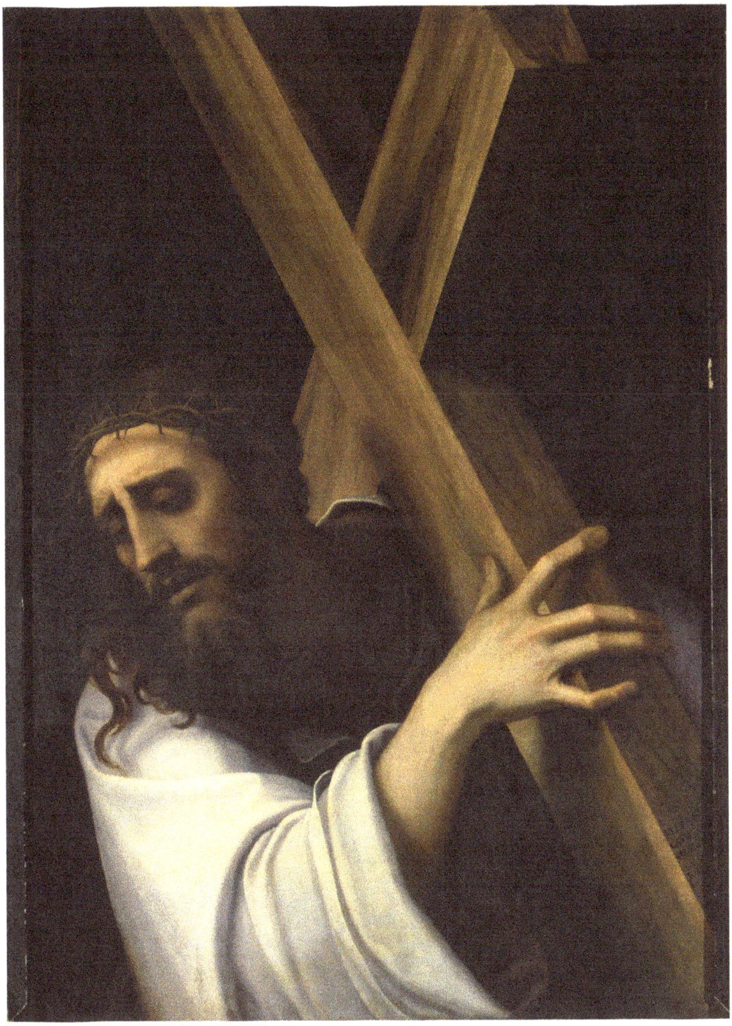

PLATE 11 *Sebastiano del Piombo,* Christ carrying the Cross, *oil on slate, 1537, The State Hermitage Museum, St Petersburg* (PHOTO: ©THE STATE HERMITAGE MUSEUM. PHOTO BY LEONARD KHEIFETS).

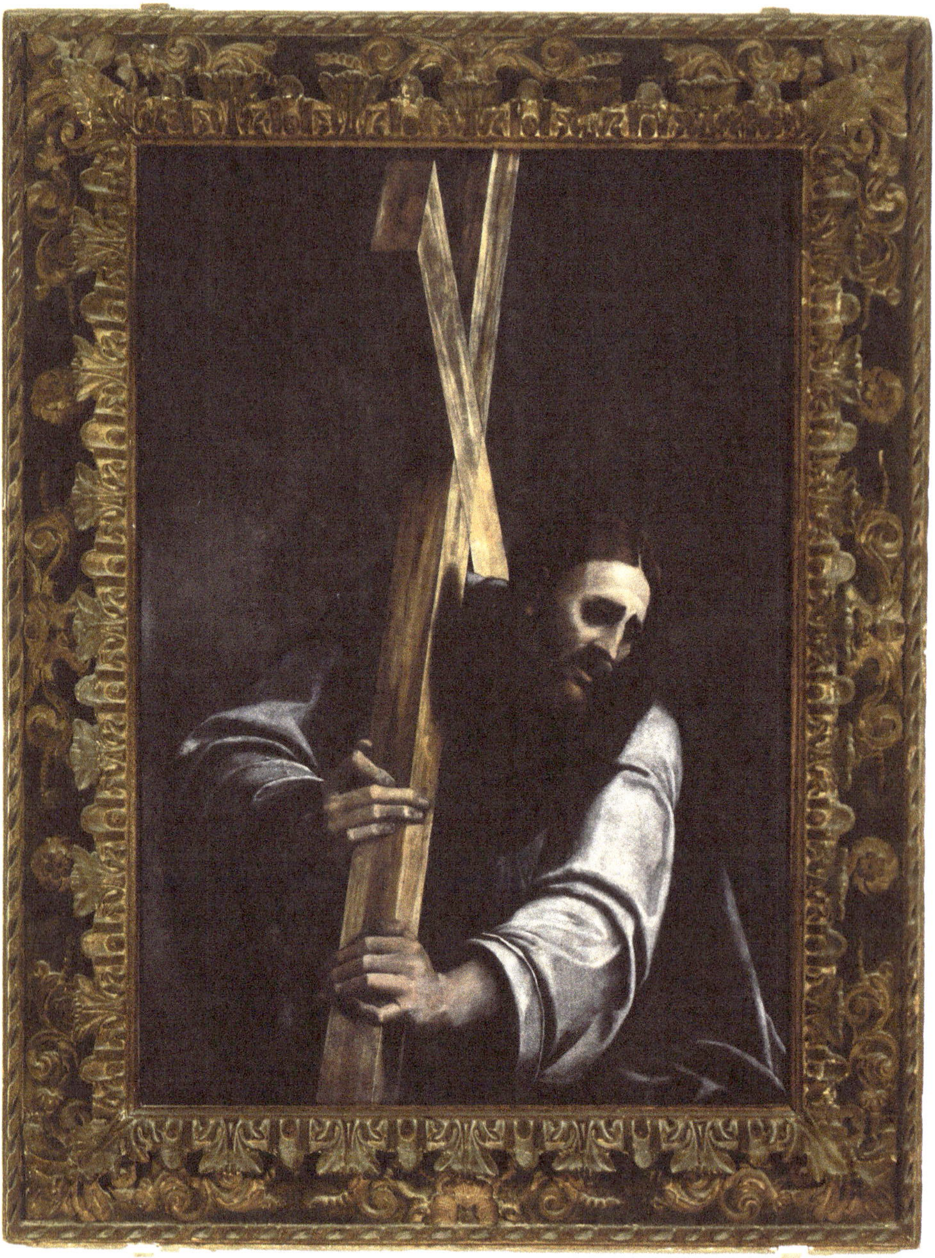

PLATE 12 *Sebastiano del Piombo,* Christ carrying the Cross, *oil on slate, ca. 1535-1540, Budapest, Szépmüvészeti Museum* (PHOTO: © SZÉPMÜVÉSZETI MUSEUM, BUDAPEST).

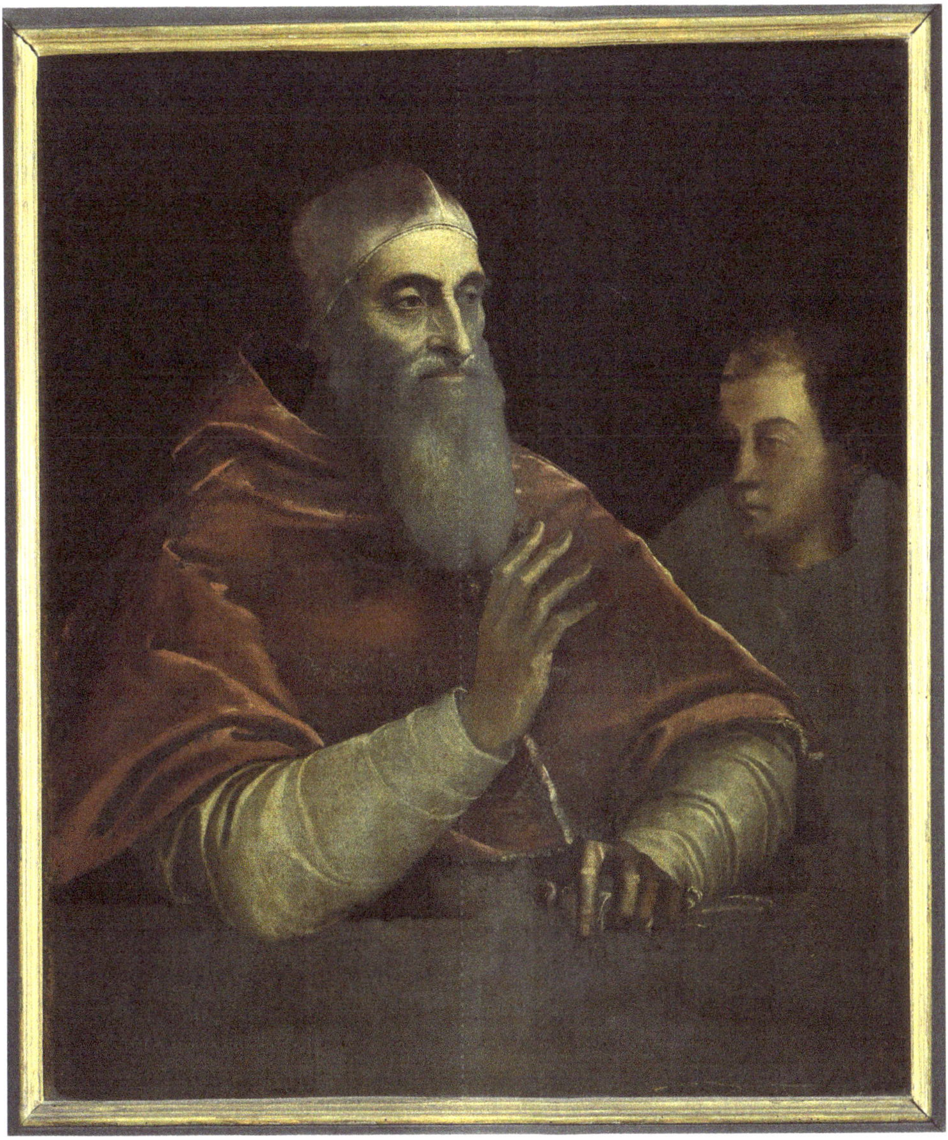

PLATE 13 *Sebastiano del Piombo,* Paul III and a Nephew, *oil on slate, ca. 1535, Parma, Galleria Nazionale* (PHOTO: ©SCALA, FLORENCE, 2015).

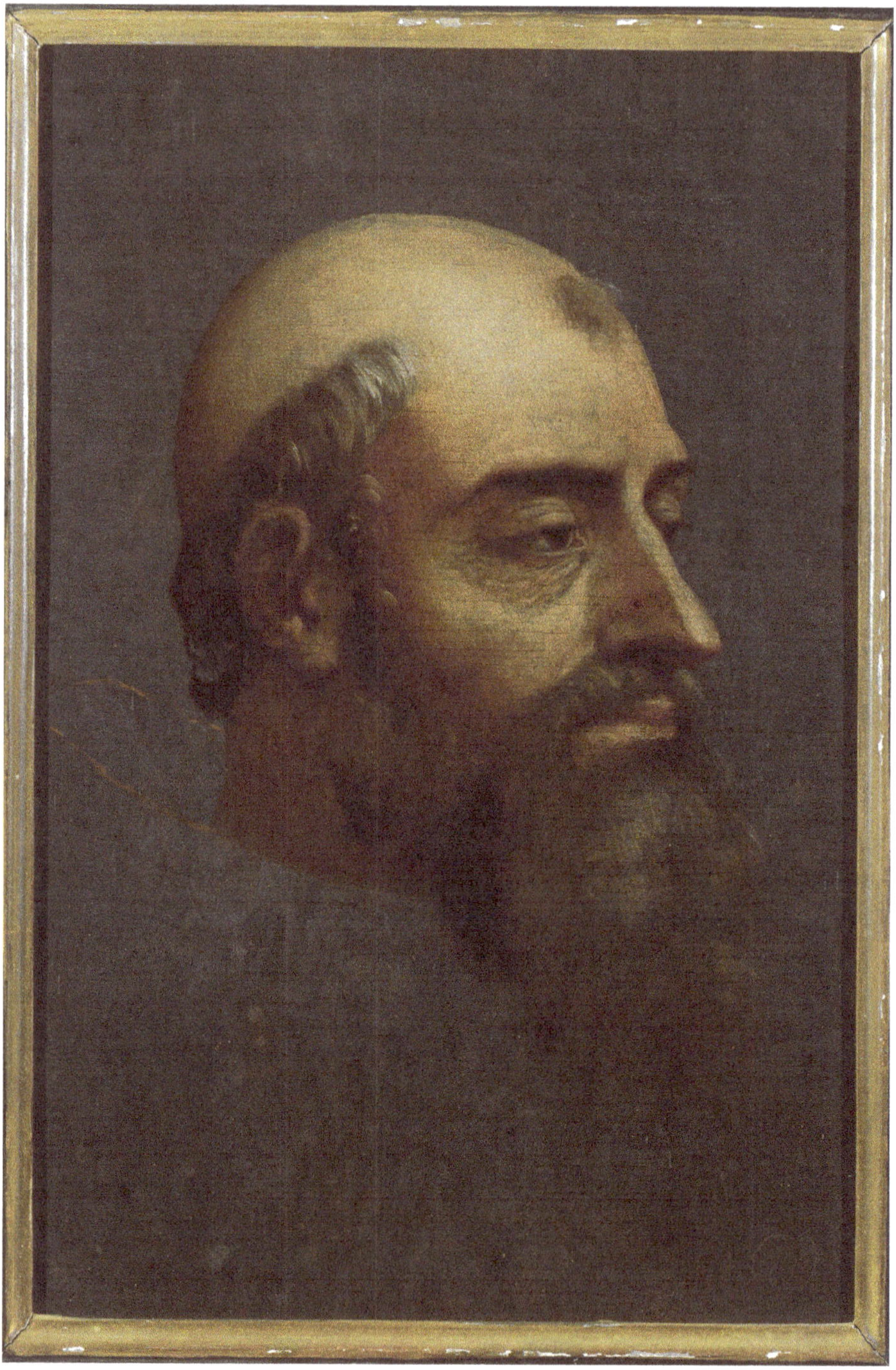

PLATE 14 *Sebastiano del Piombo,* Clement VII, *oil on slate, ca. 1531, Naples, Museo di Capodimonte* (PHOTO: ©SCALA, FLORENCE, 2015).

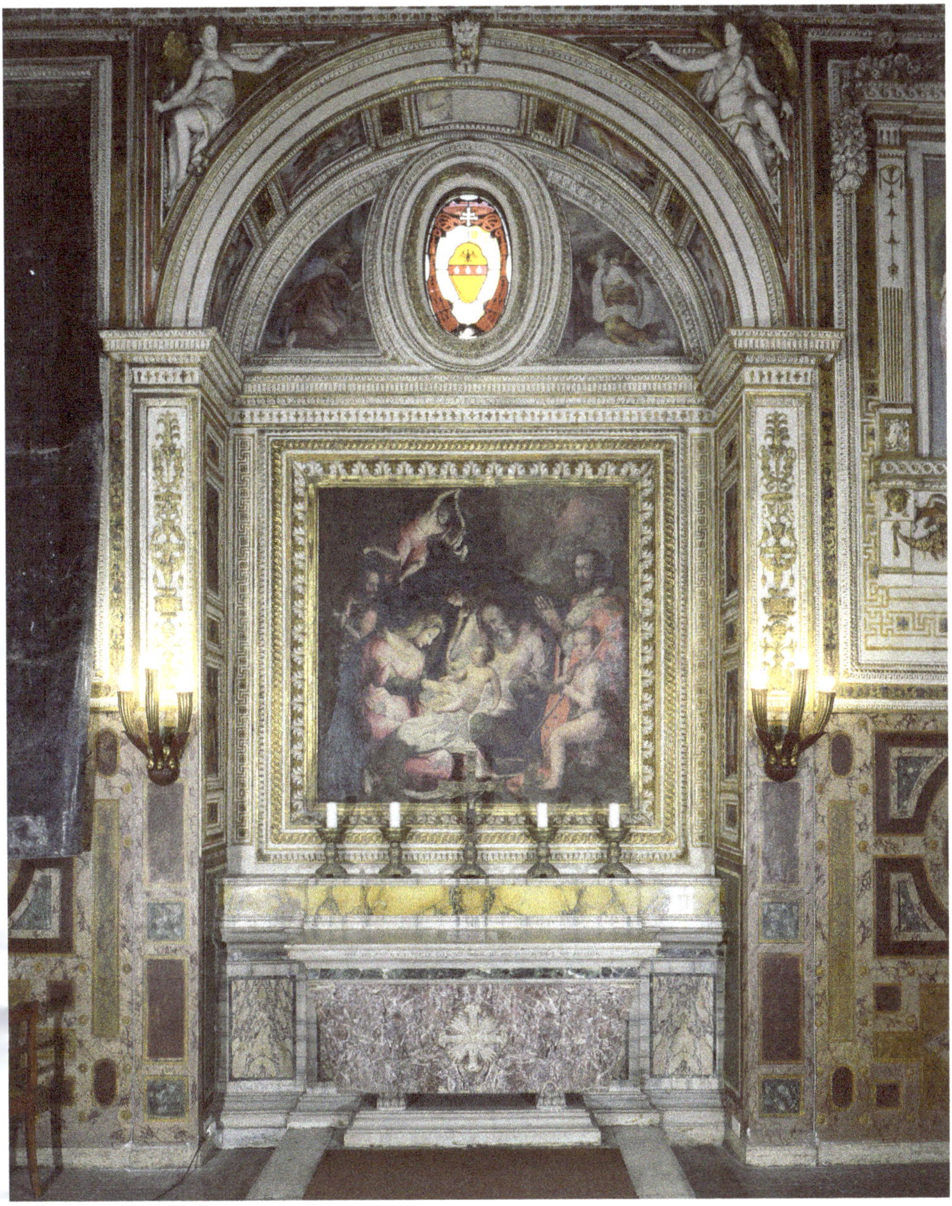

PLATE 15 *Francesco Salviati,* Cappella del Pallio, *oil on peperino, ca. 1549, Rome, Palazzo della Cancelleria* (PHOTO: VATICAN MUSEUMS).

# PART 1

## *Sebastiano del Piombo's Invention and Its Initial Influence*

CHAPTER 1

# 'Uno Nuovo Modo di Colorire in Pietra': Technical Experimentation in the Art of Sebastiano del Piombo*

*Piers Baker-Bates*

The phrase, 'a new mode of colouring in stone,' cited in this essay's title comes from the well-known letter written by Vittore Soranzo in June 1530 concerning that unspecified 'imagine di Christo' that Sebastiano del Piombo had presented to Pope Clement VII.[1] As discussed in the Introduction, this letter is the earliest surviving reference to this method of painting by Sebastiano and is also important as it describes not only the method itself but refers to Sebastiano's reasons for adopting it. Nonetheless our knowledge of technical developments in Sebastiano's art, and of their importance, remains limited, despite new research on Sebastiano's work on stone that has resulted from recent restorations.

Given the state of the question, what I will discuss in what follows represents a preliminary introduction. I have also written more broadly recently about Sebastiano's artistic practice and its innovative nature, so what I shall do here in the context of this volume is to focus on his paintings on stone specifically.[2] Chris Nygren has written recently about Titian's turn to stone and what he terms "the first generation of painted stones."[3] This chapter seeks to anchor Sebastiano del Piombo's practice in current debates around materiality, especially material religion. What follows will analyse the reasons behind Sebastiano

* I would like to thank Elena Calvillo for the original idea for this collection, for inviting me to be her co-editor and for encouraging me to think about Sebastiano's materiality more deeply in what follows. I would also like to thank Dorigen Caldwell for inviting me to present an earlier version of this research and Paul Joannides for his, as always, insightful comments throughout.

1 *Delle Lettere da diversi Re et Principi et Cardinali et altri huomini dotti a Mons. Pietro Bembo scritte* (Venice, 1560), p. 110.

2 Piers Baker-Bates, "A Painting little less than Eternal," in *Michelangelo & Sebastiano*, ed. Matthias Wivel (London, 2017), pp. 74-85.

3 Christopher J. Nygren, "Titian's Ecce Homo on Slate: Stone, Oil and the Transubstantiation of Painting," *Art Bulletin* 99, no. 1 (2017), 36-66.

 | DOI 10.1163/9789004361492_004

using such stone surfaces, in terms of their performative roles, of their aesthetic value, and of their religious dimension.[4]

For example, although the conclusions have yet to be published, the restoration of the Chigi Chapel in the Roman church of Santa Maria del Popolo has already delivered striking visual results. The heart of the chapel is Sebastiano's massive altarpiece of the *Nativity of the Virgin*, that measures nearly six meters by three and a half. [**Plate 2**]. Roughly contemporaneous with the letter of Soranzo, but shrouded for many years in soot, poorly lit, and obscured by the nearby works by Raphael and Bernini, this late masterpiece of Sebastiano's Roman years, on which work began some time after August 1530, had been unduly neglected.[5] In preparation for this altarpiece, Sebastiano had the back wall of the chapel removed and replaced by large blocks of peperino. The cleaning of the surface has revealed an extraordinarily subtle handling of paint rich in glowing colors—precisely those effects Sebastiano seems to have sought to achieve when working in oils on a stone surface.

In two senses this altarpiece represents a bravura performance on Sebastiano's part—not only in its sheer size and elaborate preparation but also in its ability to enmesh the spectator directly in the sacred drama. These are qualities inherent in Sebastiano's art but they reach their apogee in his paintings on a stone surface. In these, he is harnessing his consummate technical abilities in the service of a deeply considered religious art. Sebastiano's group of paintings on stone represent the culmination of a delight in technical experimentation that began immediately after his arrival in Rome, if not even earlier. From the beginning of his Roman career in 1511, he had proved himself to be among the leading technical innovators of the early 16th century, although this aspect of his practice has largely been overlooked.

The few studies of this phenomenon of painting on stone mentioned in the Introduction have highlighted Sebastiano's role as a pioneer but not concerned themselves with the reasons behind this development or its technical aspects.[6] Such neglect is unjustifiable, as none other of his contemporaries at Rome or

---

4 In what follows I am especially indebted to the insights that were provided by Elena Calvillo's thought provoking and ground breaking 2013 article, "Authoritative copies and divine originals: Lucretian metaphor, painting on stone, and the problem of originality in Michelangelo's Rome," *Renaissance Quarterly* 66, no. 2 (2013), 453-508.

5 John Shearman, "The Chigi Chapel in Santa Maria del Popolo," *Journal of the Warburg and Courtauld Institutes* 24 (1961), 129-60; Antonio Pinelli, "La Cappella delle tombe scambiate: Novità sulla Cappella Chigi in Santa Maria del Popolo," in *Francesco Salviati ou la Bella Maniera. Actes des colloques de Rome et de Paris (1998)*, ed. Catherine Monbeig Goguel, Phillipe Costamagna, and Michel Hochmann (Rome, 2001), pp. 253-85.

6 Anne-Laure Collomb, *Splendeurs d'Italie: La peinture sur pierre à la Renaissance* (Rennes, 2012).

beyond, with the significant exception of Sebastiano's great rival Raphael, explored the technical boundaries of the painter's practice as extensively as Sebastiano did. While he was alive, Raphael, endlessly innovative, was perhaps the leader in this exploration, but after his death Sebastiano became the unrivalled pioneer.[7] Particularly novel and innovative were Sebastiano's continuous experiments with grounds and with surfaces that brought the material into his work and which culminated in his paintings on stone.

## The Beginnings

Before turning to his actual practice of painting on stone, however, a brief recap of how he reached that precise moment may be helpful. In Sebastiano's first paintings at Rome, the Ovidian cycle in the lunettes of the *Sala di Galatea* at the Villa Farnesina, he experimented with a grey ground that reduced the luminosity of his colors.[8] The only slightly later *Christ carrying the Cross*, now dated to around 1513-14, a commission from the ambassador at Rome of Ferdinand of Aragon, Don Jerónimo de Vich y Valterra, is painted in oils on a canvas of especially fine weave which allowed him to play with the effects of light [**Fig. 1.1**].[9] Sebastiano's career has been seen as divided in two by the caesura of the Sack of 1527, and his paintings on stone placed after that date consequentially. There is however greater consistency in his artistic practice than has been acknowledged. By the 1520s, in the *Flagellation* in the Borgherini Chapel in San Pietro in Montorio, Sebastiano was using an oil-based mural technique on a colored *imprimitura* onto the wall itself [**Fig. 1.2**].

If Sebastiano was not, as Vasari states, the first artist to paint in oils on the wall, he was the first who was able to make it stick. Restoration of the Borgherini Chapel, the results of which were published in 2010, represented a start in research on his mural practice.[10] Furthermore, in this chapel Sebastiano was in head to head competition with Raphael as the primary technical innovator

7 For Raphael's innovations see most recently Ana González Mozo, "Raphael's Painting Technique in Rome," in *Late Raphael*, ed. Tom Henry and Paul Joannides (Madrid, 2012), 319-49.

8 Almamaria Mignosi Tantillo, "Restauri alla Farnesina," *Bollettino d'Arte* 57 (1972): 33-43; Costanza Barbieri, "*Tam foelix pictor vate, ut pictore Poeta*: The Iconography of Sebastiano del Piombo's Lunettes in the Loggia della Galatea," *Artibus et Historiae* 72 (2015), 125-52.

9 *Sebastiano del Piombo y Espana*, ed. Manuela B. Mena Marqués and Fernando Benito (Madrid, 1995), pp. 95-6.

10 Giorgio Vasari, *Le Vite de'più eccellenti pittori scultori e architettori nelle redazioni del 1550 e 1568*, ed. Rosanna Bettarini and Paola Barocchi, 9 vols. (Florence, 1967-1987), 5:89-90;

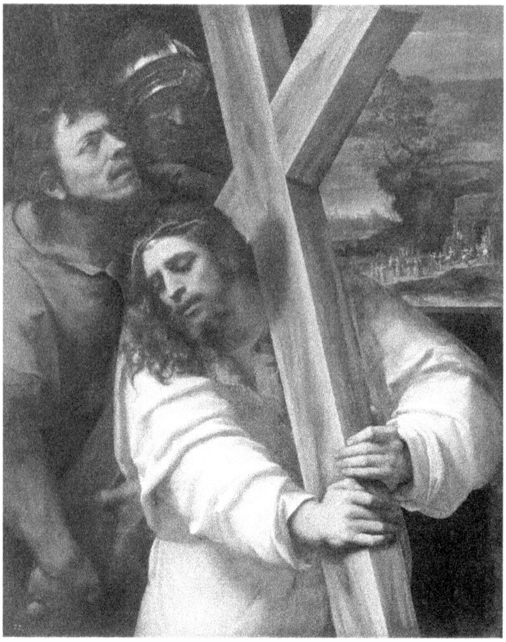

FIGURE 1.1
*Sebastiano del Piombo,* Christ carrying the Cross, *oil on canvas, ca. 1513-14, Madrid, Museo del Prado* (PHOTO: © MUSEO DEL PRADO, MADRID).

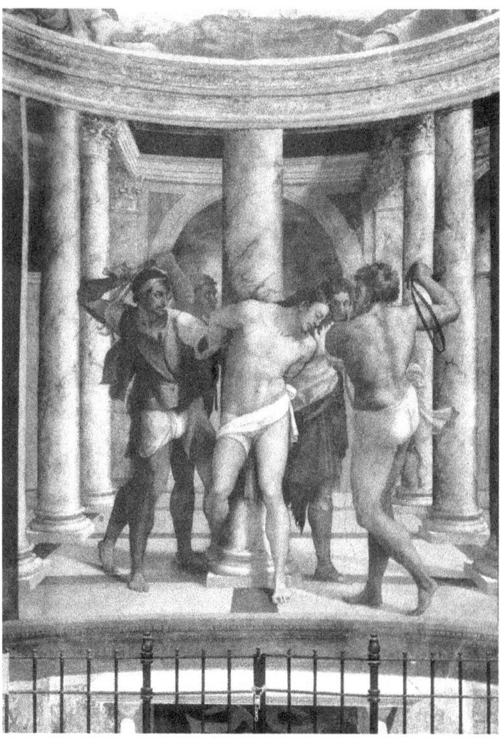

FIGURE 1.2
*Sebastiano del Piombo,* Flagellation, *oil mural, 1520-1524, Rome, San Pietro in Montorio* (PHOTO: ©SCALA, FLORENCE, 2015).

in Rome. Shortly before his death, Raphael was contemporaneously attempting to use oil on the wall in the *Sala di Constantino*. Sebastiano had painted the semi-dome of the Borgherini chapel in conventional fresco, but for the lower register, centered on the *Flagellation*, he turned to painting in oils on the wall, although which artist inspired the other is now impossible to tell.

Sebastiano explicitly acknowledged his rivalry with Raphael in a letter to Michelangelo of 6 September 1521: "I have painted it with oils on the wall which I think will please you in that it won't run down the wall as that in the *Palazzo* did."[11] After Raphael's death his experiments were abandoned and the *Sala di Constantino* was painted instead in conventional fresco by his pupils. It was Sebastiano who became known for his success in exploiting such new techniques, including before too long that of painting on stone. In the contract for the Chigi Chapel altarpiece of August 1530 he is already specifically charged to "paint it in oils in that new manner and invention that he has found thanks to long labour and experience."[12] Not that these technical innovations always pleased Michelangelo; according to Vasari, it was Sebastiano's preparation of the wall for the *Last Judgement* for painting in oils that was to cause a final rupture between the two artists. Michelangelo is reported as scornfully stating later that "painting in oils was an art for women and for leisurely and idle people like Fra Sebastiano."[13]

It was not, though, a leisurely process. As will be discussed, Sebastiano's paintings in oil on a variety of surfaces was both complicated and time consuming.[14] It is possible that Sebastiano was experimenting with a stone surface even before 1527, indeed the wording of the Chigi chapel contract of August 1530 would appear to imply exactly that. Recently, an oval bust portrait, painted on slate and an almost exact derivation from Sebastiano's three-quarter length

---

*Sebastiano del Piombo e la Cappella Borgherini nel contesto della pittura Rinascimentale*, ed. Santiago Arroyo Esteban, Bruno Marrocchini, and Claudio Seccaroni (Florence, 2010).

11 *Il Carteggio di Michelangelo*, ed. Paola Barocchi and Renzo Ristori, 5 vols. (Florence, 1965-83) 2:315. There are of course two possible instances that Sebastiano could be talking about here, either the *Sala di Costantino* in the Vatican Palace or it could be that he is referring to Leonardo's *Battle of Anghiari*, in the Palazzo Vecchio: see Rona Goffen, *Renaissance Rivals: Michelangelo, Leonardo, Raphael, Titian* (New Haven and London, 2002), pp. 149-56.

12 The full contract was published in Michael Hirst, "The Chigi Chapel in Santa Maria della Pace," *Journal of the Warburg and Courtauld Institutes* 24 (1961), 161-85 (for contract see pp. 183-85).

13 Vasari, *Le Vite*, 5:101-02.

14 Vasari unfortunately describes only Sebastiano's method for preparing wall surfaces for paint and not his preparation for painting on stone.

portrait of *Andrea Doria*, dateable to 1526, has been discovered in the private apartments of the Palazzo Doria Pamphili.[15] The only difference between the two paintings is that the head in the slate oval has a slightly longer beard. Why Sebastiano would have chosen to paint the Genoese admiral after the Sack is uncertain, the Admiral had not only left Rome but was also now in the service of the Pope's enemy, Charles V. All this implies that this slate version of the *Andrea Doria* may be earlier in date, although Sebastiano several times made copies and versions of his paintings some years after the original. Whatever the case here, after Sebastiano's return to Rome in 1529, there survive some eight to ten paintings by Sebastiano from the early to mid-1530s that use an independent stone surface.

His invention found an immediate popularity, just a year after Soranzo's first mention of it, for, in the letter to Michelangelo of 22 July 1531, Sebastiano relates that he had been asked to make a copy on stone of his canvas portrait of Clement.[16] Here the choice of surface is intimately bound up with the subject-matter, as Calvillo and Nygren also note in their contributions to this volume. Far too little is still known however of the supports on which Sebastiano painted, although they are generally described as *lavagna*, that is slate, or about how he prepared them. Their preparation is likely to have been an intricate process of trial and error that belies descriptions of Sebastiano's indolence after the Sack. He was criticized for being a slow worker but his lifelong quest for technical innovation refutes these charges and seems consistent with a process of continual experimentation.

Research in this area has also not been helped by the widely scattered locations of these stone panels—from Madrid to Saint Petersburg—few, if any, of which can now travel; just to list then the paintings that will be discussed here. Of the paintings on stone that survive by Sebastiano the majority are religious in subject matter. The standard *corpus* consists of the *Nativity of the Virgin* in Santa Maria del Popolo [**Plate 2**], three versions of *Christ carrying the Cross* (in the Prado, the Hermitage, and Budapest) [**Plates 10, 11, 12**], the *Pietà* from Úbeda in the Prado [**Plate 1**], and the *Madonna del Velo* in Naples [**Plate 3**]. There are also a handful of portraits, that of *Baccio Valori* in the Pitti [**Fig. 1.3**], two of Clement VII, one in Naples [**Plate 14**] and one (if it is by Sebastiano) at

15 Andrea De Marchi, "Il Ritratto di Andrea Doria: Una nuova versione," in *Sebastiano del Piombo e la Cappella Borgherini nel contesto della pittura Rinascimentale*, ed. Santiago Arroyo Esteban, Bruno Marrocchini, and Claudio Seccaroni (Florence 2010), pp. 123-25; Andrea De Marchi, *Collezione Doria Pamphij: Catalogo Generale dei Dipinti* (Milan, 2016), pp. 342-43.

16 *Il Carteggio*, 3:318.

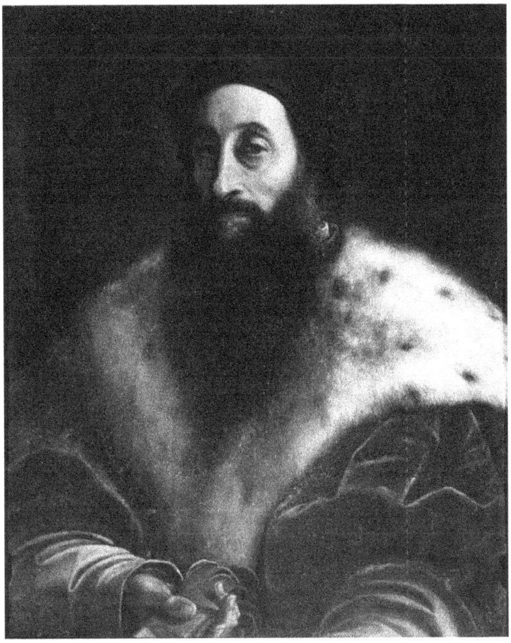

FIGURE 1.3
*Sebastiano del Piombo,* Baccio Valori, *oil on slate, 1531, Florence, Palazzo Pitti* (PHOTO: ©SCALA, FLORENCE, 2015).

the Getty [**Plate 4**], and an unusual and unfinished double portrait in Parma, that is now described as *Paul III and a Nephew* [**Plate 13**].

Sebastiano's practice of painting on stone was more extensive than these surviving works suggest; the surface of slate paintings has often been misidentified historically. While his own slates continue to surface, the Getty painting was only re-discovered in 1987. As well as the *Andrea Doria*, another head on slate attributed to Sebastiano and called *Portrait of a Man in Armour* appeared on the art market in 2017 [**Fig. 1.4**].[17] The sitter is thought to be Cardinal Ippolito de'Medici, given its relation to the portraits by Titian in the Pitti and by Sebastiano in the Johnson Collection. Other now lost portraits on stone are recorded in Vasari and other contemporary accounts, notably those of Cardinal Pirro Gonzaga and of his cousin Giulia Gonzaga, neither of which appear to survive.[18] I have argued elsewhere the role of the portraits of Clement VII as

17 Alessandro Ballarin, "Un nuovo ritratto su lavagna di Sebastiano del Piombo," *Nuovi Studi. Rivista di Arte Antica e Moderna* 20, no. 21 (2015), 71-80; Angela Cerasuolo, "Osservazioni sulla tecnica di Sebastiano del Piombo," *Nuovi Studi. Rivista di Arte Antica e Moderna* 20, no. 21 (2015), 81-86. I am grateful to Filippo Benassi, Edoardo Roberti, and David Pollack for facilitating my own study of this painting.

18 There are versions on slate of the *Giulia Gonzaga* in Munich and Wiesbaden and the latter has recently been attributed to Sebastiano himself: Costanza Barbieri, "Rodomonte Gonzaga, il sacco di Roma e un riscoperto ritratto di Sebastiano del Piombo," in *I Gonzaga e i*

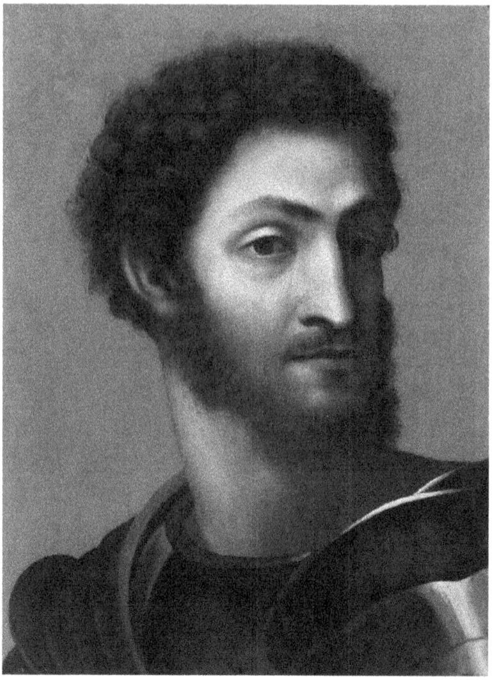

FIGURE 1.4
*Sebastiano del Piombo (?),* Portrait of a Man in Armour, *oil on slate, Private Collection* (PHOTO: PRIVATE COLLECTION, PINO DELL'AQUILA).

Clementine icons, and they at least should tentatively be included among Sebastiano's works of a sacred nature.[19]

There are also a couple of paintings closely associated with Sebastiano's compositions on a stone surface, copies or versions of his own works, that will play a role in the discussion that follows. Despite their number, however, until very recently, only those paintings on a stone surface in the Prado had been restored and examined technically, for the 1995 exhibition.[20] The restored portrait of *Baccio Valori* is now on public display again at the Palazzo Pitti in Florence.[21] Sebastiano's technique can be glimpsed with the naked eye on the *Madonna del Velo* and the *Clement VII*, both in Naples, and the Parma double portrait all of which remain unfinished and perhaps deliberately so. In these, thanks to the exposed surface, the artist can be seen exploiting the stone to the

*papi Roma e le corti padane fra Umanesimo a Rinascimento (1418-1620): Atti del convegno Mantova-Roma 21-26 febbraio 2013*, ed. Renata Salvarani (Vatican City, 2013), pp. 331-51.

19 Piers Baker-Bates, "Beyond Rome: Sebastiano del Piombo as a Painter of Diplomatic Gifts," *Renaissance Studies* 27, no. 1 (2013), 51-72, esp. 65ff.

20 *Sebastiano del Piombo y España*: unfortunately, however, no detailed restoration information was published in the accompanying catalogue.

21 *Sebastiano del Piombo e la pittura su pietro: il Ritratto di Baccio Valori, restauro e ricerche*, ed. Alessandra Cecchi, Marco Ciatti, and Oriana Sartiani (Florence, 2014).

fullest effect; some have argued that this *non finito* quality reflects the artist's purposeful exploitation of the stone surface.[22]

## Soranzo's Report

There is no evidence for how Sebastiano originally came to discover a preparation for painting on slate, or even became interested in the tricky chemical experiments it would have involved.[23] It was Kathleen Weil Garris Posner who speculated over forty years ago, although her insight has never properly been followed up since, that it may well have been through Sebastiano's contact with Leonardo da Vinci.[24] Leonardo was in Rome from 1513 until at least August 1516, and their three years together could have first inspired Sebastiano to experiment; according to Vasari part of Leonardo's time in Rome was spent as "he tried the strangest methods in seeking out oils for painting, and varnish for preserving works when painted."[25] Despite Leonardo's evident influence on Raphael, which has received considerable attention in the literature, few scholars have addressed the impact of his presence at Rome on Sebastiano.[26] Sebastiano was immediately and deeply influenced by Leonardo's presence in the city after 1513—an influence that is particularly visible in the *Raising of Lazarus*—on which Sebastiano was at work after 1516. Indeed, into the later 16th century, paintings on a slate surface were being attributed to Leonardo himself. In 1591 the Florentine author, Francesco Bocchi, saw what he described as "una notte su la lavagna di Lionardo da Vinci maravigliosa" in Vasari's own house in Florence.[27]

---

22 See Ana González Mozo in this volume, pp. 91-92.

23 Ana González Mozo also proposes in the next essay that Sebastiano may even have learned the technique in Venice; see pp. 75 and 83 in this volume.

24 Kathleen Weil Garris Posner, *Leonardo and Central Italian Art: 1515-1550* (New York, 1974), pp. 14-15.

25 Vasari, *Le Vite*, 4:35

26 Weil Garris Posner, *Leonardo*, pp. 11-15; Marcia Hall, "Sebastiano del Piombo fra innovazione e sperimentazione," in *Notturno Sublime Sebastiano e Michelangelo nella Pietà di Viterbo*, ed. Costanza Barbieri (Rome, 2004), pp. 42-7, esp. p. 43; Carmen Bambach, "Leonardo and Raphael in Rome in 1513-16," *Late Raphael: Proceedings of the International Symposium Madrid, Museo Nacional del Prado, October 2012*, ed. Miguel Falomir (Madrid, 2013), pp. 26-37.

27 Francesco Bocchi, *Le Bellezze della città di Fiorenza*. ed. Giovanni Cinelli (Florence, 1977), p. 306.

As noted, the earliest actual reference to Sebastiano's painting on a stone panel describes a work which does not appear to survive; that is the passage in the letter from the papal secretary, Vittore Soranzo, to Pietro Bembo, dated 8 June 1530: "Using this method Sebastiano has made an image of Christ and shown it to our Lord."[28] In analysing Sebastiano's painting on a stone surface there is much to unpack in Soranzo's letter, starting with the material itself. First, it is a curiosity that Soranzo states specifically that this image of Christ was painted on marble, although none of Sebastiano's surviving stone paintings use this material as a surface.

Soranzo could have been misinformed, of course, but Vasari in his second, 1568 edition went further; he described how Sebastiano "worked on peperino-stone, white and variegated marble, porphyry, and slabs of other very hard kinds of stone."[29] There is actual evidence of Sebastiano's practice, however, only in the inventory of the contents of his studio made after his death a few years before, in June 1547. Here, among the mass of unfinished work listed, are included no less than thirty-seven stone supports of varying types, mostly marble, more or less worked up.[30] There is evidence that later artists chose certain stone surfaces on the basis either of their material value, or of their scarcity. Titian used marble in his works for the Emperor Charles, as Chris Nygren has discussed recently, while Murillo, in the only known instance of his employment of stone, used a rare Mexican obsidian.[31] Those of Sebastiano's paintings on stone that survive, however, utilize only the humbler peperino and slate.

Peperino was a local Roman stone and would have been easy for Sebastiano to obtain but Italian slate comes only from the coastal area between Genoa and La Spezia. Indeed, the term that is used most commonly in Italian, *lavagna*, is that of the town almost equidistant between the two ports where this material is mostly to be found, while another term used at the period was therefore 'Pietra di Genova.'[32] Given problems of transport, however, this would have been both difficult to obtain and expensive as a material. Vasari in his introduction to the *Vite* describes this particular stone and the reasons why it was used:

28 *Delle Lettere*, p. 110.

29 Vasari, *Le Vite*, 5:98-99.

30 Michael Hirst, *Sebastiano del Piombo* (Oxford, 1981), pp. 155-57.

31 Isabel Horovitz and Joan Reifsnyder, "The use of slate or stone as a painting support," in *Conservation of Easel Paintings*, ed. Joyce Hill Stoner and Rebecca Rushton (Abingdon and New York, 2012), pp. 97-100.

32 For an example see Pierre de Nolhac, "Une Galerie de peinture au XVIe siecle: Les Collections de Fulvio Orsini," *Gazette des Beaux Arts*, 29 (1884), 427-36, esp. pp. 431ff.

> They have found a suitable kind off the sea coast of Genoa ... which are very well fitted for this purpose, for the reason that they are compact and of fine grain, and take an even polish.[33]

While in his fellow Florentine Raffaello Borghini's *Il Riposo* of 1584 the benefits of this slate are described specifically:

> Those who prefer to use color on stone will find certain slabs very good that are found on the coast of Genoa. On these it is only necessary to put the priming and then work, colouring with care.[34]

Nonetheless, despite this weight of literary evidence, further investigation is required on the precise nature of the materials on which Sebastiano worked.

The date of Soranzo's letter, June 1530, also raises important questions. It is three years after the Sack of Rome, and a year after Sebastiano's return to Rome from Venice. After Sebastiano's return to Rome the greater part of his paintings were on a stone surface, or wall paintings, and, although he continued to use other supports, these are far fewer in number.[35] I would therefore argue that this group of paintings on stone were begun in a single burst of creativity between about 1530 and 1535. This would support Vasari's assessment of Sebastiano as having almost abandoned his artistic career, perhaps in my opinion as early as 1535. Furthermore, at this period, Sebastiano reconsiders versions of earlier subjects, such as *Christ carrying the Cross*, on the new surface. He had reproduced his own work before the Sack, as in the variant of the Borgherini chapel *Flagellation* in Viterbo, but now the same compositions assumed a new aesthetic, more somber and arguably more devout in appearance.

Still the completion of several dragged on interminably, like the Úbeda *Pietà*.[36] In many cases they were never finished, as happened with the Santa Maria del Popolo, *Nativity of the Virgin* or, among moveable paintings, the Naples *Madonna del Velo*. Sebastiano was never deterred however when work-

33 Vasari, *Le Vite*, 1:137-38.

34 Raffaello Borghini, *Il Riposo*, ed. Mario Rosci, 2 vols. (Milan, 1967), 1:176; Raffaello Borghini, *Il Riposo*, ed. and trans. Lloyd H. Ellis jr. (Toronto, 2007), p. 125.

35 A census of Sebastiano's non-slate paintings after 1527 is not aided by a lack of consensus on his chronology. It is arguable that he painted only three paintings on panel for definite after that date, the portraits of *Francesco Arsilli*, *Ippolito de'Medici* and *Cardinal Giovanni Salviati*.

36 Michael Hirst, "Sebastiano's Pieta for the Commendador Mayor," *Burlington Magazine* 114, no. 834 (1972), 585-95.

ing with this troublesome and weighty material, by questions of destination, or of size. The Úbeda *Pietà* was a diplomatic gift, and as such probably always intended for a remote location in Andalusia, a fact of which Sebastiano would have been well aware from the outset.[37] When it was eventually completed in 1539 a striking illustration survives of the difficulties involved in its transport, in a letter from Nino Sernini, the Gonzaga agent at Rome, to the patron, Ferrante Gonzaga. The painting "could not be sent by land because mules could not be found to transport it safely. I have found a frigate that will take it ... but sending it without someone to ensure its safe arrival is pure folly, because of its fragile nature."[38]

## Depicting the Divine

A particularly important aspect of Soranzo's letter is the subject matter, that is an 'image of Christ,' although what exactly Soranzo meant by this is impossible to tell. I will concentrate my argument in what follows on Sebastiano's religious paintings on stone, without, however, aligning myself with the arguments that attempt to connect these paintings to a specific notion of reform current in the Catholic church in the 1530s, before religious divisions had hardened.[39] Rather, using Federico Zeri's own subtitle for his classic study of the relationship between art and religious reform, Sebastiano's religious paintings on a stone surface become I would argue an *arte senza tempo*.[40]

In a number of ways Sebastiano's subject matter and style indeed become timeless, increasingly anachronic in the recent theoretical use of that term, the longer he spent at Rome.[41] For example, in these late religious paintings on stone Sebastiano concentrates on a restricted subject matter of the *Pietà*, the *Christ carrying the Cross*, or the *Madonna and Child*. The first two subjects in particular were *andachtsbild*, to use Erwin Panofsky's term: "a devotional picture inviting the beholder to lasting and empathic meditation rather than a

37 Baker-Bates, "Beyond Rome," pp. 52-54 and 63-5; Calvillo, "Authoritative Copies and Divine Originals," pp. 483-4.

38 Hirst, "Sebastiano's Pieta," pp. 591-2.

39 Federico Zeri, *Pittura e Controriforma L'"arte senza tempo" di Scipione da Gaeta*, 3rd ed. (Vicenza, 2001); Maria Calí, *Da Michelangelo all'Escorial: momenti del dibattito religioso nell'arte del Cinquecento* (Turin, 1980); Massimo Firpo, *Artisti, Gioiellieri, Eretici: Il Mondo di Lorenzo Lotto tra Riforma e Controriforma* (Rome and Bari, 2001), pp. 333-38.

40 Zeri, *Pittura e Controriforma*.

41 Alexander Nagel and Christopher S. Wood, *Anachronic Renaissance* (New York, 2010), pp. 7-19.

narrative arousing his interest in a specific event."[42] As Calvillo has discussed in relation to the *Pietà* and Nygren has argued in relation to Titian's paintings on a stone surface, through these images Sebastiano is consciously inserting himself and asserting his own place in a tradition of Christian image making at Rome.[43]

Furthermore, through such images, the viewer was now invited to share in the sacred drama of Christ's Passion, not as a passive spectator but as an engaged participant. A number of these paintings, such as the *Pietà* and at least one of the *Christ carrying the Cross*, were intended for Spanish clients and Sebastiano's devout works painted on a stone surface enjoyed a particular success in that country. Into the next generation the reception the works on stone of the German artist, Sigismondo Laire, based at Rome, found with Spanish clients proves this success was no flash in the pan.[44] For these patrons, it wasn't so much the technical accomplishment of the paintings themselves as it was the visceral depiction of Christ's sufferings in his Passion that found favor.

As further evidence for these developments in Sebastiano's art, besides those paintings mentioned already, a handful of other religious images associated with Sebastiano around 1530 fall into this same category. Michael Hirst discussed an *Ecce Homo* on panel in the Pitti in 1981, and related it to Soranzo's letter; Hirst described this painting as bearing "all the signs of recording his [Sebastiano] own invention" [**Fig. 1.5**].[45] The Pitti painting has gone by various titles since Hirst wrote, and it is telling that it represents more a generic image of the Passion rather than one specific moment. Whoever its author, there is indeed a closeness to Sebastiano's style in the face and hands. Hirst's insight was supported when a previously unknown drawing, undoubtedly by Sebastiano and dateable to after 1530, now in a New York private collection,

42 Erwin Panofsky, "Jean Hey's 'Ecce Homo': Speculations about its author, its donor and its iconography," *Bulletin des Musées Royaux des Beaux Arts*, 5, nos. 3-4 (1956), 95-138, esp. p. 95; the use of this term is specifically applied to the Ubeda *Pietà* in Panofsky's, "Die Pietà von Ubeda. Ein kleiner Beitrag zur Lösung der Sebastianofrage," in *Festschrift für Julius Schlosser zum 60. Geburtstage*, ed. Arpád Weixelgärtner and Leo Planiscig (Zurich, 1927), pp. 150-61.

43 Calvillo. "Authoritative Copies and Divine Originals," pp. 480-84; Nygren, "Titian's Ecce Homo," pp. 45-51.

44 Germán Ramallo Asensio, "Toda la obra conservada en España y hasta ahora conocida, del pintor Segismundo Laire. Alemán en Roma," *Archivio Español d'Arte*, 79, no. 315 (2006), 243-261; Anne-Laure Collomb, "La peinture sur pierre entre Rome et Madrid: Sigismondo Laire parmi ses contemporains," *Paragone* 57, no. 69 (679) (2006), 75-87.

45 Hirst, *Sebastiano del Piombo*, p. 125.

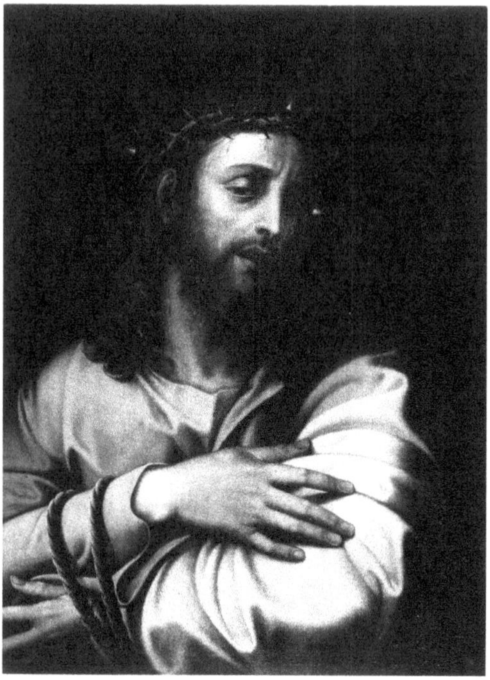

FIGURE 1.5
*After Sebastiano del Piombo,* Ecce Homo, *oil on panel, Palazzo Pitti* (PHOTO: FIRENZE GABINETTO DEI DISEGNI E DELLE STAMPE DEGLI UFFIZI).

appeared at Sotheby's three years later [**Fig. 1.6**].[46] On the *recto* is an image of a bearded man that was described in the Sotheby's catalogue in 1990 as a *Christ as the Man of Sorrows* and as having "the intensity of feeling found in his late religious works." Nonetheless, in the Pitti *Ecce Homo* the exaggerated pietism presented to the viewer in both in the crown of thorns and the bound hands is quite unlike Sebastiano's other works.

Recently, however, further evidence has emerged of this circulation of a religious imagery that is related directly to Sebastiano's late compositions. There is a further drawing in the Musée Condé, Chantilly, *Tête de Christ exprimant le Douleur*, that has been attributed by some scholars to Sebastiano and which I do believe is his own work.[47] Two paintings on panel, related to both the New

46 Sotheby's London, 2 July 1984, p. 8, lot 10; Sotheby's London, 2 July 1990, pp. 38-39, lot 39; *Sebastiano del Piombo 1485-1547* (Milan, 2008), pp. 326-7. I am most grateful to Charles and Jessie Davis for their kindness in facilitating my access to this drawing.

47 Oskar Fischel, "A new approach to Sebastiano del Piombo as a Draughtsman," *Master* Drawings 14 (1939-1940), 21-33, esp. p. 33, had related this drawing to the Pitti *Ecce Homo*. It was most recently discussed in *Dessins Italiens du Musée Condé de Chantilly III, Vénétie, Lombardie, Piémont, Émilie, XVe-XVIe siècle*, ed. Olivier Villard (Paris, 1998), pp. 66-8. Professor Paul Joannides has told me orally that he too believes this drawing to be by Sebastiano, and that Professor David Ekserdijian holds the same opinion.

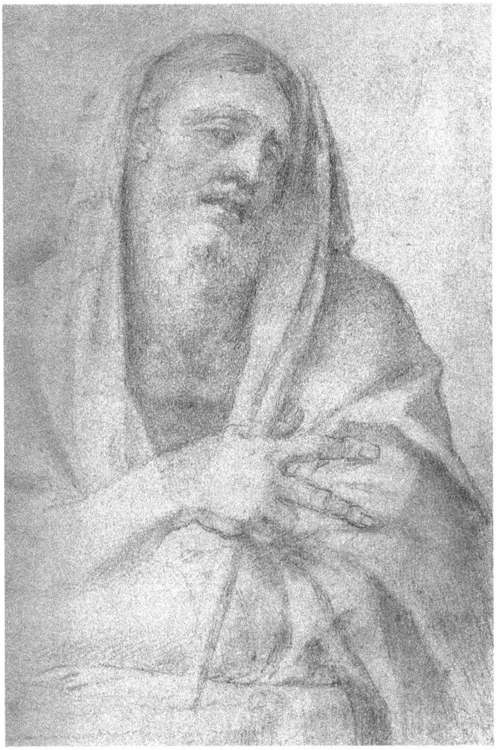

FIGURE 1.6
*Sebastiano del Piombo,* Drawing, *Private Collection, United States* (PHOTO: PRIVATE COLLECTION).

York and Chantilly drawings, were sold at Sotheby's within the last ten years.[48] Although they were marketed as autograph paintings, I do not think that either of these two can be Sebastiano's own work. Two characteristics are, however, worth highlighting in relation to the argument here: the gold background of the first image and the almost photographic closeup of the second that are consistent with other devout imagery of the period. This closely related pair do illustrate in a variety of ways the kind of iconic imagery that was circulating within Sebastiano's Roman ambit in these years of the late 1520s early to mid-1530s. The difference in supports between these paintings and Sebastiano's religious work in the 1530s suggests however that they cannot be attributed to him.[49]

48 Sotheby's New York, June 5 2009; Sotheby's London July 16 2014.

49 That said, the converse is also true, there are a handful of versions of portraits on panel that are transferred to a stone surface. Notable among these are a version of the Radnor *Portrait of a Lady* in the Louvre and the *Giulia Gonzaga* at Wiesbaden and Munich, already mentioned. None of these in my opinion are Sebastiano's own work, although a portrait of Giulia on slate is recorded as being sent to France after the artist's death, along with one of Clement VII. In that case, all three turn to a stone surface where Sebastiano's original

## Eternity and the *Paragone*

Whatever the case here, Sebastiano's paintings on a stone surface, whether for religious art or for portraiture, invoke a sense of the eternal and this invocation of the eternal appears in the very first descriptions both by Soranzo and Vasari. The stone surface plays a role in an increasing shift from narrative to icon; particularly so in the context of Sebastiano's religious art, a shift that had gradually taken place throughout his Roman career. As discussed, his use of a stone surface was intimately connected to his quest for a deeper religious meaning and this too is a thread that can be traced from the beginning of his Roman career.

A quest for the preservation of works of art, seen as transient objects, had motivated artists such as Leonardo before although none took it to such lengths as Sebastiano. The stone surface reveals Sebastiano's influential role in the debates that began with Leonardo, concerning the *paragone*, in this case of the relationship between and comparative hierarchical positioning of painting and sculpture.[50] While Chris Nygren has recently and compellingly questioned the modern conception of this term, it is nonetheless a concept that is still fruitful when considering Sebastiano's paintings on a stone surface.[51] Sebastiano's style had become increasingly sculptural throughout his Roman career and if his actual paintings could be made as durable as sculptures they would demonstrate a clear superiority.

As early as 1546, the year before Sebastiano's death, this claim for his art is made by Benedetto Varchi in his *Della Maggioranza delle Arti*.[52] However, in another Florentine's writings, Anton Francesco Doni's, *Disegno*, published only a few years later in 1549, the sculptor Silvio makes the claim that sculpture

---

(although that is uncertain in the case of the Giulia) was on panel. The number of such versions, particularly of the Giulia Gonzaga, testify to the significance of the originals but why the copyist, whoever that may be, in each case should move to a slate surface is unclear.

50 Claire J. Farago, *Leonardo da Vinci's* Paragone*: A critical interpretation with a new edition of the text in the* Codex Urbinas (Leiden, 1992); Suzanne B. Butters, *The Triumph of Vulcan: Sculptors' Tools, Porphyry, and the Prince in Ducal Florence*, 2 vols. (Florence, 1996), 1:108-11; Ben Thomas, "'The lantern of painting': Michelangelo, Daniele da Volterra and the *Paragone*," *Apollo* 154, no. 474 (2001), 46-53; Goffen, *Renaissance Rivals*, pp. 31ff; Monica Azzolini, "In Praise of Art: Text and Context of Leonardo's *Paragone* and its critique of the Arts and Sciences," *Renaissance Studies* 19, no. 4 (2005), 487-510; Sefy Hendler, *La guerre des arts: le paragone peintre-sculpture en Italie XV-XVII siècle* (Rome, 2013).

51 Nygren, "Titian's Ecce Homo," pp. 51-5.

52 Benedetto Varchi, "Lezzione nella quale si disputa della maggioranza delle arti e qual sia più nobile, la scultura o la pittura," in *Trattati d'Arte del Cinquecento*, ed. Paola Barocchi, 3 vols. (Florence 1960-62), 1:3-82, esp. p. 41.

remains "la ... madre." As with painting on stone the pigments cannot truly unite with the surface material and therefore it cannot in fact become eternal.[53] Later in the century, it is precisely this argument in favor of sculpture that is put into the mouth of Baccio Valori's nephew, also Baccio, as one of the participants in Borghini's, *Il Riposo*.[54] Indeed, Valori uses almost the same words here as those of Soranzo and of Vasari; for Valori, it is now sculpture that is 'quasi eterna,' thus proving its superiority over painting. According to Christiane Hessler, the role in the dialogue given to Valori by Borghini is therefore quite deliberate.[55] Furthermore in Sebastiano's case, this desire for preservation was double edged.

Not only was it, as Vasari has it, to preserve these paintings from such earthly problems as worms or fire, or following Borghini and Doni to match them with sculpture. Given the subject matter, Sebastiano's intention was to make them manifest symbols of the divine, especially of Christ's Passion. There is an evocative description by the Prior of the Escorial, José de Sigüenza, of how Philip II came to use an earlier *Christ carrying the Cross*. Perhaps Siguenza is referring to the painting for the Count of Cifuentes painted on stone by Sebastiano. It had become an altarpiece in the King's private oratory, where "The pious King Philip passed a good number of the night hours contemplating just how much is owed to Our Saviour who carried so heavy a Cross upon his Shoulders for the sins of Mankind and his own."[56] The symbolic potential of the stone surface reaches a logical conclusion in the vast Budapest *Christ carrying the Cross*. This is probably that described by Vasari as intended for the Patriarch of Aquileia and is virtually life-size, measuring just over one point five meters by one point two. Here the figure of Christ is reduced to the bottom third of the picture by the enormous Cross that he bears; the weight of the material on which this iconic image is painted is surely a deliberate reflection of His burden and substantiates Siguenza's comments.[57]

---

53 Anton Francesco Doni, *Disegno Fac Simile della Edizione del 1549 di Venezia*, ed. Marco Pepe (Milan, 1970), ff.20v-21r.

54 Borghini, *Il Riposo*, 1:30ff

55 Christiane J., Hessler, "The Man on Slate: Sebastiano del Piombo's Portrait of Baccio Valori and Valori the Younger's Speech in Borghini's Il Riposo," *Source* 25, no. 2 (2006), 18-22, esp. p. 21.

56 José de Siguenza, *Historia de la Orden de San Jerónimo* ed. Juan Catalina García, 2 vols. (Madrid, 1907-09), 2:631.

57 This Renaissance association between weight and sanctity has recently been discussed and explored in relation to Michelangelo's contemporary *Last Judgement*; Erin Sutherland Minter, "Virtuous Prelates, Burdensome Relics and a Sliver of Gold in the *Last Judgement*," in *Michelangelo in the New Millennium: Conversations about Artistic Practice, Patronage and Christianity*, ed. Tamara Smithers (Boston, 2016), pp. 147-78, esp. pp. 168ff.

### Style and Technique

Sebastiano's technique is tied to longevity in very practical ways also, as the stone support of these works has guaranteed these paintings' very survival. The original location of the *Pietà* was the chapel of El Salvador in the Andalusian town of Úbeda, where it was only saved from destruction at the outbreak of the Spanish Civil War in 1936 by its stone support. The adjacently displayed statue of a young *Saint John the Baptist*, supposedly by Michelangelo, was smashed into myriad fragments, the framing *retablo* by Alonso Berruguete was burnt, but the *Pietà* only blunted the bayonets of the anti-clerical mob, whose marks it preserved until its 1995 restoration.[58]

Also, as Sebastiano did not realize, chemical reaction with the black ground of the stone changed the paints he used over time, altering drastically the tonal range he sought. The restoration of the Santa Maria del Popolo altarpiece has helped to demonstrate the extent this deterioration can reach, just how far Sebastiano's paintings on stone can discolor over time and lose their original effect. The chromatic difference in the portrait of *Baccio Valori* and the subtlety of the painting of hair and fabric uncovered by the restoration completed in 2014 (the second restoration in the last fifty years) in particular is revelatory.[59] Furthermore, in the case of the Santa Maria del Popolo altarpiece, the use of more than one block of peperino has caused increasingly significant problems to the paint surface as the joins begin to part.

The use of this impermeable stone surface for painting in oils would once have magnified the chiaroscuro effect, which Sebastiano had already used so strikingly in the *Raising of Lazarus*. Especially in dimly lit spaces such as the Borgherini or Chigi chapels Sebastiano's range of colors became increasingly restricted by the 1530s. The varied, Venetian-inflected palette of the *Raising of Lazarus* is no longer present, but the colors Sebastiano does include stand out clearly as is evident in the Santa Maria del Popolo *Birth of the Virgin*. The restoration has revealed for the first time in living memory how much Sebastiano gained from using oils when painting on a stone surface, as it displays the subtle play of lights and shadows that he had intended. It has now been discovered that Sebastiano altered the architectural structure of the chapel further

---

The fragility of the slate surface has however caused problems, including a long crack at the top left (it is ironic that Sebastiano's quest for the eternal has not necessarily proved successful).

58 *Sebastiano del Piombo y Espana*, pp. 127-133.

59 *Sebastiano del Piombo e la pittura su pietro*, pp. 47ff.

in order to allow light to enter and to strike the altarpiece at certain points in particular.[60]

Until these recent restorations, there had been little research on the actual techniques used by Sebastiano. Technical analysis in the early 1990s had, however, already allowed a small *Christ carrying the Cross* from the Prado's collection to be re-attributed to Sebastiano himself. The slate support worked on was small in size and remarkably thin, it must have taken considerable skill just to cut and prepare this panel; and this explains the contemporary references to the Úbeda *Pietà* as fragile.[61] The slate surface is smoothed for painting but with a certain roughness to hold the paint. On this was laid with regular brushstrokes an oil-based ground of grey priming whose exact nature has not yet been determined; this is what ensured the stability of the picture surface. This layer is remarkably thin so the nature of the slate itself could be manipulated and even at points remain visible. Finally, the surface was ready for painting with an extraordinary attention the finest details. The very recent restorations of the *Baccio Valori* and of the *Portrait of a Man in Armour* have only confirmed these conclusions.[62]

The slate surface has further allowed Sebastiano to reduce deliberately the brightness of his colors and renders his religious subjects in particular in an appropriately somber hue. Those colors stand out all the more as the oil paint does not sink in to the stone as it does with other surfaces. The 17th-century Florentine author, Filippo Baldinucci, wrote in his *Vocabulario Toscano* of 1681 that the great advantage of *lavagna* is that "the colours applied ... do not sink in as much as on panel or canvas."[63] This gives an added richness and depth to the subject matter depicted, recalling the medieval view that the richness of the material used should equate to the importance of the subject matter. While the large fields of empty space that Sebastiano increasingly left in his paintings, for which slate was particularly suitable, was a further way of exploiting this feature to the maximum. The slate support produced a far darker background than a rendering on wood or canvas would have done, and emphasized the desolation and isolation of the subject depicted.

60 Personal communication on 27 May 2016 with Simonetta Antellini, Direttore dell'Ufficio Esportazione di Rome, who had overall responsibility for the restoration; the full results of which should eventually be published.

61 Ana González Mozo will discuss this technique in greater detail in the next essay.

62 *Sebastiano del Piombo e la pittura su pietro*, pp. 45-82; Cerasuolo, "Osservazioni sulla tecnica," pp. 83-5.

63 Filippo Baldinucci, *Vocabulario Toscano dell'Arte del Disegno* (Florence, 1681), p. 80.

Finally, in terms of the paintings' materiality, an important aspect not mentioned in Soranzo's letter and unfortunately not surviving for any of these paintings is the *ornamento*, or frame. Frames are, however, consistently referred to elsewhere and a concern for the decent setting of his religious works can probably be traced back to the beginning of Sebastiano's Roman career. For the *Raising of Lazarus*, there is detailed correspondence from 1518 in which Sebastiano seeks Michelangelo's intervention with Cardinal Giulio de'Medici to ensure that it is framed correctly.[64] According to both Vasari and Borghini, a concern with framing became an integral part of Sebastiano's use of stone supports. As early as Vasari's first edition of 1550 he makes mention of Sebastiano's "ricignendole di ornamenti di altre pietre mischie belle."[65] These stone frames not only functioned practically but also meaningfully, acting as a mediator between the subject depicted and the viewer. They also call attention immediately to the material of the support.

These are not the only evidence for this practice though as, in the correspondence between Ferrante Gonzaga and Nino Sernini, there are several discussions concerning Sebastiano's wish for a suitable stone frame for the Úbeda *Pietà*.[66] The inventory of Sebastiano's studio after his death also records a number of such frames incomplete utilizing a wide variety of precious materials are involved, as the materiality of Sebastiano's work came to extend beyond the painting itself.[67] Unfortunately, little or no trace of these elaborate frames survives. There is only one small sketch of a cornice on the reverse of a drawing. Interestingly enough, this is the same drawing in an American private collection that has the sketch of the bearded Christ on the front which may suggest to us Sebastiano's original work on stone.[68] Given this sketch is dated to sometime in the 1530s, it could even represent the design for such a frame.

## By Way of Conclusion

Besides the innovation itself, there is one further important and neglected aspect to Sebastiano's painting on a stone surface. Vasari names only one pupil of Sebastiano, the Sicilian, Tommaso Laureti; and Sebastiano is generally seen

64 *Il Carteggio*, 2:32-33ff.

65 Vasari, *Le Vite*, 5:98; Borghini, *Il Riposo*, 1:454.

66 Hirst, "Sebastiano's Pieta," pp. 589-92.

67 Hirst, *Sebastiano del Piombo*, pp. 155-57.

68 *Sebastiano del Piombo 1485-1547*, pp. 326-7.

as being an artist without direct successors.[69] Besides the impact of his work in Spain, his influence at Rome may have been more diffuse than Vasari suggests. Vasari himself could have learned from Sebastiano's technical innovations, with a small painting on slate of the unusual subject of *Christ in the House of Martha and Mary*, dateable to 1546 precisely around the years when he was in contact with Sebastiano.[70] When this was at Ickworth House in Suffolk it was often attributed to Sebastiano, although it has since been shown to be a rare work of Vasari himself on slate; its present whereabouts are unfortunately unknown. Whatever the case, that Vasari himself produced such a work lends greater credence to his description of the technique at the beginning of the *Vite*.

In particular, it was Sebastiano who was the direct progenitor of that very Roman phenomenon of painting on stone for large-scale altarpieces.[71] I shall end with a discussion of the *Assumption of the Virgin* in the Bandini Chapel off the left transept of San Silvestro al Quirinale, which has also recently been recently restored to an approximation of its original state [**Fig. 1.7**].[72] This was painted by Scipione Pulzone in 1585, in his only use of this method, and was restored between 2006 and 2007.[73] The chapel is otherwise well-preserved but the magnificent setting that it provides for the altarpiece has never received its full due; this commission brings our argument directly back to where this essay began.

As well as the altarpiece itself, the complete ensemble designed by Ottavio Mascherino is a succinct version of the Chigi Chapel in Santa Maria del Popolo.[74] With the octagonal plan, the statues in niches, the tondi, the altarpiece whose

---

69 Vasari, *Le Vite*, 5:103.

70 Edmund Pillsbury, "Three unpublished paintings by Giorgio Vasari," *Burlington Magazine* 112, no. 803 (1970), 94-101, esp. pp. 97-98; Sotheby's London December 9 1992.

71 For a partial listing of these see Marcia Hall, *Color and Meaning Practice and Theory in Renaissance Painting*, (Cambridge 1992), p. 254, n. 16; Fabio Fernetti, "Verso un Censimento dei Dipinti Cinquecenteschi su Pietra o Lavagna,", in *Sebastiano del Piombo e la Capella Borgherini nel contesto della pittura Rinascimentale*, ed. Santiago Arroyo Esteban, Bruno Marrocchini and Claudio Seccaroni (Rome, 2010), pp. 54-57

72 *Restauro della Cappella Bandini in San Silvestro al Quirinale*, ed. R. Bernini, C. Bigari, L. Carloni, P. Giacomazzi (Rome, 2006).

73 Cristiana Bigari, "Note sul restauro dell'*Assunzione della Vergine* in San Silvestro al Quirinale," *Scipione Pulzone e il suo tempo*, ed. Alessandro Zuccari (Rome, 2015), pp. 233-47.

74 *Leonardo da Vinci Michelangelo and the Renaissance in Florence*, ed. David Franklin, (New Haven and Ottawa, 2005), pp. 244-47; *Scipione Pulzone: Da Gaeta a Roma alle Corti Europee*, ed. Alessandra Acconci and Alessandro Zuccari (Rome, 2013), pp. 328-33 and 404-5.

surface is composed of a number of blocks of stone, every element is present. Pulzone's slate altarpiece like those of Sebastiano suffered over time and was not clearly legible until the most recent restoration. Since the patron Pierantonio Bandini, a Florentine banker was the equivalent of Agostino Chigi for the late 16th century, the richest banker in Rome, one might argue that the imitation here is entirely deliberate. There is one major difference, illustrating a post Tridentine shift. Sebastiano had had a relatively free hand as to subject matter and material in the Chigi Chapel. Bandini and his son, Monsignor Ottavio, both men closely connected to the Curia, sought advice indirectly on the depiction of the subject matter from no less a figure than Cardinal Gabriele Paleotti.

The author of the *Discourse on sacred and Profane Images* of 1582 (and therefore a very recent publication), the Bolognese Paleotti was a proselytizer for the proper role and function of sacred art after the Council of Trent.[75] There survives a revealing, lengthy and minutely detailed three way correspondence in April of the following year, 1583, between Paleotti and the humanist, Carlo Sigonio, in Bologna and the future Cardinal Silvio Antoniano in Rome as to the proper way of depicting the Assumption of Mary.[76] Unfortunately, the one thing this lengthy correspondence does not reveal is whether Pulzone's unusual choice of surface was motivated directly by Paleotti too. It is telling that there has been little if any discussion of this aspect to the commission, especially given the attention paid to the Oratorian fathers and the choice of slate for Rubens altarpiece in the *Chiesa Nuova*.[77]

Whatever the case, the Bandini chapel commission moves the argument of this volume forward and in a new direction; it demonstrates the importance and novelty that this aspect of Sebastiano's practice had for Rome. Sebastiano had very specific reasons for the development of painting on stone, as analysed above and in the other essays in this volume. That he became known for this practice in particular demonstrates that its subsequent impact was profound.

---

75 Paolo Prodi, *Il Cardinale Gabriele Paleotti*, 2 vols. (Rome, 1959-67); Fabrizio d'Amico, "Scipione Pulzone, l'Assunta di S. Silvestro al Quirinale, e un aspetto del rapporto fra pittura e controriforma," in *L'Arte in Roma nel Secolo XVI*, ed. Daniela Gallavotti Cavallero, Fabrizio d'Amico, and Claudio Strinati, 2 vols. (Bologna, 1992), 2:231-43; Gabriele Paleotti, *Discourse on Sacred and Profane images*, ed. Paolo Prodi and trans. William McCuaig (Los Angeles, 2012).

76 Paolo Prodi, "Sulla teorica delle arti figurative nella riforma cattolica," *Archivo Italiano per la Storia della Pietà* 4 (1965), 121-212, esp. pp. 170-71 and 191-93.

77 *The Letters of Peter Paul Rubens*, ed. and trans. Ruth Saunders Magurn (Cambridge, 1955), pp. 42-43; Michael Jaffé, *Rubens and Italy* (Oxford, 1977), pp. 85-99.

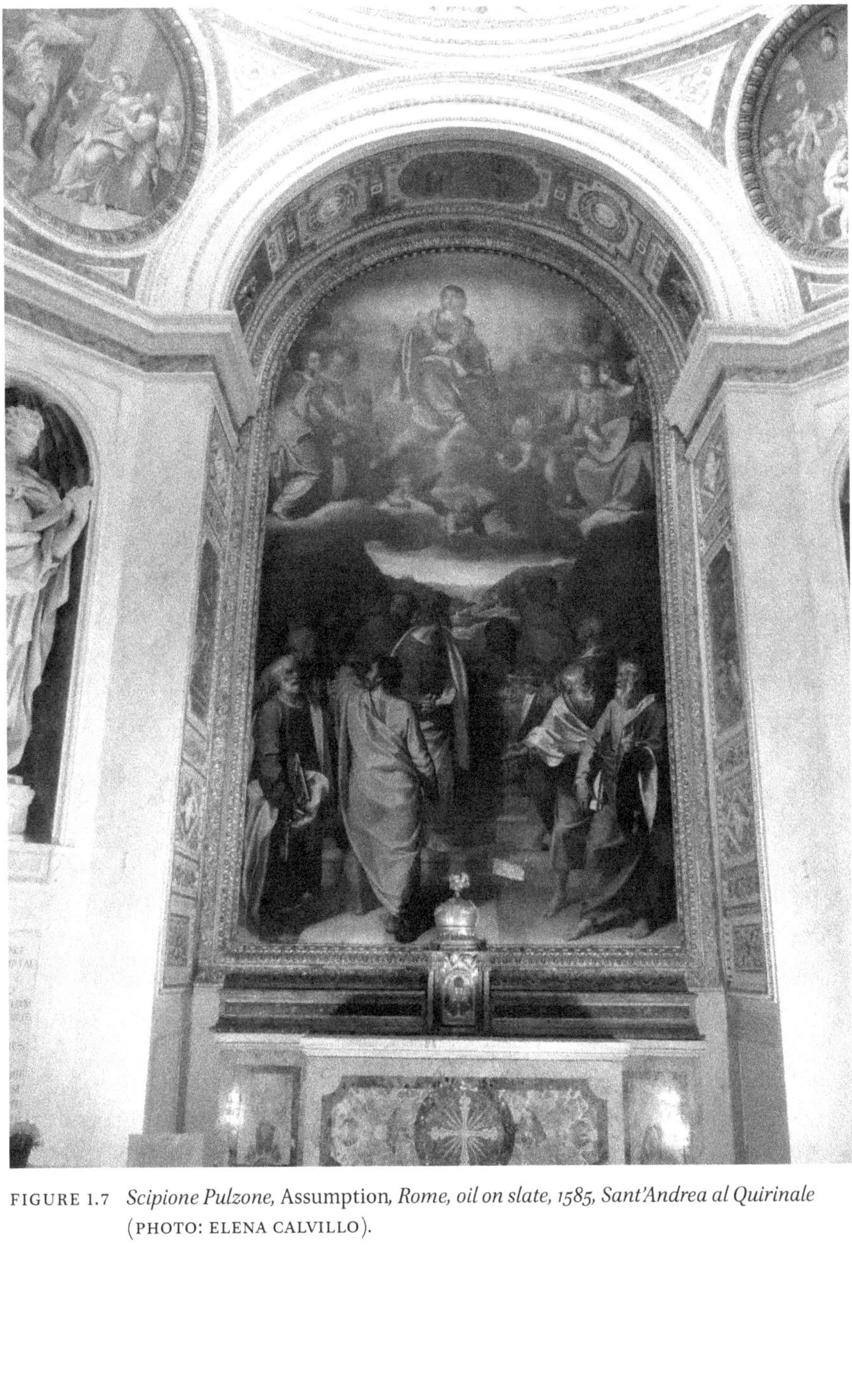

FIGURE 1.7 *Scipione Pulzone,* Assumption, *Rome, oil on slate, 1585, Sant'Andrea al Quirinale* (PHOTO: ELENA CALVILLO).

## Bibliography

Azzolini, Monica. "In Praise of Art: Text and Context of Leonardo's Paragone and its critique of the Arts and Sciences." *Renaissance Studies* 19, no. 4 (2005), 487-510.

Baker-Bates, Piers. "Beyond Rome: Sebastiano del Piombo as a Painter of Diplomatic Gifts." *Renaissance Studies* 27, no. 1 (2013), 51-72.

Baker-Bates, Piers. "A Painting little less than Eternal." In *Michelangelo & Sebastiano*, edited by Matthias Wivel, 74-85. London, 2017.

Baldinucci, Filippo. *Vocabulario Toscano dell'Arte del Disegno*. Florence, 1681.

Ballarin, Alessandro. "Un nuovo ritratto su lavagna di Sebastiano del Piombo." *Nuovi Studi. Rivista di Arte Antica e Moderna* 20, no. 21 (2015), 71-80.

Bambach, Carmen. "Leonardo and Raphael in Rome in 1513-16." In *Late Raphael: Proceedings of the International Symposium Madrid, Museo Nacional del Prado, October 2012*, edited by Miguel Falomir, pp. 26-37. Madrid, 2013.

Barbieri, Costanza. "Rodomonte Gonzaga, il sacco di Roma e un riscoperto ritratto di Sebastiano del Piombo." In *I Gonzaga e i papi Roma e le corti padane fra Umanesimo a Rinascimento (1418-1620): Atti del convegno Mantova-Roma 21-26 febbraio 2013*, edited by Renata Salvarani, pp. 331-51. Vatican City, 2013.

Barbieri, Costanza. "*Tam foelix pictor vate, ut pictore Poeta*: The Iconography of Sebastiano del Piombo's Lunettes in the Loggia della Galatea." *Artibus et Historiae* 72 (2015), 125-52.

Bigari, Cristiana. "Note sul restauro dell'*Assunzione della Vergine* in San Silvestro al Quirinale." In *Scipione Pulzone e il suo tempo*, edited by Alessandro Zuccari, pp, 233-47. Rome, 2015.

Bocchi, Francesco. *Le Belleze della cittá di Fiorenza*. Florence, 1677.

Bocchi, Francesco. *Le Bellezze della città di Fiorenza* (1591). Farnborough, 1971.

Borghini, Raffaello. *Il Riposo*. Edited by Mario Rosci. 2 vols. Milan, 1967.

Borghini, Raffaello. *Il Riposo*. Edited and translated by Lloyd H. Ellis jr. Toronto, 2007.

Butters, Suzanne B. *The Triumph of Vulcan: Sculptors' Tools, Porphyry, and the Prince in Ducal Florence*. 2 vols. Florence, 1996.

Calí, Maria. *Da Michelangelo all'Escorial: momenti del dibattito religioso nell'arte del Cinquecento*. Turin, 1980.

Calvillo, Elena. "Authoritative Copies and Divine Originals: Lucretian Metaphor, Painting on Stone, and the Problem of Originality in Michelangelo's Rome." *Renaissance Quarterly* 66, no. 2 (2013), 453-508.

Cerasuolo, Angela. "Osservazioni sulla tecnica di Sebastiano del Piombo." *Nuovi Studi. Rivista di Arte Antica e Moderna*." 20, no. 21 (2015), 81-86.

Collomb, Anne-Laure. "La peinture sur pierre entre Rome et Madrid: Sigismondo Laire parmi ses contemporains." *Paragone* 57, no. 69 (679) (2006), 75-87.

Anne-Laure Collomb, *Splendeurs d'Italie: La peinture sur pierre à la Renaissance*, Rennes and Tours, 2012

D'Amico, Fabrizio. "Scipione Pulzone, l'Assunta di S. Silvestro al Quirinale, e un aspetto del rapporto fra pittura e controriforma." In *L'Arte in Roma nel Secolo XVI*, edited by Daniela Gallavotti Cavallero, Fabrizio d'Amico, and Claudio Strinati, vol. 2, pp. 231-43. Bologna, 1992.

*Delle Lettere da diversi Re et Principi et Cardinali et altri huomini dotti a Mons. Pietro Bembo scritte*. Venice, 1560.

De Marchi, Andrea. *Collezione Doria Pamphij: Catalogo Generale dei Dipinti*. Milan, 2016.

De Nolhac, Pierre. "Une Galerie de peinture au XVIe siecle: Les Collections de Fulvio Orsini." *Gazette des Beaux Arts* 29 (1884): 427-36.

De Siguenza, José. *Historia de la Orden de San Jerónimo*. Edited by Juan Catalina García. 2nd ed. 2 vols. Madrid, 1907-09.

*Dessins Italiens du Musée Condé de Chantilly III, Vénétie, Lombardie, Piémont, Émilie, XVe-XVIe siècle*. Edited by Olivier Villard. Paris, 1998.

Doni, Anton Francesco. *Disegno Fac Simile della Edizione del 1549 di Venezia*. Edited by Marco Pepe. Milan, 1970.

Farago, Claire J. *Leonardo da Vinci's* Paragone: *A critical interpretation with a new edition of the text in the* Codex Urbinas. Leiden, 1992.

Fernetti, Fabio. "Verso un Censimento dei Dipinti Cinquecenteschi su Pietra o Lavagna." In *Sebastiano del Piombo e la Capella Borgherini nel contesto della pittura Rinascimentale*. Edited by Santago-Arroyo Estaban, Bruno Marrocchini, and Claudio Seccaroni, pp. 54-7. Rome, 2010.

Firpo, Massimo. *Artisti, Gioiellieri, Eretici Il Mondo di Lorenzo Lotto tra Riforma e Controriforma*. Rome and Bari, 2001.

Goffen, Rona. *Renaissance Rivals: Michelangelo, Leonardo, Raphael, Titian*. New Haven and London, 2002.

González Mozo, Ana. "Raphael's Painting Technique in Rome," In *Late Raphael*, edited by Tom Henry and Paul Joannides, pp. 319-49. Madrid, 2012.

Hall, Marcia. *Color and Meaning: Practice and Theory in Renaissance Painting*. Cambridge, 1992.

Hall, Marcia. "Sebastiano del Piombo fra innovazione e sperimentazione." In *Notturno Sublime: Sebastiano e Michelangelo nella Pietà di Viterbo*, edited by Costanza Barbieri, pp. 42-7. Rome, 2004.

Hendler, Sefy. *La guerre des arts: le paragone peintre-sculpture en Italie XV-XVII siècle*. Rome, 2013.

Hessler, Christiane J. "The Man on Slate: Sebastiano del Piombo's Portrait of Baccio Valori and Valori the Younger's Speech in Borghini's Il Riposo." *Source* 25, no. 2 (2006), 18-22.

Hirst, Michael. "The Chigi Chapel in Santa Maria della Pace." *Journal of the Warburg and Courtauld Institutes* 24 (1961), 161-85.

Hirst, Michael. "Sebastiano's Pieta for the Commendador Mayor." *Burlington Magazine* 114, no. 834 (1972), 585-95.

Hirst, Michael. *Sebastiano del Piombo*. Oxford, 1981.

Horovitz, Isabel, and Joan Reifsnyder. "The use of slate or stone as a painting support." In *Conservation of Easel Paintings*, edited by Joyce Hill Stoner and Rebecca Rushton, pp. 97-100. Abingdon and New York, 2012.

*Il Carteggio di Michelangelo*. Edited by Paola Barocchi and Renzo Ristori. 5 vols. Florence, 1965-83.

Jaffé, Michael. *Rubens and Italy*. Oxford, 1977.

Joannides, Paul. "The Portraits of a Lady." In *Artistic Practice and Cultural Transfer in Early Modern Italy: Essays in Honour of Deborah Howard*, edited by Nebahat Avcioğlu and Allison Sherman, pp. 233-41. Farnham and Burlington, 2013.

*Leonardo da Vinci Michelangelo and the Renaissance in Florence*, ed. David Franklin, pp. 244-47. New Haven and Ottawa, 2005.

Mignosi Tantillo, Almamaria. "Restauri alla Farnesina." *Bollettino d'Arte* 57 (1972), 33-43.

Minter, Erin Sutherland, "Virtuous Prelates, Burdensome Relics and a Sliver of Gold in the *Last Judgement*." In *Michelangelo in the New Millennium: Conversations about Artistic Practice, Patronage and Christianity*, edited by Tamara Smithers, pp. 147-78. Boston, 2016.

Nagel, Alexander, and Christopher S. Wood. *Anachronic Renaissance*. New York, 2010.

Nygren, Christopher J. "Titian's Ecce Homo on Slate: Stone, Oil and the Transubstantiation of Painting." *Art Bulletin* 99, no. 1 (2017), 36-66.

Paleotti, Gabriele. *Discourse on Sacred and Profane Images*. Edited by Paolo Prodi and translated by William McCuaig. Los Angeles, 2012.

Panofsky, Erwin. "Die Pietà von Ubeda. Ein kleiner Beitrag zur Lösung der Sebastianofrage." In *Festschrift für Julius Schlosser zum 60. Geburtstage*, edited by Arpád Weixelgärtner and Leo Planiscig, pp. 150-61. Zurich, 1927.

Panofsky, Erwin. "Jean Hey's 'Ecce Homo': Speculations about its author, its donor and its iconography." *Bulletin des Musées Royaux des Beaux Arts, Brussells* 5, nos. 3-4 (1956), 95-138.

Perguidi, Stefano. "Le Portrait de Clement VII par Sebastiano del Piombo du J. Paul Getty Museum." *Revue de l'Art* 160, no. 2 (2008), 55-9.

Pillsbury, Edmund. "Three unpublished paintings by Giorgio Vasari." *Burlington Magazine* 112, no. 803 (1970), 94-101.

Pinelli, Antonio. "La Cappella delle tombe scambiate: Novità sulla Cappella Chigi in Santa Maria del Popolo." In *Francesco Salviati ou la Bella Maniera. Actes des colloques de Rome et de Paris (1998)*, edited by Catherine Monbeig-Goguel, Phillipe Costamagna, and Michel Hochmann, pp. 253-85. Rome, 2001.

Prodi, Paolo. *Il Cardinale Gabriele Paleotti (1522-1597)*. 2 vols. Rome, 1959-1967.

Prodi, Paolo. "Sulla teorica delle arti figurative nella riforma cattolica." *Archivo Italiano per la Storia della Pietà* 4 (1965), 121-212.

Ramallo Asensio, Germán. "Toda la obra conservada en España y hasta ahora conocida, del pintor Segismundo Laire. Alemán en Roma" *Archivio Español de Arte* 79, no. 315 (2006), 243-61.

*Restauro della Cappella Bandini in San Silvestro al Quirinale*. Edited by R. Bernini, C. Bigari, L. Carloni, and P. Giacomazzi, Rome 2006.

*Scipione Pulzone: Da Gaeta a Roma alle Corti Europee*. Edited by Alessandra Acconci and Alessandro Zuccari. Rome, 2013.

*Sebastiano del Piombo 1485-1547*. Milan, 2008.

*Sebastiano del Piombo e la Cappella Borgherini nel contesto della pittura Rinascimentale*. Edited by Santiago Arroyo Esteban, Bruno Marrocchini, and Claudio Seccaroni. Florence, 2010.

*Sebastiano del Piombo e la pittura su pietro: il Ritratto di Baccio Valori: Restauro e ricerche*. Edited by Alessandra Cecchi, Marco Ciatti, and Oriana Sartiani. Florence, 2014.

*Sebastiano del Piombo y Espana del 1 de marzo al 30 de abril 1995*. Edited by Manuela B. Mena Marqués and Fernando Benito. Madrid, 1995.

Shearman, John. "The Chigi Chapel in Santa Maria del Popolo." *Journal of the Warburg and Courtauld Institutes* 24 (1961), 129-60.

*The Letters of Peter Paul Rubens*. Edited and translated by Ruth Saunders Magurn. Cambridge, Mass., 1955.

Thomas, Ben,. "'The lantern of painting': Michelangelo, Daniele da Volterra and the *Paragone*." *Apollo* 154, no. 474 (2001), 46-53.

Varchi, Benedetto. "Lezzione nella quale si disputa della maggioranza delle arti e qual sia più nobile, la scultura o la pittura." In *Trattati d'Arte del Cinquecento fra manierismo e controriforma*, edited by Paola Barocchi, vol. 1, pp. 1-82. Florence, 1960.

Vasari, Giorgio. *Le Vite de'più eccellenti pittori scultori e architettori nelle redazioni del 1550 e 1568*. Edited by Rosanna Bettarini and Paola Barocchi. 9 vols. Florence, 1967-1987.

Weil Garris Posner, Kathleen. *Leonardo and Central Italian Art: 1515-1550*. New York, 1974.

Zeri, Federico. *Pittura e Controriforma L''arte senza tempo' di Scipione da Gaeta*. 3rd ed. Vicenza, 2001.

CHAPTER 2

# Painted Stone: Idea and Practice in Italian Renaissance

*Ana González Mozo*

… blue will be more beautiful when there is greater darkness below[1]

## Background and Arguments

Years after his arrival in Rome in 1511, Sebastiano del Piombo laid the foundations of the method for painting on stone supports. Although there have been differing opinions among scholars about the motivation and influence of his method, this essay proposes that Sebastiano's visual and technical background in Venice greatly influenced the solutions that he adopted. The success of the invention is testified to by Titian's emulation in 1545 and later by Daniele da Volterra's adaptation, which indicates a reinterpretation according to Roman aesthetics.

The technique and appearance of polychromy on stone were described by the authors of classical treatises and philosophers in antiquity,[2] always linked to sculpture, and perfectly suited to the debate on the *paragone*, which was a topic of interest in the Roman environment in the mid-16th century. The art of painting and sculpture were pitted one against the other, as were mural and oil painting, each trying to justify their supremacy according to the principles of durability, or persistence over time, and the difficulty of the particular medium. Stone seemed the place to represent this dispute in easel painting, but an attentive look at the practice, through a theoretical lens, suggests that there were other substantive reasons linked to the artistic challenges in question. Stone provided a particular support that seemed stable and able to reproduce

1 Aristotle, *Meteorologica*, 2.239. The black in antiquity was the negation of color, invisible, *Melan*, darkness, related to the idea of the blue. See Homer, *Iliad*, 14.191, and *Odyssey*, 11.93; Virgil, *Aeneid*, 6.134, 237, and 267.

2 Pliny, *Natural History*, 35.61, 149, and 133, and Pausanias, *Description*, 1.22.6-76.

 | DOI 10.1163/9789004361492_005

nuances inspired by a new perception of the world and its *propietà*,[3] concepts that had their highest level of excellence in the Veneto. The recovery of the ancient world played an important role in this process.

Decisions in painting are seldom spontaneous; requiring favorable conditions to register ideas and give shape to experience. In the first decades of the 16th century, it was in Venice where a series of factors converged to inspire Sebastiano's development: on the one hand, the coexistence of polychrome stone and murals on Venetian facades, and, on the other hand, the city's position as an important center of the collecting of antiquities and as a receptacle of materials and visual sources coming from the East.[4] The printing in the *Serenissima* and in nearby Ferrara of many classic texts and of ancient works, whose descriptions suggest a pictorial mode of thinking, would have awakened interest in the technique of Greek artists.[5]

The physical properties of slate and marble, the focus of this essay, favored a special oil absorption. These surfaces allowed the chromatic and lighting effects codified by Aristotle or Plotinus,[6] and especially by Pliny in his *Natural History*, to be reproduced. Their qualities made them an ideal material for realizing certain Greco-Roman ideals of painting: the diffused bodies, chiaroscuro, transparent and homogeneous dark, halftones control, and glare. These ancient concerns were adapted to Venetian taste and its sensual conception of art, although neither Sebastiano, the principal innovator of this method, nor Titian painted on stone for the Republic of Venice. Of considerable importance were the relations between Venice, Padua, Mantua and Ferrara where

3 A term coined by Paolo Pino in reference to the material nature of the represented object. Paolo Pino, *Dialogo di Pittura*, ed. Susanna Falabella (Rome, 2000), pp. 116ff.. This concept concerns the texture of the surfaces and the relationship between colors, in the sense of the nuances reached by the artists through their handling of color materials. It also concerns the effect produced by the painted surfaces according to how the light is received and reflected. Boschini, in his *Breve Instruzione*, held Giorgione up as an example, writing that "in coloring, he invented the impasto of the soft brush." In this regard, see Moshe Barasch, *Light and Color in the Italian Renaissance Theory of Art* (New York, 1978), p. 97.

4 Promoted by Ciriaco of Ancona and his travels. Many pieces were acquired by Ludovico Trevisani and Pietro Barbo (later Pope Julius II).

5 In Ferrara, Guarino da Verona published in 1433 the *Natural History by Pliny*. The first edition of Pausanias' *Description* was done there in 1516 (Printing House of Aldus Manutius). The two texts considered as a prelude to the Venetian art theory: *De Sculptura* and *Hypnerotomachia Poliphili* were written by scholars who had a very close contact with the classical world: the first by Pomponio Gaurico, who had, according to Marcantonio Michiel many works of ancient sculpture and maintained relationship with Mantegna, and the second by Francesco Colonna, who demonstrates his passion for the "hoary and venerable Antiquity."

6 Aristotle, *Meteorologica*, and Plotinus, *Enneads*, 1.6.

the classical past was received and interpreted in a suggestive way. Tinged with the passage of time, the marble works, Aegean and Roman pieces, which entered collections in the region, seduced the local artists and provided them with formal and aesthetic archetypes. If Roman *Cinquecento* painting captured the formal strength of classical statuary, Venetian art sensed the diffuse appearance, lightly tinted, that was bequeathed to its stone. Still, the shape and volume of sculpture provided the basis for discussion between the arts of painting and sculpture;[7] the image of the achromatic statuary described by Ciriaco d'Ancona after his trip to Greece and its conceptual clarity, the *intacta pureza* of the ancient works that had lost their polychrome,[8] began to provide the measure by which artists were judged. This shift, evidence suggests, partly based the discussion or *paragone* between the arts on foundations that were, if not false, manipulated. In fact, the ancient texts described the color of the sculptures, and, paradoxically, in Greece 'white' was synonymous with a disorder that the color had to organize.[9]

In a changing environment marked by a depletion of the technical solutions, the competition between painting and sculpture seems to have become an endless subject in the theoretical art writings of the 16th century, which concluded by acknowledging the supremacy of drawing.[10] A reinterpretation of the classical texts was necessary in order to provide an explanation for the origin of both disciplines; the quotes are endless. This fueled the debate on

7 Philippe Jockey, *Le mythe de la Grèce blanche. Histoire d'un rêve Occidental* (París, 2013).

8 For polychrome architecture, see Melucco Vaccaro, *Policromia*, 19. For Ciriacus of Ancona, see Edward W. Bodnar and Charles Mitchell, *Cyriacus of Ancona's journeys in the Propontis and the Northern Aegean, 1444-1445*, Memoirs of the American Philosophical Society, vol. 112 (Philadelphia, 1976). Their travelogue was lost in the fire at the Sforza Library, in the middle of 16th century. For polychromy and ancient sculpture, see Michelangelo Cagiano Di Azevedo, "Il colore nella antichità," *Aevum* 28 (1954), 151-67. He talks about restorations made in the 16th and 17th centuries in the polychrome sculptures, but does not mention the source. The ancient sources discussed in Eurípides, *Hipsipila*; Plato, *Republic*, 4.420c-d; Pliny, *Natural History*, 35.133; and Plutarch, *De Gloria Atheniensium'*, 6.348e.

9 Jockey, *Le mythe*, p. 12.

10 Lucian, *Somnium*, 7; Quintilian, *Institutio Oratoria*, IX, 4.147; Philostratus, Introduction to Book I of *Imagines*, 0-2; Aristotle, *Poetics*, 1450a, 14ff; Horace, *Ars Poetica*, 4.361-65; Leonardo da Vinci, *Traité de la peinture, trans.* Andrè Chastel (Paris, 1964), p. 102. Lodovico Dolce, *Dialogo della pittura di M. Lodovico Dolce, intitolato l'Aretino* (Florence, 1735), 2:29r and ff. "Letter from Michelangelo Buonarrotti to Benedetto Varchi," in Paola Barocchi ed., *Trattati D'Arte del Cinquecento. Fra Manierismo e Contrariforma*, 3 vols. (Bari: Giuseppe Laterza e figli, 1960-1962), 1:82. "Letter from Benvenutto Cellini to Benedetto Varchi," in Barocchi, ed., *Trattati*, 1:80-81; and Giovanni Paolo Lomazzo, *Idea del tempio della pittura*, trans. Robert Klein (Florence, 1974), pp, 61, 91, and 151.

questions of mimesis and eternity and determined the evolution of the painting method discussed here, inspired, as always, by the descriptions of Pliny.[11] Durability and difficulty were arguments that stimulated the use of stone supports, but they were not the only ones.[12] Only Daniele da Volterra, twenty years after Sebastiano had painted his first *lavagna*'s, competed clearly with sculpture. In the context of the *paragone*, Giovanni della Casa, in order to illustrate his treatise on painting proposed one of the most intriguing challenges to painters, commissioning Daniele da Volterra to create his *David and Goliath* [**Plates 6a and 6b**], painting it on both sides of a slab of slate, a kind of rock that cannot be carved, and thereby placing the two disciplines at a point of greater inequality.[13] This painting also engaged other topics related to the *paragone*, among which was the multiplicity of views that sculpture allowed: up "to eight," or so stated Benvenuto Cellini in a letter to Benedetto Varchi 1546.[14] Francesco Doni claimed, along the same lines, that the *grandissima* difficulty of sculpture was found in its endless views, which should be joined gracefully. At this point the comparison was in vain. Nevertheless, the challenge inspired

11 "Today they exist in Ardeade, in temples, in paintings older than those of Rome [...] that even so, without being protected by a roof, after so long, they keep their freshness." Pliny, *Natural History*, 35.17.

12 "... e fanno delle pitture, che non meno dalle ingiurie del tempo si difendono, che le statue, come le pitture nel marmo, & musaici." Raffaello Borghini, *Il Riposo di Raffaello Borghini in cui della pittura, e della scultura si fauella, de' più illustri pittori, e scultori, e delle più famose opere loro si fa mentione; e le cose principali appartenenti a' dette arti s'insegnano* (Florence, 1584), 1:40. Inspired by the myth of Pygmalion, Petrarch in *De rebus familiaribus*, 5.17-18, compared painting with sculpture according to their proximity to nature: "the sculptor's works are more alive than the painter's ones" but Boccaccio, in the *Decameron*, 6.5, was the first who related the visual arts with the imitation of nature. See his famous passage about the painting of Giotto. "Hence the human sense of sight has often been deceived by its works, taking as real what was only painted" refers to the competition between Parrhasius and Zeuxis. All statements on this topic were collected by Benedetto Varchi in *Leccion que hizo Benedicto Varqui en la Academia Florentina ... año 1546 sobre la primacia de las artes y gual sea mas noble, la escultura o la pintura; con una carta de Michael Angelo Buonarroti*, trans. Felipe de Castro (Madrid, 1753).

13 *C.* 1550, Louvre Musée. Cécile Scaillierez et al., "L'ardoise double face de Daniele Volterra, figurant David et Goliat: étude et restauration d'une oeuvre d'exception," *Techné* 25 (2007), 5-11.

14 Barocchi, ed., *Tratatti*, pp. 519-20, "A painting is nothing more than one of the eight principal views required of a statue [...]. These views are not only eight, but more than forty, because even if the figure is rotated no more than an inch, there will be some muscle showing too much or not enough, so that each single piece of sculpture presents the greatest variety of views."

one of the most evocative pictures of the Roman Renaissance, which has survived in perfect condition.[15]

Another argument relevant to the *paragone* between painting and sculpture was that of durability or preservation, that is to say the superior durability of sculpture and by extension stone objects, an argument made most extensively by Benedetto Varchi and Francesco Sangallo. For them, it was more dignified to be more durable and resistant to atmospheric and natural dangers, and thus susceptible only to time itself.[16] Other artists such as Pontormo and Zuccaro were more pessimistic;[17] their kind of disillusionment may also be seen in a letter from Galileo to Cigoli:

> The argument of eternity is worthless, since it is not the sculpture that makes the marble eternal, the reality is that marble makes the sculptures eternal. This privilege belongs only to the hard rocks, perhaps painting and sculpture are both destined to perish.[18]

Galileo's judgment that the permanence of the works depended not on the genius of the artist but rather on the choice of materials, hence reproducing many centuries later—and again in a rhetorical key—the same lamentation of Cicero and Seneca.[19] In the Venetian sphere, Doni also thought that whatever the chosen support was, the problem of preservation remained.[20] His comments suggest that even at that time the practice of oil painting on stone was not satisfactory, an opinion already present in a letter from Niccolò Sernini to Ferrante Gonzaga in 1537, one of the first testimonies of the conservation problem.[21]

---

15 Jacques Touret (School of Mines in Paris) detected the presence of finely dispersed rare micro-crystals of pyrite in the rock mass, which increases their quality. Scaillierez et al., "L'ardoise," pp. 8 and n, 10.

16 Barocchi, ed., *Tratatti*, p. 516.

17 Barocchi, ed., *Tratatti*, pp. 506, 698-69, and Federico Zuccaro, *L'Idea L'idea de' pittori, scvltori, et architetti, del Cavalier Federico Zvccaro; Diuisa in due Libri* (Turin, 1607), L II, VII, 35-35. Vasari himself demonstrates skepticism about its strength.

18 Lauriane Fallay d'Este and Nathalie Bauer, *Le Paragone: le parallèle des arts*, trans. Lauriane Fallay D'Este and Natalie Bauer (Paris, 1992), p. 228.

19 Cicero, *Brutus*, 17, Seneca, *Letters*, 86.6s, and Vitruvius, *Ten Books on Architecture*, 7.5.7-8.

20 Anton Francesco Doni, *Il disegno del Doni partito in piu raggionamenti, ne quali si tratta della scultura e pittura, de colori, de getti, de modegli, con molte cose appartenenti a quest'arti & si termina la nobilta dell'una et dell'altra professione* ... (Venice, 1549), p, 20.

21 Biblioteca Estense di Módena, Aunto. Campori, lettere 32, published by Michael Hirst, in "Sebastiano's Pietà for the Commendador Mayor," *Burlington Magazine* 114, no. 834 (1972), 589.

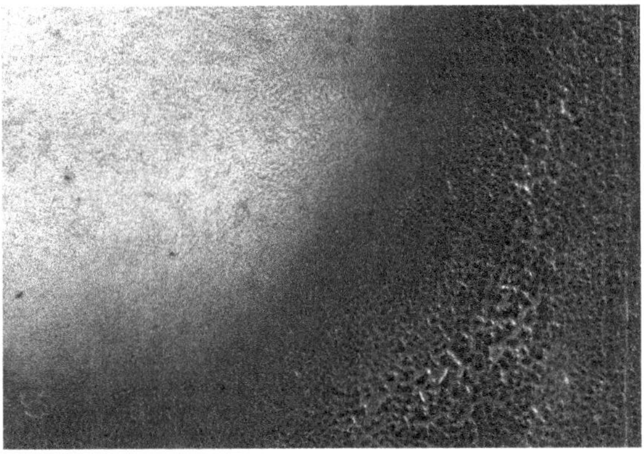

FIGURE 2.1
*Titian,* Ecce Homo, *Oil on slate, 1547, Madrid, Museo Nacional del Prado. Detail of the cracks in the surface* (PHOTO: © MUSEO DEL PRADO, MADRID).

Who managed to create paintings for eternity? Oil on the stone is not preserved much better than on the canvas or wood, and artists had been aware of this since the technique was implemented. One common problem or alteration was the premature contraction of the layers, caused by the oily priming of polished stones. In the latter, neither the priming nor the painting, dried well, forcing the addition of drying products to oil matter, making the densest parts crack quickly [**Fig. 2.1**]. If the preparation was defective, something that often happened, it compromised the adhesion of the paint, which easily peeled away from the support, even at the time of execution.[22] What is more, some supports—at least until 1604—came from the material looted from Roman buildings of late imperial period,[23] materials that had been exposed for centuries to environmental changes and exuded moisture. This condensed into the resinous ground, which was tarnished and lost transparency by varying its index of refraction, causing the visually distracting blanching.[24] The weakness of slate, however, is due to its principal character, friability, a tendency to disintegrate because of humidity [**Fig. 2.2**]. Slate's fragility was known at the time of its use,[25] as demonstrated by Sernini's reference to the *natura frangibile* of the

22 Salviati had technical problems in the Pallio Capella della Cancelleria, painted in oil on *peperino*. He had to repair his own work before finishing.

23 "Ferdinando fece venire -1597- da Roma a Firenze, 350 pezzi di pórfido, frammenti di opere antiche e 5 casse di marmi lavorati per servicio della capella di S. Lorenzo; i quali sino dai tempi di Cosimo erano stati comprati da Giorgio Vasari." Archivio della R. Guardaroba (f. 222). Antonio Zobi, *Notizie Storiche riguardanti l'imperiale e reale stabilimento dei lavori di comesso in pietre dure di Firenze* (Florence, 1841).

24 Raffaella Rossi Manaresi, "Pietre porose: alterazione e conservazione," in "Materiali lapidei. Problemi relativi allo studio del degrado e della conservazione," supplement, *Bollettino d'Arte* 41, no. S2 (1987): p. S133.

25 It is an exfoliable material by the effect of atmospheric agents or mechanical actions. The water chemically attacks the clay minerals in the slate, causing division in subtle stone

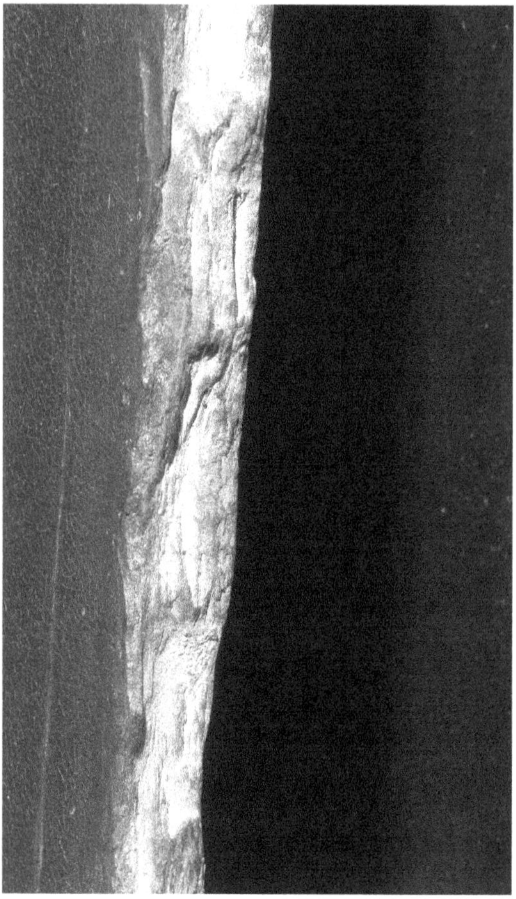

FIGURE 2.2
*Titian,* Ecce Homo, *oil on slate, 1547, Madrid, Museo Nacional del Prado. Detail of the border in which the structure of the slate can be seen* (PHOTO: © MUSEO DEL PRADO, MADRID).

*Pietà* by Sebastiano destined to Spain.[26] Years later, the big slab of stone selected by Daniele da Volterra for his *David and Goliath* was broken while he was painting it. Perhaps it was this problem, in addition to the difficulty of finding pieces that fit his formats and the subsequent great weight, that caused Titian to rule out stone as a support after painting his *Our Lady of Sorrows* [**Plate 7**],

blades, increasing the porosity and the progressive loss of water-repellent capacity. Raffaella Rossi Manaresi, "L'Ardesie ligure: deterioramento e conservazione, primi risultati di una ricerca," in *Del Dipingere e Scolpire in Pietra: Genova, Fiera Internatiionale de Genova 8-23 dicembre 1984: Piazza della Meridiana 5 dicembre 1984-12 gennaio 1985*, ed. Piero Boccardo and Ida Maria Bott (Genoa, 1985), p. 125.

26 Hirst, "Sebastiano's Pietà," p. 585.

as it was not suitable for those pieces that had to travel.[27] Despite these weaknesses, his and Sebastiano's works are well preserved, especially those of the latter. It is reasonable to think that stone is eternal; as raw materials go, and in non-extreme conditions, depending on its crystalline configuration, its structure is more stable than the processed materials, but there were certainly other concerns that motivated the use of this technique.

## Precedents and Beginnings of the Technique in the Italian Renaissance

It is unclear when the practice of painting in oil on two-dimensional stone supports began. Pliny seems to describe it in the introduction to the book 35 of his *Natural History*, and Cennino mentions it in the late 14th century, suggesting a more ancient medieval origin.[28] Nevertheless, no one gave as much importance to it as Vasari.[29] The Aretine describes in detail the method in his *Vite* and associates it with the murals of Sebastiano del Piombo in Rome.[30] There, he had found the appropriate atmosphere to develop his Venetian experiences. At the same time, Raphael seems to have been thinking of ways to use oil paint on walls to expand his wide chromatic register to large areas. It is significant that the big change in Raphael's technique (1518-20) coincided with the possible preparation of the wall in the *Stanza di Constantino* to paint in oil.[31] I suspect that this was due to the inability to reproduce in fresh lime the

27 Titian waited a long time until he reached the plaque, which makes me think that could come from an excavation.

28 Franco Brunello, in his introduction to *El Libro del Arte*, explains that Cennino, who probably wrote his treatise in the Veneto, transmits a little surprise in the art, because maybe it was not a great discovery [Cennino Cennini, *El libro del arte*, ed. Franco Brunello and trans. Fernando Olmeda Latorre (Madrid, 1988)]. Classical sources mention it, and Theophilus talks about it in *Diversarum artium Schedula*, 1.22.

29 "Our painters are becoming increasingly daring, after having used the oil painting on the wall, they have managed to use the stone." Vasari, "Della Pittura" in *Vite*.24.

30 In 1521, he painted in oils on the wall his *Flagellation* in the Chapel Borgherini. In 1526-30 Chigi made him responsible for two works in oil on *peperino*: *The Rising of Christ* for Santa Maria della Pace and *The Nativity* for Santa Maria del Popolo. The marble of this work, according to Vasari, was extracted from the Roman ground.

31 According to Vasari, the *Stanza di Constantino* was ready to be painted in oil before the death of Raphael. Sebastiano, in a letter dated July 3, 1520, writes down that the *bottega* made a figure in oil as an evidence months after the death of Sanzio in Vincenzo Golzio, *Raphael Nei Documenti, Nelle Testimonianze Dei Contemporanei e Nella Letteratura del Suo Secolo* (Westmead, 1971), p. 129.

*colori cangianti*, the metallic shades and the deep blacks he was able to obtain in his woodworks in oil. It is not clear what osmosis occurred between Raphael and Sebastiano, who, despite his criticism of the painter from Urbino, was constantly matched with him.

When Sebastiano was called by Agostino Chigi to participate in the decoration of the *Loggia della Farnesina* (1511-12), he ran into difficulties with the fresco, a new technique for him, by using an *intonaco* prepared in a Venetian way. However, despite this failure, the shades that Piombo obtained on the surface of marbled appearance, using a colored base, must have drawn the attention of Sanzio. Shortly after his work in the Farnesina, it seems that Raphael began his first attempt at using oil for a mural. Although Sebastiano said that Raphael's assistants first worked in it after the master's death, the news of their innovation might have inspired him to use this technique for his, which combined both substances. This success could have been the trigger for subsequent decisions. Gradually the idea was brewing in his head, and it is remarkable that his first slates are somewhat after his trip to Bologna (1529), where he could see a great Madonna painted on a black stone by Andrea del Sarto.[32] While this account might give the impression that Sebastiano's innovations were conceived in the shadow of Raphael, even while trying to surpass him, the latest invention always has the advantage, as in any *technē*.[33] Stone and mural oil painting allowed him to work in media that had no comparisons, and he became the executor of his pictorial testament.[34]

## Venice and Ancient Technical Precedents

Secret techniques, new mechanisms that would ensure the durability of artists' works, were commonplace in discussions between writers and artists. Unknowingly, artists had in fact used materials and techniques not very different from those used in the Paleolithic era and Antiquity. In the first cave paintings found to this date (Cuevas del Castillo, Cantabria, 40,800 BC), the pigments bound with animal fats made body with the support by carbonation and thus have

32 Mentioned by Pietro Lamo in *Graticola di Bologna: ossia descrizione delle pitture, sculture e architetture di detta città fatta l'anno 1560* (Bologna, 1844), p, 32, placed in the *Palazzo* of the Counts of Bentivoglio.

33 Thucydides, *History of the Peloponnesian War*, trans. Philippe Remacle. 1.71.3 [3]. <http://remacle.org/bloodwolf/historiens/thucydide/livre1.htm>.

34 "To overcome Raphael, Bastiano wanted to paint some things about on stone" Vasari, "Vita di Sebastiano Veneziano," in *Vidas*, pp. 711, 714, and 716.

been preserved this far. In short, artists, without having direct knowledge of these techniques, arrived at similar solutions over time, through practice, by trial and error and the use of the workshop treatises. Returning to the Renaissance, in 16th-century Italian workshops customs converged with the need and desire to find pictorial stable processes. As I previously pointed out, Venice was in a position to set a precedent. At the time of its Republic (11th century to 17th century) architects were concerned with preserving the polychrome façades of buildings, exposed to constant humidity and the salinity of its water, which penetrated the walls by capillarity.[35] After centuries of experimentation, they found the formula of the *stucco veneciano o marmorino*,[36] a product impermeable to extraneous water in a place where the steam rising from the channels was common. Its superficial layer with Istrian ground stone and crude linseed oil,[37] polished in a particular way, acquired the look of a marble plaque.[38] This new product was also easy to polychrome. They had found a way to realize stronger murals, while maintaining the aesthetics of light and the scenographic taste of the city.

Sebastiano del Piombo knew this technique and translated it to Rome, introducing some variants, given that the weather there was not so extreme. The solution was identical for easel paintings: to isolate the support from the damp and prepare it so that it does not lose its granularity. The oil must be caught in its structure and not only in a possible ground layer that could come off. All the sources, at any time, are in agreement about the particular suitability of porous and rough stones, which better absorbed the hot oil and retained

35 Silvia Collepardi and Mario Fogliata, "Stucco e marmorino della Repubblica Venezia a nostri giorni," *Enco Journal* 4 (1997) <http://www.encosrl.it/enco-journal/archivio-articoli/ej/archivio/antico/4.html>; Vincenzo Scamozzi, *L'Idea Della Archittetura Universale*, ed. Stefano Ticozzi (Milan, 1838), vol. 1, bk. 2, ch. 5; Emanuele Armani, "Materiali e tecniche di esecuzione degli intonaci dell'architettura veneziana," in"Intonaci, colori e coloriture dell'edilizia storica. Atti Convegno, Roma, 1984," supplement, *Bollettino d'arte* 35-36, no. S1 (1986): pp. S69-S75; and Michelangelo Muraro, *Pitture murale nel veneto* (Vicenza), pp. 50 and 62.

36 Consisting on six layers. The most internal compounds by mixing water and hydrated lime, fine inert: sand, natural *pozzolana*, or *cocciopesto, Opus Siginum* Roman: finely ground brick constituted by a thin aerial lime mortar.

37 To *Intonaco* surface oil was added flax oil to increase its plasticity and prolong absorption of paint, reducing the risk of crackles by moisture loss. The final polishing was done with animal or vegetable waxes diluted in turpentine to give greater brightness in the colors. Adhesion and hardening of the mixture is ensured by carbonation of lime in contact with air.

38 Raw linen oil was added in order to increase its plasticity and prolong the absorption time of the paint layer, reducing the risk of cracking due to the loss of moisture.

the color. Regarding the procedures, what is recommended by the treatises do not differ much from what has been found when examining sculpture or painting of all periods. Since these techniques have been studied by Anne-Laure Collomb and Angela Cerasuolo,[39] I will only briefly analyze a practice that was transmitted almost unchanged for centuries until the Renaissance. There is nothing mysterious about the ingredients used, and the study of classical polychromy reveals that the working methods of the ancient painters of sculptures were very close to those of Sebastiano or Daniele da Volterra.[40] The stones extracted from the quarries were worked with metal tools, then they grooved after that, they were sanded down with pumice and finally polished.[41] Although well-prepared surfaces were not smooth, Sebastiano made a *sgraffito* on them to emphasize their texture and ensure perfect adhesion of the paint.[42] The two works by Titian at the Prado Museum exhibit a similar treatment, though with more disordered marks [**Figs. 2.3a, 2.3b, 2.3c**].

The area that received the paint was prepared similarly: with heat, a compound of oil and molten resins—Greek pitch or Colophony and mastic- was thickly applied onto the wall resulting in a textured layer, and thinly layered on the slate, *pietra di paragone*,[43] or *peperino* supports; that is the preferred stones

39 Anne-Laure Collomb, *Splendeurs de l'Italie: la peinture sur pierre à la Renaissance* (Rennes and Tours, 2012), and Angela Cerasuolo, "Un nuovo modo di colorire in pietra. Vasari e la fortuna dell'invenzione di Sebastiano," in *Sebastiano del Piombo e la Cappella Borgherini nel contest della pittura rinascimentale: Atti Convegno Roma, Real Academia de España en Roma, May 2009*, ed. Santiago Arroyo Esteban, Bruno Marrocchini, and Claudio Seccaroni (Florence, 2010).

40 Philippe Jockey, "Praxitèle et Nicias, le débat sur la polychromie de la statuaire antique. Le Sculpture antique et la question de la couleur: de sa reception à l'époque moderne aux recherches contemporaines," in *Praxitéle*, ed. Alain Pasquier and Jean-Luc Martinez (Paris, 2007), pp. 62-81.

41 Collomb describes a line system to prepare the stone, led by a teacher, a group of artisans preparing the plaques. Then a *stuccatore* or *Scarpellino*, polished and sometimes carved plaques in specific dimensions. Finally, they were assembled on the wall, paved by a *muratore*.

42 The backgrounds stands the *Portrait of Baccio Valori* (h.1530), *Portrait of Pablo III with a Nipote*, the *Portrait of Clemente VII bearded* (1532-34) and the *Madonna of the Veil* (*c.* 1532) are treated with small scratch of irregular direction, sometimes simulating the texture of a fabric, as it did Volterra in his *Portrait of a Young Man* at Capodimonte Museum. Angela Cerasuolo, "I dipinti di Sebastiano del Piombo del Museo di Capodimonte. Note sulla tecnica," in *La "Pietà" di Sebastiano a Viterbo: storia e tecniche a confronto*, ed. Costanza Barbieri, Enrico Parlato, and Simona Rinaldi (Rome, 2009), pp. 129-48.

43 Also called touchstone, is extended in different workshops in Verona and Bassano. It is a lidite, a compact variety of quartz, black color, and sometimes veined. Well-polished, it

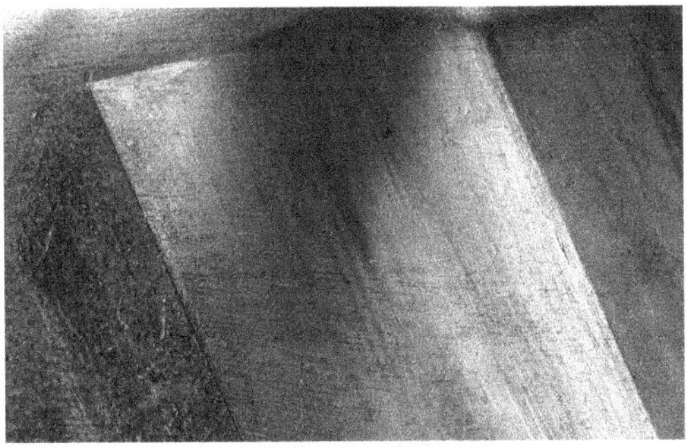

FIGURE 2.3a
*Sebastiano del Piombo,* Christ carrying the Cross, *oil on slate, 1532-35, Madrid, Museo Nacional del Prado. Detail* (PHOTO: © MUSEO DEL PRADO, MADRID).

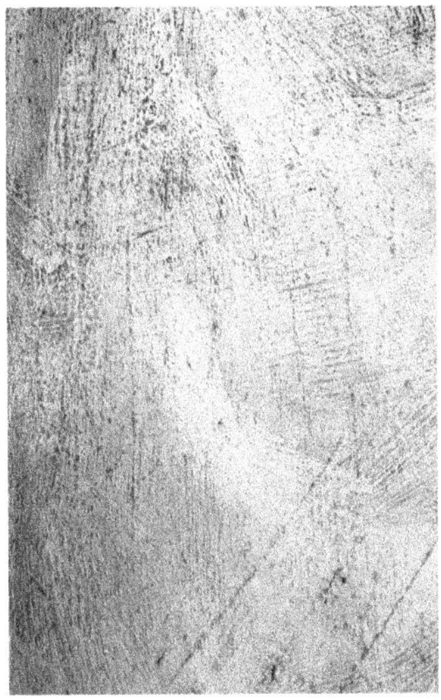

FIGURE 2.3C
*Titian,* Our lady of Sorrows, *oil on marble, 1555, Madrid, Museo Nacional del Prado. Detail (photo: © Museo del Prado, Madrid)*

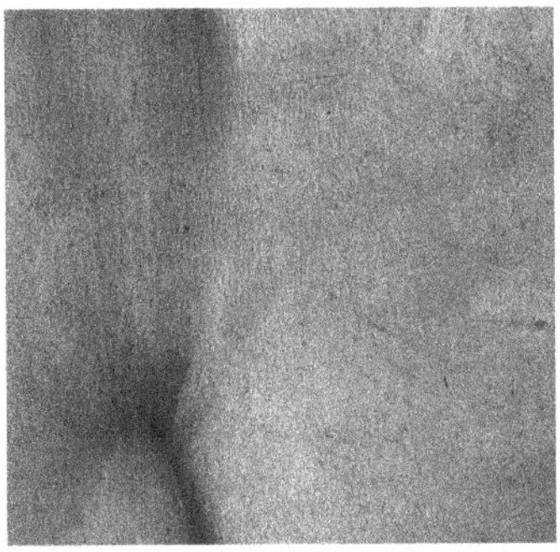

FIGURE 2.3b
*Titian,* Ecce Homo, *oil on slate, 1547, Madrid. Museo Nacional del Prado. Detail taken with raking light (photo: © Museo del Prado, Madrid).*

for monochromatic backgrounds, at least until the mid-16th century.[44] White marble only seems to have been chosen by Titian, and only once.[45] Lead white, the Holy Grail of this technique since time immemorial, was added to the water-repellent materials. Another common practice, and an important one, was its application mixed with oil as under-layer.[46] Remains of Hellenic polychromy testify that this ground, which was made with this pigment—pure or mixed (*psimythion* or *cerussa*)—on a previously eroded surface, was common

was used to check the purity of gold. It is mainly extracted from a quarry at the foot of the mountain Gironda di Salo, from where it was shipped across Lake Garda to Verona and from Adigio to Venice and its surroundings. It is also mined in Egypt and Greece.

44 Vasari also mentions a large number of stones used in paint, chapter 24 of the *Vite*. Also, Scamozzi in *Delle Architetture*, vol. 1, bk. 2. ch. 5, pp. 194-5.

45 Miguel Falomir Faus, ed., *Tiziano: 10 de junio-7 de septiembre 2003* (Madrid, 2003), pp. 226-28.

46 Linen, nuts, poppy.

during antiquity.[47] This preparation had, as in painting on stone, multiple objectives: to waterproof, to smooth the heterogeneous surface of the marble, to improve adherence, to give greater elasticity to the paint, and to enhance the brightness of the color. Eraclius gives the medieval recipe in the verse 25 of his book III *De Coloribus et de artibus romanorum*,[48] and it was recommended by *Le Livre des Metiers* (1268).[49] Evidence of its use has been documented in the porticoes and murals from several medieval cathedrals.[50] Cennini advised artists to prepare the ground stone colors with linseed oil, giving a coating paint,[51] evoking the classical and medieval tradition to polychrome columns.[52] Filarete, who worked in Venice in 1464, also speaks of *biacca* as a primer for *intonaco*, which has been found in the Dome of Sala di Psiche in the Palazzo del Té painted in oils by Giulio Romano.[53] Both Sebastiano and Daniele da Volterra followed this same procedure to prepare their slates.

47 Philippe Jockey, "Delos, Carrefour international des couleur," in *Diversamente bianco. La policromía della scultura romana*, ed. Paolo Liverani and Ulderico Santamaria (Rome, 2014), pp. 109-29. Brinkman, however, says that the color is applied directly onto the polished marble.

48 Mary P. Merrifield, *Original Treatises on the Arts of Painting in Oil, Miniature, Mosaic and on Glass*, 2 vols. (New York, 1967), 1:182-257.

49 Marcel Aubert, *Trionfo del Gotico* (Milan, 1964), pp, 71 and 74.

50 Portico of the Cathedral of Ferrara (1135), South Portico of the cathedral of Bourges, Choir of the Cathedral of Angers (early 14th century), and Chapel of St. Stefano of Westminster (1350-63). See Ulderico Santamaría and Paola Santopadre, "Il Maestro Oltremontano: studio della tecnica pittorica mediante indagini chimiche," in *Il Cantiere Pittorico della Chiesa Superiore di San Francesco in Assisi*, ed. Giuseppe Basile and Pasquale Magro (Assisi, 2001), pp. 37-42. Analyses in the sculptures of the façades of the cathedrals of Ferrara (from 1135 to 13th century), Bourges (1170-75/1228-55), Chartres (Portrail Royal, half of the 12th century); Lausanne (1216-20), Strasbourg (13th century). In the Basilica of San Marco in Venice (Central Porch and arches, 13th century) have characterized a superposition of layers of color in the sculpted forms, almost all on a base of white lead, sometimes with protein binders.

51 Cennino in the XC chapter of his *El Libro del Arte* explains how to start oil painting on a wall and on the preparation of the pigments to paint over it in chapter 93. He does not mention the stones on which it is advisable to paint, and some authors think that he might refer to the polychrome sculpture.

52 Heraclius. *Si vis aliquam columnam vel laminam de petra pingere* in Merrifield, *Original*, 2:230 and Charles Lock Lock Eastlake, *Methods and Materials of Painting of the Great Schools and Masters*, 2 vols. (New York., 1960), 1:34.

53 Report concerning the restoration in Fabio Talarico and Giuseppina Vigliano, "Contributo alla conoscenza della tecnica di esecuzione ed allo stato di conservazione dei dipinti murali della Sala di Psiche," in "L'Istituto Centrale del restauro per Palazzo Te," special volume, *Bollettino d'arte, Serie speziale* 2 (1994), 75-84. Also, Andrea Mantegna, in the

The fascination with the method reinforced the myth of Greek and Roman encaustics, although there is no physical evidence in the 15th and 16th centuries of the practices with wax. In Sebastiano's mural of the *Flagellation*, the receiving layer is an *impasto* painting resin, beeswax and linseed oil, which, curiously, was the final treatment given to *marmorino*. Vitruvius and Pliny describe a finishing touch with Punic wax in the plastered walls: melted and combined with a small amount of oil, to preserve the color to be applied on them, a technique that Plutarch mentions in regard to statues.[54] But it is an addition to the mixture, with the goal of a sound preservation, and in my opinion, this technique cannot be considered encaustic. Ultimately, the key to the preservation of some works was the careful preparation of the stone surface, rather than the addition of unusual or 'secret' ingredients. Sebastiano's contemporaries talked about his *segreto*, which after the publication of the inventory of his property by Girolamo Amati in 1865,[55] was interpreted as acorn oil.[56] Technical analysis has not provided yet conclusive data on this identification.[57] Angela Cerasuolo proposes that 'Sebastiano's secret' was the

---

scene of the "Camino" in the *Camera Picta* (Mantua) used lead white mixed with oil—linen and walnut- as ground, in order to increase the organic color refraction index and their power of covering. He also used oil as binder with the ultramarine and some kind of green. See Michele Cordaro, et al., *Mantegna. La Camera degli Sposi* (Milan, 1992), 242-7.

54 Plutarch, *Etymologies*, C98, and *Roman Questions*, 287B.

55 Scholar, collector and Roman merchant, Amati joined collections of ancient documents, antiquarian curiosities and transcribed documents and literary works (by the middle of the 19th century). He is an ambiguous figure, and his ties to the antiquarian market have put his publications under suspicion by some scholars. The manipulated inventory was published under the pseudonym of Momo, *Lettere Romane* (Rome, 1872), p. 65.

56 Described in the study "a bank for adoperar a far l'olio, […], an instrument da fare olio di Cerqua […]." The antiquarian thought that was used to extract oil from acorns and concluded that this was the secret of Sebastiano. Also in 1865, another transcription of the inventory made by the brothers F. and B. Gasperoni was published. Partially published in the monograph written by Hirst, which integrates the Sebastiano's inventory published by Amati (*Arti e Lettere, nel Quaderno è pubblicato intitolato Villa Giulia L' inventory di Sebastiano del 1547*) and confirms that the inventory was lost. Gaetano Milanesi, included the citation in the *Vita* of the painter in 1880.

57 Francesca Modugno et al., "Indagine chimiche per la caratterizzazione dell'olio di ghiande come legante pittorico," in *Sebastiano del Piombo e la Capella Borgherini: Atti Convegno Roma, Real Academia de España en Roma, May 2009*, ed. Santiago Arroyo Esteban, et al, (Florence, 2010), pp. 61-62. Acorns oil viscosity and extensibility is a suitable binder to work on smooth surfaces like stone. It would be good to know why Sebastiano might have chosen. Roberto Bellucci has found news about this oil in texts disciplines which aren't linked to the painting, like the medical treaty of Leonardo Fioravanti (1570), which

combination of *mastic* and rosin and its application by heat both for murals and stone supports.[58]

Difficulty was another aspect of the discourse concerning the technique; in the case of sculpture and fresco painting, the difficulty lay in the artist's inability to correct errors. The artist's need for secure and accurate execution, which demonstrated wisdom in Aristotelian terms, wisdom that linked to the practice of sculpting.[59] In truth, it is not easy to rectify oil on stone and keep a subtle stratigraphy or expose the support: matter permeates its pores and is very difficult to remove. This means that oil painting on stone was a practice that answered the call for the *perfezione di giudizio* required of sculpture.[60] This same need for caution was extended to fresco, although many are full of corrected errors that distance helps to conceal. The stony character of the visible image, as conceived Sebastiano and Volterra, is achieved with subtle layers without corrections and avoiding accumulation of oil. This requires clear thinking before starting to paint, as evidenced by Sebastiano's portrait of Baccio Valori, in which one can trace an underdrawing of sharp outlines, similar to the ones that Volterra projected on his slate supports. In the infrared reflectography used for Sebastiano's work, these drawings are barely perceptible, and it is incised in very specific areas, as we can see in the cross of the *Cristo Portacroce* at Prado Museum [**Fig. 2.4** and **Plate 9**]. Perhaps the color of slate and an occasional dark priming obscure the response to the infrared radiation of the black material from which it is drawn. Alternatively, as in the *Portrait of Clemente VII* (Capodimonte), Sebastiano outlined the head with a brown-red pigment, which the infrared devices often fail to register and which is only noticeable in the unfinished parts. In addition, the underdrawing was made in freehand, and since it did not fully conform to the contours, it stands out better. Such is the case in his *Ecce Homo*, where some forms are corrected with brush strokes in the intermediate stages of the working process. The works on slate by Daniele da Volterra are more revealing: the primer layers are clear and the drawing, transferred by tracing in both cases, is seen clearly in the scientific images. We must consider, however, that the ultimate reason for the execution of the Louvre painting was different than the other works noted above, and the artist could not make any mistakes. Since the paint layers

---

provides evidence of its use in the arts, in a somewhat later period than that of Sebastiano and almost contemporary to the second edition of the *Vite*.

58 Cerasuolo thinks it was a screw press made of oak wood, in "Un nuovo modo." pp. 47-53.

59 Aristotle, *Nicomachean Ethics*, 1141a 9.

60 Vasari, Proemio of the *Vite* and Varchi, "Due lezzioni …," in Barocchi, *Trattati*, pp. 532-33.

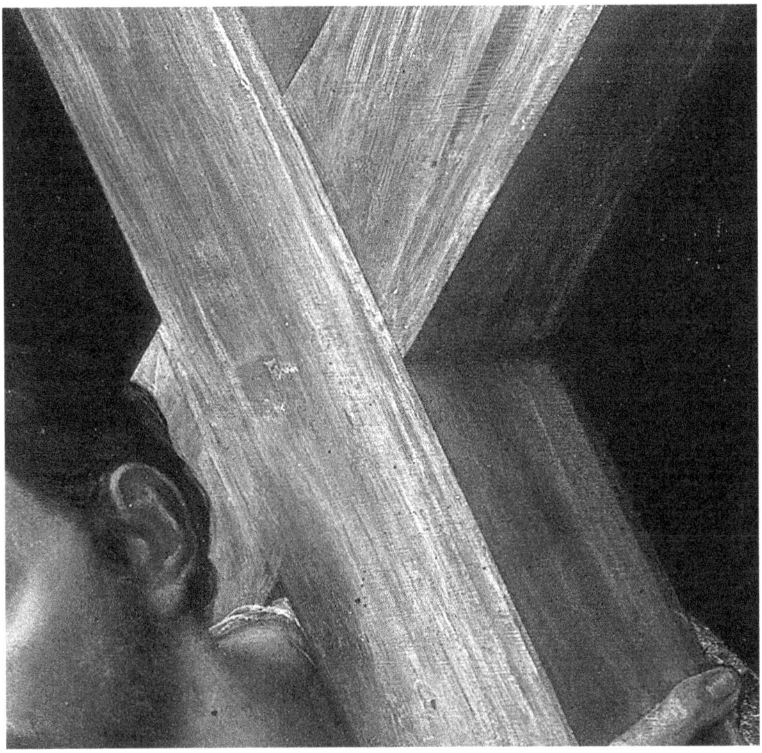

FIGURE 2.4 *Sebastiano del Piombo,* Christ carrying the Cross, *oil on slate, 1532-35, Madrid, Museo Nacional del Prado. Infrared Reflectogram. Detail* (PHOTO: © MUSEO DEL PRADO, MADRID).

should be extremely subtle, involving an accurate underdrawing, the preparative work was a painstaking and complex process, including *modelli* in clay, drawings, little canvas and even a small *modello* on slate,[61] all of which allowed Daniele a very safe execution without any *pentimenti.*

There is nothing strange in the pigments of these works. In general, in all the analysis the same materials used in other epochs on other supports have been detected: those were similar to the ones used in antiquity on painted marble sculptures. I have not found references to any pigments whose use was unadvisable for oil painting on stone or on wall. Instead, all those that were

61 Andrea De Marchi, *Daniele Volterra e la prima pietra del paragone* (Rome, 2014) and Ben Thomas, "The Lantern of Painting. Michelangelo, Daniele da Volterra and the paragone," *Apollo* 154, no. 2 (2001), 46-52.

unsuitable for frescoes can be used bonded with oil, such as cinnabar,[62] one of the regular pigments in classical and Renaissance palette, incompatible with lime, as azurite or lead white.[63] Despite the speculation concerning unusual pigments and secret recipes for the paintings from his Roman period,[64] scientific surveys have confirmed that his palette was typically *cinquecentesca* in those years.[65] The pigments of Daniele da Volterra's work conserved in Paris are similar to those used for the portrait of Baccio Valori.[66] Some of the practices that Sebastiano used for other supports are also evident in his works on stone and in mural. In the *Pietà* of Viterbo, painted on wood, there is varnish mixed with oil.[67] Other examples indicate that Sebastiano did this for his murals and works on stone, according to Vasari's advice of mixing the colors with some resin to prevent excessive brightness with the final varnishing. The oil that binds the pigments is enough to gloss them, and if the slate is covered by thin layers, and even partially visible, the varnish would intensify this property, canceling out the characteristics of the stone. This may be why Daniele da Volterra's work was not varnished. An exaggerated gloss would hamper the legibility of the image by reflections that would occur in the background, which

62 Fine pigment, easy to apply as a uniform base. In the Antiquity derived from natural cinnabar (mercury sulphide). Pliny calls it Kinnabaris: Dragon's Blood.

63 Peruzzi in the Dome of the Galatea Room at the Farnesina, executed in fresco, used the black color to fresco under the azurite, bonded with dry glue. Aldo Angelini, "La Loggia della Galatea alla Villa Farnesina a Roma: L' incontro delle scuole Toscana, Umbra e Romana (1511-1514)," in *Tecnica e stil: Esempi di pittura murale del Rinascimento italiano*, edited by Eve Borsook and Fiorella Superbi Gioffredi (Milan, 1986), pp. 95-110.

64 Michael Hirst, *Sebastiano del Piombo* (Oxford, 1981), pp. 42-48, and Marcia Hall, *Colour and Meaning: Practice and Theory in Renaissance Painting* (Cambridge, 1992), pp. 94, 104, 168-9.

65 In the analyses made in the Venetian tables of Sebastiano: *Sacra Conversazione* (1505-06); *St. Louis of Toulouse and Saint Sinibaldo* (*c.* 1509); *Visitazione* (1506-10) and the *Pala of San Giovanni Crisostomo*, (1510-11)), the characterized palette is typical of the Cinquecento. Only emphasizes the use of gold foil dress of Magdalena in Pala. See Antonella Casoli et al., "Sebastiano del Piombo a Venezia. Materiali e tecnica Pittorica di alcune sue opera," in *Sebastiano del Piombo e la Cappella Borgherini nel contesto della pittura rinascimentale: Atti Convegno Roma, Real Academia de España en Roma, May 2009*, ed. Santiago Arroyo Esteban et al. (Florence, 2010), pp. 117-22. The study by Marika Spring in the *Rising of Lazarus* has revealed another rare material, but not unusual in the palette of that time: galena particles in blue-green corset of the figure who holds Lazarus. Only a work of Venetian stage, casein was detected with the binder emulsion.

66 Sartiani, "Il Ritratto di Baccio Valori," pp. 57-82.

67 Cesare Brandi, "The Restoration of the *Pietà* of Sebastiano del Piombo," *Museum* 3 (1950), 207-19.

Sebastiano wanted to avoid at all costs, this was one of the reasons he used to explain why he painted on slate.

## New Effects in the Visible Image

With few exceptions, the stratigraphy of Sebastiano's works is simple, regardless of the support that he used; this is not the case with Titian's works.[68] Still, as I noted at the beginning of this essay, Sebastiano's paintings on slate in Rome have distinctly Venetian origins. In the region, from the 15th century onward, the painting began to be conceived in a less linear and a more tactile way, and artists ceased to think only in terms of formal representation. In the structure of the paintings, support, ground and painting began to participate in a kind of material *concinnitas*, in which all strata were part of the same act and had the same goal: the visible image. The rugged supports and subtle preparations interacting with oil help to achieve this unity. This practice, initiated by Mantegna and Cosmè Tura, strengthened the debate on the imitation of nature and emulation of the appearance of materials. The nature of the surface in the paintings attempted by Sebastiano and Titian lies in the perfect integration of the oily substance with the stony support, when pigments remain clearly embedded in the rock resulting in the interesting symbiosis that is produced between them [**Fig. 2.5**]. Moreover, the attributes of the black stones favored *non-finito*, common in the work of Piombo,[69] but not always for the same reasons, exposing some parts of the support meant that he could not make a mistake and needed to be a very confident painter, or work very carefully. We could deduce from Vasari's accusations concerning Sebastiano's indolence, and his own confession about his disinterest to continue to paint after the Sack of Rome,[70] that the fact of leaving some works on slates unfinished could be due to a state of apathy, but it was not. On the other hand, the study of these paintings, does not suggest that he wanted to save time or that it was a reflection of an innate *sprezzatura*, in my opinion, it was a deliberate technical issue. In works such as the *Madonna del Velo*, conceived as a grisaille,

68 Only in the *Rising of Lazarus*, the overlap layers is complicated. Jill Dunkerton and Marika Spring, "Il colore nella Resurrezione di Lazaro," in *Sebastiano del Piombo e la Cappella Borgherini: Atti Convegno Roma, Real Academia de España en Roma, May 2009* ed. Santiago Arroyo Esteban, et al. (Florence, 2010), pp. 71-4.

69 His inventory registers enough pictures with different scales of unfinished. Hirst, *Sebastiano*, pp. 152-7.

70 Vasari, "Vita di Sebastiano Veneziano" in *Vidas*, pp. 711, 715 and 717.

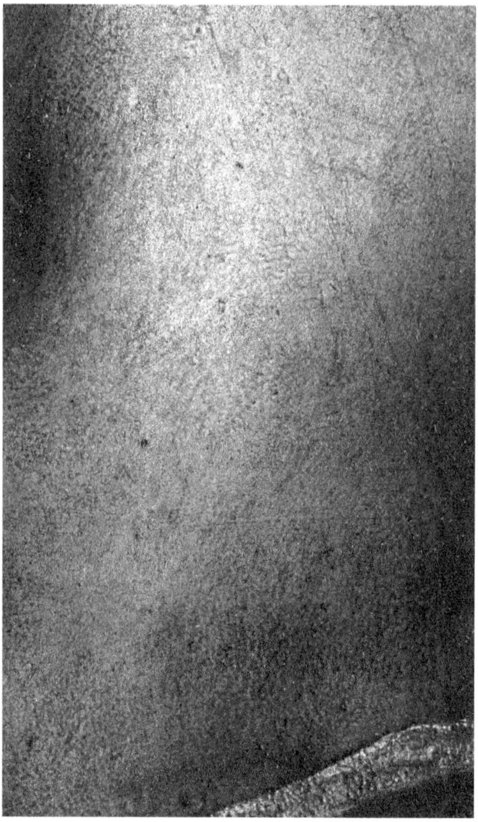

FIGURE 2.5
*Titian,* Ecce Homo, *oil on slate, 1547, Madrid. Museo Nacional del Prado. Detail* (PHOTO: © MUSEO DEL PRADO, MADRID).

the explanation of the *non-finito* may lie in the matter itself. The quality of the stony base actually strengthens the unity of the unfinished work rather than visually detracts from it. And there is the possibility that it was a demonstration of virtuosity, of showing he was painting on it. That he accorded a special importance to this property in the fleshtones is corroborated by the *ricordo* that he made of the head of Clemente VII on slate; in work, it appears as if he cared as much about the qualities rendered by the support to the painted skin as much as he did about the characterization of the portrayed. This also seems to be the case with Daniele da Volterra's *Portrait of a Young Man,* although maintaining greater independence between the layers. In Sebastiano's and more clearly in Titian's works, all the matter seems to have melted down; the bodies emerge from the stone, look like they are made of stone, without pretending to be sculptures. An infrared reflectogram of Titian's *Ecce Homo* is very revealing in this regard. A phrase of Soranzo in his famous letter to Pietro Bembo reveals part of the mystery: "the dry colors are joined to marble so that they are almost petrified." Sebastiano never used marble, except in Santa Maria del Popolo.[71] The less dense passages of Titian's *Mater Dolorosa* provide a better opportunity to observe the apparent petrifaction of pigments absorbed by the carbonate of the marble [**Plate 8**]. The effect evoked by its surface is very similar to the fresco, when dissolved in water the pigments are trapped in the plaster by a carbonation process. The material impression that Titian's marble and slate produces is different, yet in both cases it transmits the same sense of substantial unity that characterizes his works from the 1540s. The pore of the rock, added to its rugged relief, gives to the skin of Christ the same lumpy texture as the thick woven textiles in diagonal pattern or herringbone. But the dense, continuous and enveloping darkness simulation achieved with stone, would not have played in the fabric, the reflection of light in both materials, applying fine preparations is not the same. Inevitably, the texture of the stone forced him to change the gesture in painting.

The prestige achieved by the works of Titian and Sebastiano not only lay in the unusual use of slate as a support but also in the enigmatic quality of its background, which enhanced the appearance of light and the dark transparency of the paint, which is difficult to get in other supports without the necessity of multiple layers. Cennino, curiously took stone as an example to attract attention to a new and desirable perception of the world, which would allow him to recreate a more truthful representation. He writes, "if you want to acquire a good style to represent mountains that seem natural, pick up some large, rough and uncleaned stones and copy them from nature."[72] Cennino suggests that he was not encouraged to reproduce the shape of the mountain as much as its mineral structure and appearance. Interpreted in this way, the maxim would provide one of the keys to the estimation of the slate as a pictorial support: It was beautiful in itself, capable of producing natural and unexpected effects: mattes, brilliants, coarses, and glosses. Slate incorporated in its substance the light and the darkness and therefore, according to the classical tradition, the entire chromatic range, virtue shared under the *atramentum* of Apelles, the most attractive substance mentioned in ancient texts.[73] The

71 The *Natività della Vergine*, commissioned by Sigismondo Chigi to Sebastiano del Piombo is the only known work by the painter made on *peperino* marble. See Barbieri, Costanza, "La Natività della Vergine di Sebastiano del Piombo nel contesto della Cappella Chigi" in *Santa Maria del Popolo, Storia e restauri*, a cura di Ilaria Miarelli Mariani and Maria Richielo. Istituto Poligrafico e zecca dello satato, 2009, pp- 481-488

72 Cennini, *El Libro dell'Arte*, p. 88.

73 Pliny *Natural History*, 33.5. Perhaps ivory powder or burnt wood, dissolved in a semitransparent binder, wax or resin. On this matter, see also Kathleen Weil Garris Posner, *Leonardo and Central Italian Art: 1515-1550* (New York, 1974), pp. 17-22.

search for the harmonious fusion of all tones of an image was the origin of this translucent screen, inducing softness in the treatment of color, associated with Titian by Dolce, a kind of color dimly lit or, as Aristotle stated, comparable to seeing bodies through the water.[74] This ideal aesthetic found in oil glazing, the dark preparations, texturized fabric of canvases, and stone supports the ideal means for its incarnation. Paradoxically, while *atramentum* emphasized the brightness of the colors and protected the painting, through of an estimation of the lighting system it prevented that glow would be excessive for an observer located at a distance.[75] Directly or indirectly, it is an effect that had been evoked by the Venetian paintings since Giorgione through the general blurring of the image, an effect that Sebastiano and especially Titian reproduced in an unrepeatable way in their paintings and slates.

The visual turbidity of the final image could be rendered in either the finishing touch or initially projecting it using dark and grainy backgrounds, whichever support was used. If this dark and grooved 'screen' is placed under colors, and these are not very opaque, the effect is the same as glazing the image: this comes from the interaction between a pigmented transparent surface and a semi-translucent tinted layer. Well controlled glazing on these backgrounds allow for a work with fewer and purer colors. It is interesting to remember that Greek painters avoided mixing colors, considering that each pigment that was added dirtied the original dye. For them to mix meant to destroy, separate-*phthorà*- and were particularly concerned to avoid it in the fleshtones[76]. The theoretical solution is suggested by Aristotle, who defined the genesis and differentiation of tones as altering the transparency of bodies with light, and glazing as the type of combination that required artistic experience.[77] The first evidence of this practice is found in Greece,[78] where Protógenes alternated

74 Aristotle, *De sensu*, 440a.7-12.

75 Pliny *Natural History*, 35.97. Paradoxically, the *atramentum* mitigated the effect of colors *Floridi*, Pliny said that Apelles avoided its use.

76 Plutarch, *Quaestiones Romanas* and *Moralia*, 5.26. For Aristotle and his theory on color, see Raffaello Scognamillo, "La mescolanza dei colori nel trattato Del senso e i sensibili di Aristotele e la técnica dei pittori greci," in *Da Aristotele alla Cina. Sei saggi di storia dell'arte universal*, ed. Stefano Marconi (Rome, 1994), pp. 5-34.

77 Aristotle, *Meteorologica*, III, 372 a, 29 and 372b, 11; 374b, 8-15, 18ss and III, 4, 375 a, 23-28 and *De Sensu*, 440a, 7-12.

78 Licia Vlad Borrelli, "Qualche schede sulla técnica della pittura greca," sulla técnica della pittura greca," *Bollettino Istituto Centrale del Restauro* 2 (1950), 55-57. She proposes that Polygnoto was the inventor of the transparent varnish. See also, Ranuccio Bianchi Bandinelli, *Storicità dell'arte classica* (Bari, 1973).

tempera colored strata with other subtle stratas of wax,[79] a procedure that was used in the archaic sculpture (7th-5th centuries BC).[80] As discussed at the beginning of the essay, many of these ancient formulas for painted sculpture are relevant to our study of the early modern practice. The application of an underlying color, *sublinere,* to attenuate or to extol the tone of some pigments was not discovered in the period of Renaissance in Italy. The murals of Tarquinia (5th century BC) have black bases, and there are different colors in Pompeii under the red areas. Pliny recommended the use of a double primer –black on white, under the *chrysocolla* of green or pink or blue to emphasize the intensity of cheaper purple.[81] This optical rule relied upon the special incidence of light on reflectors as the slate allows; in practice, the artist would simplify the palette and stratigraphy. On slate, the black colors acquire a diffuse glow and look like blue,[82] while others reach different colors, that are more saturated and deeper without making multiple combinations. Moreover, glazing the black with clear layers produces an azure, pearly and ethereal effect, such as the fleshtones of the Virgin in the Capodimonte *Madonna del Velo* a kind of misty glow that Raphael had anticipated in the fresco in the *Liberation of St. Peter.*[83] The transparent darkness found the right medium of representation and rendered, in stone, the mysterious density of the portraits as well as the religious, nocturnal scenes since the mid-16th century. But it was a density slightly faded in the limits of the bodies, an attractive resource that exploited Sebastiano especially in anatomy, to make the paint grazed the streaks of the slate, without putting it into those channels [**Figs. 2.6a** and **2.6b**]. They were visible, thus creating a certain discontinuity in the image and therefore, the appearance of dissolution of the substance. Figures such as the Christ of Titian's *Ecce Homo* or the *Madonna del Velo* by Sebastiano, by contrast, are

79 Pliny *Natural History,* 35.45.

80 Vinzenz Brinkmann, "La polychromie de la sculpture archäique en marbre.," *Pact: Revue du Groupe européen d'études pour les techniques physiques, chimiques et mathématiques appliquées à l'archéologie* 17 (1987), 35-70.

81 Selim Augusti, *I colori Pompeiani* (Rome, 1967). Tarquinia murals also have black bases (SV BC). Noel Heaton, "The Mural Paintings of Knossos, an investigation into the methods of their production," *Journal of the Royal Society of Arts* 58 (1910), 206-12; Vlad Borelli, "Qualche schede," pp. 55-57; and Pliny *Natural History,* 35.5 and 45: an underlying layer of *atrametum* and *Paraetonium* is prepared and both are very tenacious fixatives, and bring out the color [...] The *atrametum*—black- for the whiteness of *Paraetonium* do not pale the crhysocolla, is interposed between the two.

82 Black and blue were overlapping in Greek terminology. John Gage, *Colour and Culture. Practice and Meaning from Antiquity to Abstraction* (London, 2001), p. 29 and n. 4.

83 The use of white glazes in mural painting is documented in 1400 BC at Knossos.

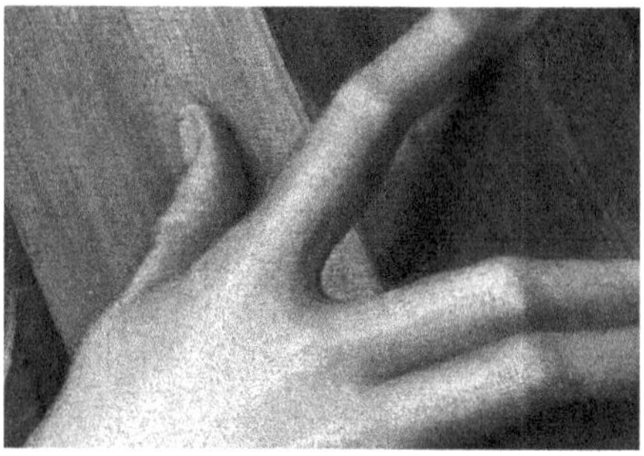

FIGURE 2.6a
*Sebastiano del Piombo,* Christ carrying the Cross, *oil on slate, 1532-35, Madrid, Museo Nacional del Prado, cat. P-348. Detail* (PHOTO: © MUSEO DEL PRADO, MADRID).

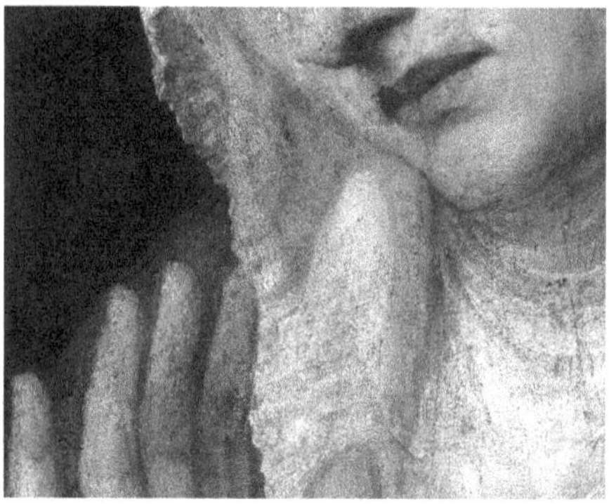

FIGURE 2.6b
*Titian,* Our lady of Sorrows, *oil on marble, 1555. Madrid, Museo Nacional del Prado. Detail* (PHOTO: © MUSEO DEL PRADO, MADRID).

painted with more constant, strokes grittier in the latter, both also making one with the surface. Even so, they visually emerge from a rough and shadowy background that blurs their contours, which are largely defined by the adjacent luminous masses, as if they appeared through the fog that involves the images on black and white films.

*Atramentum*, glazing, dark backgrounds, and *ganosis* were substances and mechanisms that gave nuance to the painted image, offsetting a formal clarity not found in nature.[84] The indefinite black of the slate was the best way to experience the synthesis of light and relief with the four-color palette of Apelles, as evaluated by Pliny, who lamented the chromatic opulence of his time.[85] We have no evidence that Sebastiano was interested in these issues,

84 *Ganosis* is the waxy and delicate *patina* that, from the 5th century BC, was used for the polychrome sculpture, to accentuate the transparency of the matter and give it the "heat of living things." See Paolo Liverani, "La policromia delle statue antiche," *Escultura Romana en Hispania* 5 (2008), 65-85.

85 This critique appears in Vitruvius, *Ten Books*, 7.7-8, and Pliny *Natural History*, 35.41.30. The antithesis is clear in the *Poetics* of Aristotle 1450ab and Plato, *Policy* 277b-c.

but Titian himself was attracted by the Greek painter,[86] as well as Titian's supporters Aretino, Doni and Dolce,[87] who regularly compared the Venetian to Apelles. Regardless of the discussion about the sobriety of the palette, there were, of course, other reasons, and other artists who reduced and darkened it, and it is true that there are artists who did it at particular times in their careers. This allowed them to control the shapes, lighting effects and shadows. Titian and Sebastiano both did this in their maturity, as Dürer, another *alter ego* of Apelles, claimed he did in his own work.[88] Some of the paintings by Sebastiano, such as the *Madonna del Velo*, the Úbeda Pietà or the small *Cristo Portacroce* at the Prado, are barely executed in four colors: white, black, blue and earth. Such is also the case with the *Ecce Homo* by Titian, whose palette is even more limited. The play with dark background creates other tones, especially with the blue ones, which are saturated and look like cleaner. Simultaneously, the flesh tones built on these dark and rough stones acquire changeable hues, without it being necessary to modulate skin by overlapping them.

The painters of the subsequent generation, such as Bassano, Palma il Giovanne, Pasquale Ottino, or Marcantonio Bassetti created "gimmicky" works directed at other customers. Their favorite stone, the paragone, when well-polished, produced a black mirror that was used to recreate the effects of night and increase the sense of depth, extending the real space, although nullifying the space of representation. In these paintings, the figures become independent of the background, and everything happens in the foreground. The space in which the bodies should be moving does not exist; the shapes seem to be plastic sheets attached to it. The Human God represented in this way,

---

86 His *Venus Bridgewater* was inspired by the story of *Venus Anadyomene* of Apelles, see Pliny *Natural History*, 36.91.

87 In 1565 he published the *Dialogue ... nel quale if ragiona della qualità, diversità and proprietà I Colori*. It collects large part of the text of Pliny and the *Libellus of Coloribus of Anthonius Thylesius* (Venice, 1528). The catalogue of paintings of Andrea Vendramin Museum', developed by the collector in the 1620s, includes many reproductions of works attributed to Giorgione and Titian. His preface is extracted from *Gallus Hospes Romae* (1585, 1609) Louis de Montjosieu pioneer in linking the old styles with contemporary practice. Printed several times with the title *Commentarius Pictura*, including one of the earliest references to the four-color palette.

88 The main defenders of this concept were Conrad Celtis, Erasmus of Rotterdam, and Philip Melanchthon. On this matter, see: Dieter Wuttke, "Dürer und Celtis: von der Bedeutung des Jahres 1500 für den deutschen Humanismus, 'Jahrhundertfeier als symbolische Form,'" *Journal of Medieval and Renaissance Studies* 10 (1980): 73-129; Erasmus of Rotterdam, letter-preface to Stanislas Sturzo for the Froben edition of Pliny, *Secundi Historia mundi denuo emendata*, Basel, Froben, 1525.

abandoned his stony and credible character, he no longer lived in the shadows, as he was clearly superimposed on them. Because of that, his image was more direct and accessible. It is not necessary to spend much time in front these works to discover their forms.

The rift that happened at the end of the century in painting on stone with respect to the solutions offered in the of previous years is partly due to the fact that it promoted the new liturgy for art as formulated by Carlo Borromeo and Cardinal Paleotti. To a significant extent, no one was interested in the properties of the natural mineral, but instead concerned with the clarity of the message transmitted by this kind of image and the focus that the artist in the sacred figures, rather than in the secondary details. And so, shortly after, the images contained in other gems were converted into subjects of works, whose quality—if it was not for their extraordinary support-is generally poor; these are the *pierres imagées*, which filled the cabinets of curiosities, but that is another story.

## Bibliography


Angelini, Aldo. "La Loggia della Galatea alla Villa Farnesina a Roma: L'incontro delle scuole Toscana, Umbra e Romana (1511-1514)." In *Tecnica e stil: Esempi di pittura murale del Rinascimento italiano*, edited by Eve Borsook and Fiorella Superbi Gioffredi, pp. 95-110. Milan, 1986.

Armani, Emanuele. "Materiali e tecniche di esecuzione degli intonaci dell'architettura veneziana." In "Intonaci, colori e coloriture dell'edilizia storica. Atti Convegno, Roma, 1984," supplement, *Bollettino d'arte* 35-36, no. S1 (1986), S69-S75.

Aubert, Marcel. *Trionfo del Gotico*. Milan, 1964.

Augusti, Selim. *I colori Pompeiani*. Roma, 1967.

Barasch, Moshe. *Light and Color in Italian Renaissance theory of Art*. New York, 1978.

Barocchi, Paola, ed. *Trattati D'Arte del Cinquecento. Fra Manierismo e Contrariforma*. Bari, 1960.

Bianchi Bandinelli, Ranuccio. *Storicità dell'arte classica*. Bari, 1973.

Bodnar, Edward W., and Charles Mitchell. *Cyriacus of Ancona's journeys in the Propontis and the Northern Aegean, 1444-1445*." Memoirs of the American Philosophical Society, vol. 112. Philadelphia, 1976.

Borghini, Raffaello. *Il riposo di Raffaello Borghini in cui della pittura, e della scultura si fauella, de' più illustri pittori, e scultori, e delle più famose opere loro si fa mentione; e le cose principali appartenenti a' dette arti s'insegnano*. Florence, 1584.

Brandi, Cesare. *Il Restauro, teoria e prattica*. Rome, 1994.

Brandi, Cesare. "The restoration of the *Pietà* of Sebastiano del Piombo." *Museum* 3 (1950): 207-19.

Brinkmann, Vinzenz. "La polychromie de la sculpture archäique en marbre." *Pact: Revue du Groupe européen d'études pour les techniques physiques, chimiques et mathématiques appliquées à l'archéologie* 17 (1987): 35-70.

Cagiano Di Azevedo, Michelangelo, "Il colore nella antichità." *Aevum* 28 (1954), 151-67.

Casoli, Antonella, Enrico Fiorin, Lorenzo Lazzarini, and Stefano Volpin. "Sebastiano del Piombo a Venezia. Materiali e tecnica Pittorica di alcune sue opere." In *Sebastiano del Piombo e la Cappella Borgherini nel contesto della pittura rinascimentale: Atti Convegno Roma, Real Academia de España en Roma, May 2009*, edited by Santiago Arroyo Esteban, Bruno Marocchini and Claudio Seccaroni, pp. 117-22. Florence, 2010.

Cecchi, Alessandro, Marco Ciatti, and Oriana Sartiani, eds. *Sebastiano del Piombo e la pittura su pietra: il "Ritratto di Baccio Valori." Restauro e ricerche*. Florence, 2014.

Cennini, Cennino. *El libro del arte*. Edited by Franco Brunello and translated by Fernando Olmeda Latorre. Madrid, 1988.

Cerasuolo, Angela. "Approfondimenti su Sebastiano del Piombo a Capodimonte." *Konsthistorisktidskrift* 81, no. 4 (2012), 254-61.

Cerasuolo, Angela. "I dipinti di Sebastiano del Piombo del Museo di Capodimonte. Note sulla tecnica." In *La "Pietà" di Sebastiano a Viterbo: storia e tecniche a confronto*, edited by Costanza Barbieri, Enrico Parlato, and Simona Rinaldi, pp. 129-48. Rome, 2009.

Cerasuolo, Angela. "Un nuovo modo di colorire in pietra. Vasari e la fortuna dell'invenzione di Sebastiano." In *Sebastiano del Piombo e la Cappella Borgherini nel contest della pittura rinascimentale: Atti Convegno Roma, Real Academia de España en Roma, May 2009*, edited by Santiago Arroyo Esteban, Bruno Marrocchini, and Claudio Seccaroni, pp. 47-53. Florence, 2010.

Collepardi, Silvia. "Stucco e marmorino della Repubblica Venezia a nostri giorni." *Enco Journal* 4 (1997). <http://www.encosrl.it/enco-journal/archivio-articoli/ej/archivo/antico/4.html>.

Collomb, Anne-Laure. *Splendeurs de l'Italie: la peinture sur pierre à la Renaissance*. Rennes and Tours, 2012.

Cordaro, Michele, Maurizio Marabelli, Giovanni Rodella, and Giuseppina Vigliono. *Mantegna. La Camera degli Sposi*. Milan, 1992.

De Marchi, Andrea. *Daniele Volterra e la prima pietra del paragone*. Rome, 2014.

Dolce, Lodovico. *Dialogo della pittura di M. Lodovico Dolce, intitolato l'Aretino*. Florence, 1735. Originally published Venice, 1557.

Dolce, Lodovico. *Lettre di diversi eccellentissimi huomini, raccolte da diversi libri*. Venice, 1554.

Doni, Anton Francesco. *Il disegno del Doni partito in piu raggionamenti, ne quali si tratta della scultura e pittura, de colori, de getti, de modegli, con molte cose appartenenti a quest'arti & si termina la nobilta dell'una et dell'altra professione: con historie, essempi, et sentenze & nel fine alcune lettere che trattano della medesima materia.* Venice, 1549.

Dunkerton, Jill, and Marika Spring. "Il colore nella Resurrezione di Lazaro." In *Sebastiano del Piombo e la Cappella Borgherini: Atti Convegno Roma, Real Academia de España en Roma, May 2009*, edited by Santiago Arroyo Esteban, Bruno Marrocchini, and Cladio Seccaroni, pp. 71-74. Florence, 2010.

Eastlake, Sir Charles Lock. *Methods and Materials of Painting of the Great Schools and Masters.* 2 vols. New York, 1960.

Fallay d'Este, Lauriane and Nathalie Bauer. *Le paragone: le parallèle des arts.* Translated by Lauriane Fallay D'Este and Natalie Bauer. Paris, 1992.

Falomir Faus, Miguel, ed. *Tiziano: 10 de junio-7 de septiembre 2003*. Madrid, 2003.

Gage, John. *Colour and Culture. Practice and Meaning from Antiquity to Abstraction.* London:, 2001.

Garzya Romano, Chiara, ed. *De coloribus et de artibus Romanorum. I colori e le arti dei Romani e la compilazione pseudo-eracliana.* Naples, 1996.

Golzio, Vincenzo. *Raffaello Nei Documenti, Nelle Testimonianze Dei Contemporanei e Nella Letteratura del Suo Secolo.* Westmead, 1971.

Hall, Marcia. *Color and Meaning: Practice and Theory in Renaissance Painting.* Cambridge, 1992.

Heaton, Noel. "The Mural Paintings of Knossos, an investigation into the methods of their production." *Journal of the Royal Society of Arts* 58 (1910), 206-212.

Hirst, Michael. "Sebastiano's Pietà for the Commendador Mayor." *Burlington Magazine* 114, no. 834 (1972): 585-595.

Hirst, Michael. *Sebastiano del Piombo.* Oxford; 1981.

Jockey, Philippe. *Le mythe de la Grèce blanche. Histoire d'un rêve Occidental.* Paris, 2013.

Jockey, Philippe. "Praxitèle et Nicias, le débat sur la polychromie de la statuaire antique. Le Sculpture antique et la question de la couleur: de sa reception à l'époque moderne aux recherches contemporaines." In *Praxitéle*, edited by Alain Pasquier and Jean-Luc Martinez, pp, 62-81. Paris, 2007.

Jockey, Philippe. "Delos, Carrefour international des couleur." In *Diversamente bianco. La policromía della scultura romana*, edited by Paolo Liverani Paolo and Ulderico Santamaria, 109-129. Rome, 2014.

Lamo, Pietro. *Graticola di Bologna: ossia descrizione delle pitture, sculture e architetture di detta città fatta l'anno 1560.* Bologna, 1844.

Leonardo da Vinci. *Traité de la peinture.* Translated by Andrè Chastel. Paris, 1964.

Liverani, Paolo. "La policromia delle statue antiche." *Escultura Romana en Hispania* 5 (2008), 65-85.

Lomazzo, Gian Paolo. *Trattato del arte della Pittura Idea del tempio della pittura.* Translated by Robert Klein. Florence, 1974.

Melanchthon, Philip, "Letter-preface to his translation into Latin of the *Contra Aristogitonem Demosthenis Orationes* (August1527), In *Opera*, edited by Carol Gottlieb Bretschneider and Heinrich Ernst Bindsell, vol 1, n. 462, cols 888-90. Halle, 1834.

Merrifield, Mary P. *Original Treatises on the Arts of Painting in Oil, Miniature, Mosaic and on Glass*. 2 vols. New York, 1967.

Modugno, Francesca, Maria Perla Columbini, Erika Ribechini, Ugo Bartolucci, and Monica Galeotti. "Indagine chimiche per la caratterizzazione dell'olio di ghiande come legante pittorico." In *Sebastiano del Piombo e la Capella Borgherini: Atti Convegno Roma, Real Academia de España en Roma, May 2009*, edited by Santiago Arroyo Esteban, Bruno Marrocchini, and Claudio Seccaroni, pp. 61-64. Florence, 2010.

Momo (Girolamo Amati). *Lettere Romane*. Rome, 1872.

Muraro, Michelangelo. *Pitture murali nel veneto*. Vicenza, 1960.

Pausanias. *Description de la Grèce*. Vol. 1. Translated by Étienne Clavier. Paris, 1821.

Pino, Paolo. *Dialogo di Pittura*. Edited by Susanna Falabella. Rome, 2000.

Pliny the Elder. *Historia Natural*. In *Textos de historia del arte*. Edited by M. Esperanza Torrego. Madrid: Visor, 1988.

Rossi Manaresi, Raffaella. "Pietre porose: alterazione e conservazione." In "Materiali lapidei. Problemi relativi allo studio del degrado e della conservazione," supplement, *Bollettino d'Arte* 41, no. S2 (1987), S133-S144.

Rossi Manaresi, Raffaella. "L'Ardesie ligure: deterioramento e conservazione, primi risultati di una ricerca." In *Del Dipingere e Scolpire in Pietra: Genova, Fiera Internatiionale de Genova 8-23 dicembre 1984: Piazza della Meridiana 5, dicembre 1984-12 gennaio 1985*, edited by Piero Boccardo and Ida Maria Bott, p. 125. Genova, 1985. Exhibition catalogue.

Santamaria, Ulderico, and Paola Santopadre. "Il Maestro Oltremontano: studio della tecnica pittorica mediante indagini chimiche." In *Il Cantiere Pittorico della Chiesa Superiore di San Francesco in Assisi*, edited by Giuseppe Basile and Pasquale Magro, pp. 37-42. Assisi, 2001.

Scaillierez, Cécile, Claudia Sindaco-Domas, and Christine Mouterde. "L'ardoise double face de Daniele Volterra, figurant David et Goliat: étude et restauration d'une oeuvre d'exception." *Techné* 25 (2007), 5-11.

Scamozzi, Vincenzo. *L'Idea Della Architettura Universale* (1615). Vol. 2, book 2, chapter 5. Edited by Stefano Ticozzi. Milan, 1838.

Scognamillo, Raffaello. "La mescolanza dei colori nel trattato Del senso e i sensibili di Aristotele e la técnica dei pittori greci." In *Da Aristotele alla Cina. Sei saggi di storia dell'arte universale*, edited by Stefano Marconi, pp. 5-34. Rome, 1994.

Talarico, Fabio, and Giuseppina Vigliano. "Contributo alla conoscenza della tecnica di esecuzione ed allo stato di conservazione dei dipinti murali della Sala di Psiche." In "L'Istituto Centrale del restauro per Palazzo Te," special volume, *Bolletino d'arte, Serie speziale* 2 (1994), 75-84.

Theophilus, "Schedula diversarium artium." In *On divers arts*. Translated by John G. Hawthorne and Cyril Stanley Smith. New York, 1963, 1979.

Thomas, Ben. "The Lantern of Painting. Michelangelo, Daniele da Volterra and the paragone." *Apollo* 154, no. 2 (2001), 46-52.

Thucydides. *History of the Peloponnesian War*. Translated by Philippe Remacle. <http://remacle.org/bloodwolf/historiens/thucydide/livre1.htm>.

Varchi, Benedetto. *Leccion que hizo Benedicto Varqui en la Academia Florentina ... año 1546 sobre la primacia de las artes y gual sea mas noble, la escultura o la pintura; con una carta de Michael Angelo Buonarroti*. Translated by Felipe de Castro. Madrid, 1753.

Vasari, Giorgio. *Las vidas de los más excelentes arquitectos, pintores y escultores italianos desde Cimabue a nuestros tiempos*. 2nd edition. Edited by Ana Ávila. Madrid, 2005.

Vlad Borrelli, Licia. "Qualche schede sulla técnica della pittura greca." *Bolettino Istituto Centrale del Restauro* 2 (1950): 55-7.

Weil-Garris Posner, Katheleen. *Leonardo and Central Italian Art: 1515-1550*. New York, 1974.

Wuttke, Dieter. "Dürer und Celtis: von der Bedeutung des Jahres 1500 für den deutschen Humanismus, 'Jahrhundertfeier als symbolische Form.'" *Journal of Medieval and Renaissance Studies* 10 (1980), 73-129.

Zobi, Antonio. *Notizie Storiche riguardanti l'imperiale e reale stabilimento dei lavori di comesso in pietre dure di Firenze*. Florence, 1841.

Zuccaro, Federico. *L'idea de' pittori, scvltori, et architetti, del Cavalier Federico Zvccaro; Diuisa in due Libri*. Turin, 1607.

CHAPTER 3

# 'Un paragone con oro su': Material Innovation, Invention and Sebastiano del Piombo's Papal Portraiture*

*Elena Calvillo*

The papal *cameriere* Vittore Soranzo's account in 1530 of Sebastiano del Piombo's new technique of painting in oil on stone specified that his first work was of a figure of Christ, presumably now lost. This essay evolved from a close study of a slightly later painting of Christ, Sebastiano's Úbeda *Pietà* [**Plate 1**], whose slate support, I proposed, rendered the painting a touchstone by which the artist claimed an authoritative status, both for his pictorial mastery and the Roman relic prominently featured in the composition.[1] Sebastiano's technical innovation also demonstrated his intellectual capacity to engage the ontology of materials, oil and stone, to enhance the identity and meaning of his subjects and, as I argue here, his role in the papal household. One line of inquiry in this earlier study, made in the context of the facture and medium of the Úbeda painting, was provisionally addressed but might bear fruit for a broader investigation of Sebastiano's material invention and their reception in the later 16th century. What follows, then, is a further consideration of the significance of Sebastiano's method of painting on stone for his office as the keeper of the papal seal, or *piombatore*, especially because he maintained this

* I would like to thank Nicholas Baker, Piers Baker-Bates, Anne Dunlop, Judith Mann, Lorenzo Pericolo, Guido Rebecchini, and Giorgio Tagliaferro for their comments after either reading versions of this text or inviting me to present it. I am also grateful to the University of Richmond for supporting my research on this project in 2016 and the Villa I Tatti, where I was a Hanna Kiel Fellow, 2013-14.

1 Christopher Nygren has also discussed the touchstone metaphor in his excellent dissertation "Vibrant Icons: Titian's Art and the Tradition of Christian Image-Making" (PhD diss., Johns Hopkins University, 2011); and in a subsequent article, "Titian's *Ecce Homo* on Slate," in *The Art Bulletin* 99, no. 1 (2017), 36-66. My own understanding of the works as pictorial touchstones able to demonstrate the artist's virtuosity was inspired by Rudolph Preimesberger's study, "Ein 'Prüfstein der Malerei,' bei Jan van Eyck?" in *Der Künstler über sich in seinem Werk*, ed. Matthias Winner (Weinheim, 1992), pp. 85-100; see pp. 94-95, as cited in Calvillo, "Authoritative Copies and Divine Originals: Lucretian Metaphor, Painting on Stone, and the Problem of Originality in Michelangelo's Rome," *Renaissance Quarterly* 66, no. 2 (2013), 453-508.

 | DOI 10.1163/9789004361492_006

position for the remainder of his life, after Clement VII died, and through most of the papacy of Clement's Farnese successor, Paul III.

The importance of Sebastiano's status as a member of the papal household for his late career has recently been emphasized by both Costanza Barbieri and Piers Baker-Bates.[2] This essay asserts that there was a noteworthy and highly symbolic conjunction of Sebastiano's office and his artistic production.[3] Sebastiano's papal portraits and depictions of Christ and the Virgin in this new technique seem conceived, indeed charged, to marry the authority and function of the *piombatore* with that of an esteemed portraitist. The brilliant linking of these roles, a sort of professional reinvention circa 1531, would have appealed both to Clement VII and the artist's later Farnese protectors on a number of levels, metaphoric, ontological, and practical. That the Farnese understood the potential of Sebastiano's material and technical innovation to signal the authenticity and virtuosity of official portraits is demonstrated first by their patronage and/or acquisition of several oil paintings on stone supports before and after Sebastiano's death in 1547 and by their dealings with other artists who were both in contention for the position of *piombatore* and desirable portraitists.

## The Invention

At the height of the tensions surrounding the commission and painting of the Úbeda *Pietà*, the agent Niccolò Sernini wrote the patron Ferrante Gonzaga in May and June of 1537 to tell him that Cardinal Alessandro Farnese (1520-89) would intervene to help see the commission to completion.[4] Farnese's assistance in this matter is not surprising given his role in his grandfather's

2 For Sebastiano's position at the papal court, see Costanza Barbieri, "Fishing for Offices: Sebastiano del Piombo as Piombatore," *Burlington Magazine* 150, no. 1258 (2008), 35-36; Piers Baker-Bates, "'Sebastianello nostro Venetiano': The Career of Sebastiano del Piombo and the Concept of a 'Court Artist' in Renaissance Rome," *The Court Historian* 16, no. 1 (2011), 7-22. Barbieri, "Sebastiano's Portraits in *Paragone*: Sculpted Paintings, Praise through Images," in *Sebastiano del Piombo, 1485-1547*, ed. Carlo Strinati (Rome, 2008), pp. 53-57, has speculated about the relationship between Sebastiano's acquisition of the office and the success of his portraits of Clement VII.

3 See Calvillo, "Authoritative Copies and Divine Originals," pp. 489-93.

4 See Michael Hirst, "Sebastiano's Pietà for the *Comendador Mayor*," *The Burlington Magazine* 114, no. 834 (1972), pp. 585-93, 595; p. 590 for the letter indicating Cardinal Farnese's intervention.

administration and Sebastiano's official position in the papal household.[5] The Cardinal's intervention thus establishes the latest occasion for his acquaintance with the painter's new method of working on stone; however, Soranzo's much-cited letter to Pietro Bembo suggests both the novelty of and high level of interest for the technique in Rome in 1530. Soranzo reported, "dovete sapere che Sebastianello nostro Venetiano ha trovato un segreto di pingere in marmo a olio bellissimo il quale farà la pittura poco meno che eterna. I colori ... si uniscono col marmo di maniera che quasi impetriscono ... ha fatto ogni prova & è durevole. Ne ha fatto una imagine di Christo & halla mostrato a N. Sig."[6] His choice of language—"dovete sapere," "il segreto," "nostro Venetiano"—aims to bring Bembo up to date on news of another notable Venetian in Rome.[7] Giorgio Vasari's discussion of the technique in the first edition of his *Vite* suggests the immediate success of Sebastiano's invention, "avendo poi cominciato questo pittore un nuovo modi di colorire in pietra, ciò piaceva molto a' popoli, parendo che in quel modo le pitture diventassero eterne, a che né il fuoco né i tarli potessero lor nuocere."[8] These reports and the high number of works that Sebastiano produced on stone in the 1530s make it likely that Paul III and his grandson Cardinal Farnese would have been aware—and probably quite interested—in this new secret of painting well before the Cardinal had to persuade Sebastiano to finish the Pietà commissioned by Gonzaga as a gift for the Imperial minister Francisco de los Cobos.

The second version of the *Madonna del Velo* [**Plate 3**] also painted on slate and now part of the Farnese Collection at the Museo di Capodimonte in Naples might even have come into the Cardinal's possession while he was mediating between Gonzaga and Sebastiano. The references made in the exchange between Sernini and his patron to a copy of a work by Raphael refer, I propose, to this painting, which reverses the earlier version on wood panel painted in

5 Clare Robertson, 'Il Gran Cardinale': *Alessandro Farnese Patron of the Arts* (New Haven, 1992), pp. 15-25.

6 *Delle Lettere da diversi Lettere da diversi Re et Principi et Cardinali et altri huomini dotti a Mons. Pietro Bembo scritte* (Venice, 1560), f. 110v; 'You should know that our little Sebastiano the Venetian has found the secret with which to paint in oils on marble in the most beautiful fashion, which will make his paintings little less than eternal... the colors...unite with the marble as if they were turned to stone; every test has been tried, and it has proved durable. Using this method Sebastiano has made an image of Christ and shown it to our Lord.'

7 Baker-Bates, "The Career of Sebastiano del Piombo," discusses Soranzo's letter in terms of the pride shown in such a letter.

8 Giorgio Vasari, *Le Vite de' più eccellenti architetti, pittori, ed scultori nelle redazioni del 1550 e 1568*, ed. Rosanna Bettarini and Paola Barocchi, 8 vols. (Florence, 1966-87), 5:97-98 (Giuntina). See the Introduction to this volume, p. 7 n. 17, for the difference in this passage between the two editions.

the mid-1520s and now in Olomuoc.[9] The translation of this composition to the new support both endows the work with the cooler tonality typical of these paintings and the metaphoric possibility of stone. The sleeping child on this surface assumes a greater sense of his impending death and burial; of course, this connection is made more apparent by the unfinished area and the visibility of the slate support, which—theoretically at least if often not in practice—would have been invisible.

Certainly, Sebastiano's unfinished portrait of Paul III with a young boy [**Plate 13**], usually identified as his grandson Ottavio, provides proof of the Farnese appreciation and immediate appropriation of the painter's new technique.[10] One might even wonder if the painting was put aside to complete the politically sensitive work for los Cobos, although the number of unfinished works from Sebastiano's late career makes such an hypothesis difficult to prove. Works by other artists, a slightly later portrait of Vittoria Farnese (1521-1602), the Cardinal's sister, painted on slate by Jacopino del Conte, now in the Borghese Gallery, and the portraits of the Cardinal and the Pope incorporated into the peperino stone altarpiece by Francesco Salviati in the Cappella del Pallio of the Palazzo della Cancelleria, as we shall see, also demonstrate an early and sustained interest in the medium itself, beyond Sebastiano's initial production or specific authorship.[11] Whatever the circumstances of their first acquaintance with this method of painting, I would argue that Paul III and Cardinal Alessandro were inclined to appreciate the Venetian artist's invention and its metaphoric application to portraiture. In particular, their interest in *imprese* and poetically charged art rendered them well-suited to admire the original conceits and potential of Sebastiano's innovation.[12]

---

9 Hirst, "Sebastiano's Pietà," p. 587; see the letters that mention a copy of one of Raphael's works that was being made. For the painting and its bibliography see the exhibition catalogue, *Sebastiano del Piombo*, p. 234, no. 58

10 See *Sebastiano del Piombo: 1485-1547*, ed. Claudio Strinati (Rome, 2008), no. 53, p. 222; Michel Hochmann, "Les Dessins et les Peintures de Fulvio Orsini et la Collection Farnèse," *Mélanges de l'École Française de Rome* 105, no. 1 (1993), 49-91; see pp. 59-60.

11 For the attribution to Jacopino del Conte, see Federico Zeri, "*Me pinxit*. Salviati e Jacopino del Conte," *Proporzioni* 2 (1948), 180-83 (republished in *Giorno per giorno nella pittura*, 1994, pp. 73-75).

12 Cardinal Farnese's court artist, Giulio Clovio, devised a composition that embodied one of his patron's *impresa* and invoked simultaneously the public imagery promulagated by Paul III; see Calvillo, "'*Il Gran Miniatore*' at the Court of Cardinal Alessandro Farnese," in *Artists at Court: Image-Making and Identity 1350-1600*, ed. Stephen J. Campbell (Boston and Chicago, 2004), pp. 168-69.

In the fall of 1531, a year and a half after Soranzo's letter describing Sebastiano's new technique, Clement VII awarded the artist the office of the *piombatore*, which enjoyed an annual revenue of 800 scudi—although in Sebastiano's case, 300 of it was committed to Giovanni da Udine, an arrangement noted by Vasari and recently confirmed by Barbieri's work in the Roman archives.[13] The artist's correspondence with Michelangelo from the summer before this promotion records his work on at least three, and up to six, portraits of Clement.[14] After he became the *piombatore*, then, Sebastiano would have found himself in the remarkable position of overseeing the papal seal while simultaneously producing official papal portraiture, at least part of which were rendered in the new technique on stone.[15] As I have proposed, these 'almost eternal' works promulgating the pope's likeness—first that of Clement VII and later Paul III—were his painterly counterparts to the lead seals applied by his court office.[16]

The poetic and decorous concomitance of Sebastiano's official duty and pictorial invention was surely recognized by Clement VII, if not actually first conceived by him. The portrait of Clement VII on slate now in the Capodimonte [**Plate 14**] and the weaker but finished work at the Getty [**Plate 4**] are usually associated with a letter written by Sebastiano to Michelangelo in July of 1531, in which the painter noted that the pope had asked him to produce another portrait on a stone support.[17] In the last decade, scholarship on Sebastiano has begun to address the importance of the stone support for the conception of these portraits. Barbieri has noted both the context of the *paragone* between painting and sculpture and the coincidence of the serial portraits with Clement's award to Sebastiano of the lucrative office of the *piombatore*.[18]

13 Barbieri, "Fishing for Offices," pp. 35-6.

14 See Karla Langedijk, *The Portraits of the Medici*, 3 vols. (Florence, 1981-83); Piers Baker-Bates, "Beyond Rome: Sebastiano del Piombo as a Painter of Diplomatic Gifts," *Renaissance Studies*, 27, no. 1 (2013), 51-72, 65-66; Barbieri, *Sebastiano del Piombo: i Ritratti: Committenti, artisti e letterati nella Roma del Cinquecento* (Rome, 2012), pp. 92-93.

15 See Barbieri, *I Ritratti*, p. 92, for a consideration of the correct number; at least two of which were on stone.

16 For the connection between Sebastiano's paintings and papal seals, see Calvillo, "Authoritative Copies and Divine Originals," pp. 490 and 492.

17 Paola Barocchi and Renzo Ristori, *Il Carteggio di Michelangelo*, 5 vols. (Florence, 1965-1983), 3:332-33; see the Introduction of this volume (p. 4), which cites the passage. Also see Dawson Carr's catalogue entry for the Vienna version of the portrait, in which he notes the papal metaphor of the stone support (p. 102); in *Vittoria Colonna: Dichterin und Muse Michelangelos*, ed. S. Ferino-Pagden, exhibition catalogue (Vienna, 1997), pp. 98-102.

18 Barbieri, "Sebastiano's Portraits in *Paragone*," pp. 53-7; eadem, *I Ritratti*, p. 92.

Nygren has also recently identified these portraits as some of the first works to embrace stone's signifying potential through similitude between the subject—Peter the rock upon which the Church was built—and the material.[19] The present argument asserts the significance of linking the concept of the touchstone and the use of stone for portraiture, perhaps by Clement himself, to Sebastiano's papal office. If Clement did not necessarily spur Sebastiano's invention of a pictorial touchstone in the use of stone supports, he would, at least, have recognized immediately its capacity for meaning in both sacred images and papal portraits, thereby rewarding and marrying Sebastiano's faithful service to the office of the papal seal.

## The *Impresa*, the Seal, and the Portrait on Stone

The first evidence that Clement was interested in the touchstone metaphor appears in a section of Vasari's *Zibaldone* entitled, "I ritratti di tutti gli Eroi di Casa Medici," in which he describes a series of portraits in the Medici *guardaroba*. Paired with this list of portraits are corresponding *imprese* of the family members. For a portrait of Cardinal Giulio de' Medici, later Clement VII, Vasari notes "un paragone con oro su," or "a touchstone with gold upon it."[20] I have not been able to find an extant portrait with such an *impresa*, but Vasari's note at least serves to indicate Clement VII's personal interest in the concept of the touchstone; perhaps, given his identification of Cardinal Giulio de' Medici, before Sebastiano painted his first portrait on stone in 1531. Whether the portrait in *guardaroba* represented Giulio de' Medici while he was still a cardinal or after he became pope is not entirely clear, but other notices in *Zibaldone* refer to him according to his status at the moment of his depiction, thus suggesting that the *impresa* indeed predates Sebastiano's use of stone supports.[21] In any case, the fact that the concept of a touchstone can be directly connected to the first sitter to be rendered in Sebastiano's new method of painting substantiates the metaphorical function of the painter's technical innovation.

19 Nygren's understanding of the touchstone may be found in his PhD dissertation on Titian, "Vibrant Icons," pp. 348-62, in which he discusses the theological and contemporary discussions of Christ as a touchstone in the context of Titian's *Ecce Homo*. For his discussion of similitude and the materiality of stone, see his essay in this volume.

20 Vasari, *Lo Zibaldone*, ed. Alessandro del Vita (Arezzo, 1938), pp. 258-65; p. 258 for the note of the portrait in the Guardaroba and the *impresa*, and p. 264, n. 22, for the identification of Clement VII by Alessandro del Vita.

21 Cardinal Giulio de' Medici became Pope Clement VII in 1523, and Sebastiano did not begin using stone supports until around 1530.

That his representation of the Medici Pope was the most authoritative by the mid-century, or the time of Vasari's description, is testified to by another series of portraits: Bronzino's small-scale Medici likenesses painted on tin and installed in Cosimo I's Scrittoio di Calliope in 1559. In this series, the portrait of Clement VII is based on Sebastiano's works from the early 1530s, while another of Pope Leo X is based on Raphael's portrait of 1518, indicating that Bronzino selected canonical models for the duke's study.[22]

Giulio de' Medici's touchstone *impresa*, as Suzanne Butters has noted, was designed to signal authenticity, truth, and personal virtue.[23] We might also connect it to conceptions of purity, which in the context of his illegitimate birth, would align the notion of a touchstone to the Cardinal's (later Pope's) better known *impresa*, *Candor Illeseus*.[24] Though most are familiar with the figural use of the word 'touchstone,' denoting something that sets the standard by which like things are judged, the practical use of a touchstone is less well known. From antiquity, as Pliny records, these small black stones were used to test the true substance of gold and silver in order to detect debased or fraudulent coins.[25] Drawn across the surface of a touchstone, a false coin will leave a mark of a different color from that left by pure gold. Applied to the art of painting, the concept of the touchstone might demonstrate the truth of an artist's mark, as Rudolph Priemesberger suggested in his analysis of Jan van Eyck's obsidian-like surface for the Annunciation in the Thyssen Bornemisza Collection.[26] In the case of portraiture, the metaphor might convey the authenticity of the sitter's likeness and the purity of his or her character.[27] Such a

22 For this work and the series of small Medici portraits by Bronzino and his workshop, see the recent exhibition catalogue, *Bronzino. Pittore e poeta alla corte dei Medici*, ed. Carlo Falciani and Antonio Natali (24 September 2010-23 January 2011; Palazzo Strozzi, Florence), pp. 144-47; and Andrea Gáldy, "Identity and Likeness: Bronzino's Medici Portraits," in *Agnolo Bronzino: Medici Court Artist in Context*, ed. Andrea Gáldy (Newcastle on Tyne, 2013), pp. 31-50, esp. pp. 45-47.

23 Suzanne Butters, *The Forge of Vulcan: Sculptors' Tools, Porphyry and the Prince in Ducal Florence*, 2 vols. (Florence, 1996), 1:101-02.

24 For this *impresa* and its connection to Giulio de' Medici's illegitimate birth, see Sheryl Reiss' discussion in her dissertation, "Cardinal Giulio de' Medici as a Patron of Art, 1513-1523" (PhD diss., Princeton University, 1992), pp. 106-7.

25 Pliny discusses the use of the touchstone in *Natural History*. Vol. 9: *Books 33-35*. ed. G.P. Goold and trans. H. Rackham (London and Cambridge, MA, 1952; reprint edition, Cambridge, MA, 1999), pp. 94-97.

26 Preimesberger, "Ein 'Prüfstein der Malerei.'" pp. 85-100, esp. pp. 94-95.

27 In the case of a female sitter, we should think of Sebastiano's famed portrait of Giulia Gonzaga for Cardinal Ippolito de' Medici in 1532, which has been discussed profitably in terms of the poetics of female beauty and the question of the *paragone* between painting

conceit allows for a more specific application in the context of Sebastiano's papal office, whose most central concern was the application of a seal that guaranteed authenticity and signaled Petrine authority. Without—in any way—dismissing the achievement of Sebastiano's technical or intellectual innovation, it is worth considering the extent to which Clement VII might have collaborated with the painter in the production of an object that became the embodiment of one of the pope's *imprese* and a representational equivalent to the papal seal.

By infusing a recognizable likeness, the painted portrait, with an abstract concept, the *impresa*, the artist collapsed or merged two forms of representation. As noted above the stone used by Sebastiano functioned as another sign for Clement VII's identity, that of the papacy.[28] In his discussion of the portrait and the coat of arms, Hans Belting differentiates between a body 'image' and 'sign' and notes that the portrait (the body image) might have developed out of necessity, for while it was a counterpart to the coat of arms (the body sign), it was also its contradiction in the sense that they "coexist in a state of symbolic and aesthetic opposition."[29] The *impresa* and the papal seal conveyed information about Clement that is similar to that of coats of arms, indicating an aspect of character and an institution that extended beyond the individual represented, in this case an ecclesiastical office rather than a hereditary familial line. The material of the support itself assumed a function that we associate with the visual symbols of an *impresa*, the allegorical figures of a coin, or the historical/sacred figure of the papal seal. Beyond conveying a physical resemblance, these portraits on stone communicated the legitimacy and verity of Clement's

---

and the rival arts of sculpture and poetry but has not been understood in the context of this metaphor, despite the fact that this may have been his most famous work on a stone support. For a discussion of this work in the context of the *paragone*, see Barbieri (2008, 2013) and Nygren in this volume.

28 See Valentin Groebner, *Who are You? Identifiation, Deception, and Surveillance in Early Modern Europe* (New York, 2007), pp. 184-86; and Stephen Perkinson, *The Likeness of the King: A Prehistory of Portraiture in Late Medieval France* (Chicago, 2009), pp. 20-22, 89-90, whose discussion Brigitte Bedos-Rezak's work on medieval seals is particularly relevant, especially in relation to Hans Belting's chapter, "The Coat of Arms and the Portrait: Two Media of the Body," in *An Anthropology of Images: Picture, Medium, Body*, trans. Thomas Dunlap (Princeton and Oxford, 2011), pp. 62-63. Bedos-Rezak has written extensively, see in particular, "Signes et insignes du pouvoir royal et seigneurial au Moyen Age: Le témoignage des sceaux," in *Form and Order in Medieval France: Studies in Social and Quantitative Sigillography* (Aldershot, 1993), pp. 47-62; "Medieval Identity: a Sign and a Concept," *American Historical Review* 105, no. 5 (2000), 1488-1533. I want to thank Nic Baker and Anne Dunlop directing me to Groebner's and Perkinson's works, respectively.

29 Belting, *Anthropology of Images*, pp. 62-63, and 69 for citation.

likeness and character specifically, via the metaphor of the touchstone, and the institution of the papacy generally. In this volume, Christopher Nygren proposes the way in which stone was able to convey the stability of the Petrine office and signal ontological similitude; establishing a similitude between the subject, St. Peter's successor, and the matter used in representing him (the stone support of these portraits) that surpasses naturalistic representation in its construction and communication of the sitter's identity.[30] In this sense, Sebastiano's likenesses of Clement VII and later Paul III painted on stone accomplish more than the seals of each pope, which would have had an image of St. Peter on one side and the pope's name on the reverse, by rendering their recognizable and specific physiognomy in a material more closely associated with their ecclesiastical office; while the naturalistic portrait substituted for the specific name, the stone at once invoked Peter (*pietro*/*pietra*) and embodied the historical, almost eternal, Petrine office. Sebastiano's invention thus resolves the symbolic and aesthetic opposition between different forms of representation noted by Belting, while also claiming the individual sitter's place in an institution whose longevity and system of succession could not be demonstrated by physiognomic likeness, as hereditary features in a series of familial portraits might.[31]

## The *Paragone* and Antiquity

As Baker-Bates establishes in chapter one of this volume, Sebastiano had been developing or discovering new techniques in painting for well over a decade when he perfected the process of painting in oil on stone. Although the theme of physical preservation consistently appears in the 16th-century rhetoric of the *paragone* between painting and sculpture and in the few contemporary descriptions of Sebastiano's technique, the Gonzaga agent Niccolò Sernini, who attempted to manage the Úbeda Pietà commission, warned of the stone's fragility in the midst of his battle with the painter; at one point even suggesting that another work on copper be sent to Spain because it would be less likely to

30 See Nygren, "The Matter of Similitude: Stone Paintings and the Limits of Representation in Cavaliere d'Arpino's *Perseus and Andromeda* and Jacques Stella's *Jacob's Ladder*," chapter four of this volume, p. 148.

31 Belting, *Anthropology of Images*, p. 69. In the context of Belting's discussion, the function of Sebastiano's inventive form of portraiture was to convey, through the material association with the ecclesiastical office, authority and legitimacy to an individual beyond the confines of his physical presence. Baker-Bates discusses the distribution of these portraits in his "Painter of Diplomatic Gifts," p. 65.

break.[32] Sebastiano's invention was more likely conceived to function symbolically, even if the metaphors in question encompassed the idea of eternity.[33] The language used by Soranzo to describe this technique seems to marvel not only at its durability but also its 'discovery.'[34] In this sense, Sebastiano's intellect and drive to experiment placed him in the company of Leonardo and Raphael. Miguel Falomir's suggestion that the paintings on stone would have recalled mosaic icons is particularly striking, if we consider Sebastiano's work in the Chigi Chapel and its inevitable comparison with Raphael's mosaics.[35] The illusionism of Sebastiano's style, however, might have held more appeal to his sitters than the obvious craftsmanship or artistic mark of mosaics, if the passage by Paolo Giovio cited in the Introduction is any indication of the 16th-century reception of that medium. Praising Vittoria Colonna, Giovio admires the coherence of her beauty and virtue contrasting it to the shortcomings other women, whose "lot ... besides her alone" it was "to be like the images of a mosaic," which upon close inspection begin to show seams, "and both the enormous gaps and the crowded awkwardness, the almost haphazard workmanship of the crudely inserted pebbles—all discerned clearly by the

32 See Hirst, "Sebastiano's *Pietà*," p. 589. Sernini wrote, "Il lavorar in Ramo mi piacerebbe per esser cosa più sicura e durabile che che la pietra (I would be pleased with a work on copper because it is something more secure and durable than stone)." Despite this comment, made in April 1537, relatively late in the commission's tortuous history, the patron or the recipient seems initially to have been interested in having a work on stone by Sebastiano, a desire that would confirm Vasari's report that people sought out these works (*Le Vite*, 5:97-98). For a discussion of the *paragone* between painting and sculpture and the question of durability see Leatrice Mendelsohn, *Paragoni: Benedetto Varchi's* Due Lezzioni *and Cinquecento Art Theory*, ed. Donald B. Kuspit (Ann Arbor, 1982), pp. 129, 276, n. 136, who first proposed this context in her discussion of Sebastiano's portrait of Giulia Gonzaga. Also see, Barbieri, *I Ritratti*, pp. 95-108.

33 Miguel Falomir Faus, in "Sebastiano and 'Spanish Taste,'" in *Sebastiano del Piombo*, also addresses the slate supports in a way that allows for a more complex reading of Vasari's 'quasi eternal.' Also see Christianne Hessler for this technique and the portrait of Baccio Valori in her "The Man on Slate: Sebastiano del Piombo's Portrait of Baccio Valori and Valori the Younger's Speech in Borghini's *Il Riposo*," *Source* (2006), 18-22.

34 For a more extensive discussion of Sebastiano's demonstration of intellect and *ingegno* through his development of this technique, see Calvillo, "Authoritative Copies and Divine Originals," pp. 485-86, 500.

35 Miguel Falomir Faus, "Sebastiano and 'Spanish Taste,'" p. 71 n. 60. Alexander Nagel and Christopher S. Wood's recent work, *Anachronic Renaissance* (New York, 2010) has demonstrated the extent to which the term *eternal* could suggest much more complicated and ambitious conceptions of the work of art in time. Also see Anna Kim's chapter in this volume for the connection between painting on stone and mosaics as a kind of painting.

eyes—shatter one's entire opinion of the work."[36] This desirable illusionism made possible through his use of oil paint on surfaces such as walls and stone supports would not only have been tied Sebastiano to Leonardo's experimentation, a connection noted by Kathleen Weil Garris Brandt almost 40 years ago, but also to Pliny—not only for the concept of the touchstone but also for the value of naturalistic portraits and the painting of stone.[37]

Along these lines, one might read the comments of a Roman gentleman in Francisco de Holanda's *Roman Dialogues* as a commentary on Sebastiano's revival of an ancient portrait type. As frequently occurs in the fourth dialogue, Holanda's interlocutor slightly adjusts the paraphrase of Pliny's text to address a particular topic of concern; in this case, the anonymous Roman introduces the question of portraiture that opens Book Thirty-five:

> He was complaining already then that in the past the art of painting had been very noble and sought after by kings and peoples, which ennobled those whom art destined for renown in posterity; but that in their own day it enjoyed less standing because of the invention of encrusting marble with colors, which was a newer and more ambitious way of painting—a complaint that certainly pertains not to this period but to that.[38]

The care that Holanda takes at the end of this passage to distinguish his time from that of Pliny, while typical of this dialogue, is particularly striking and, I believe, meaningful here, especially because of the parenthetic phrase "a newer and more ambitious way of painting" after the "invention of encrusting marble with colors." Holanda's interpretation of the Plinian portrait clearly

36 Paolo Giovio, *Notable Men and Women of our Time*, ed. and trans. Kenneth Gouwens (Cambridge, MA, 2013), p, 505; see this volume's Introduction for the full citation, p. 20.

37 Kathleen Weil Garris Brandt, *Leonardo and Central Italian Painting, 1515-1550* (New York, 1974), p. 14 for the connection to Leonardo's experimentation in Rome; her discussion of Pliny's *atramentum* (pp. 17-22), however, also deserves to be reintroduced into discussions of Sebastiano's use of stone supports, as a potential *lapis specularis*. Ana González Mozo's essay in this volume (pp. 73-75, 93-94) emphasizes the importance of antique sources for this invention. The important theoretical connection to antiquity has been noted by Anne-Laure Collomb, "Les traits artistiques et la peinture sur pierre, XVI^e^–XVII^e^ siècles," *Histoire de l'art* 52 (2003), 111-20, esp. 111-12; and Mario Casaburo, "La Pittura su Pietra in Italia. Primi Scenari di Ricerca: Appunti sulle techniche, i materiali e le Opere," *Per la Conoscenza dei Beni Culturali* III (2010), 239-57, esp. p. 41.

38 For the most recent English translation of Holanda's dialogues, now significantly published with the preceding treatise, see Francisco de Hollanda, *On Antique Painting*, trans. Alice Sedgwick Wohl. (University Park, 2013), p. 229, for this passage.

privileges the medium of painting, as opposed other ancient and durable techniques such as mosaic. He also insists that Pliny's complaint "pertains not to this period but to that" (Pliny's), distancing the criticism of the ancient author from his own appraisal of such works. In doing so, Holanda's comments remove Sebastiano's innovation from the taint of excess clearly expressed by Pliny, who continues by indicting the ancients' privileging of material value over naturalistic likenesses. In light of this passage, I would argue that Sebastiano's invention intended, or was at least immediately understood, to provide a modern and ambitious corrective of the colored marble panels of Pliny's text.

Sebastiano's portraits were prized for their naturalism and probably also their rhetorical and poetic potential. The artist's and his work's virtuosity appear through his skill and invention; one would never argue that the stone supports of Sebastiano's paintings were intended to add material value to the works, in contrast to later works that emphasize the natural beauty and preciosity of the figurated stone. The value of Sebastiano's work was entirely artistic and intellectual. Still, Holanda's engagement of Pliny the Elder's lament is similar enough to Vittore Soranzo's description of Sebastiano's new technique, which marveled that the colors united, almost becoming stones themselves, to suggest that the 16th-century invention was immediately understood in the context of a *paragone* between ancients and moderns.

## Portraiture and the Office of the Piombo

Sebastiano's innovative engagement of Pliny's history and Clement VII's *impresa* surely would have delighted both his Medici and Farnese patrons for its ability to present an authoritative papal image issued from the office of the *piombatore* while simultaneously asserting that likeness' authenticity through its material kinship to a touchstone and its symbolic consubstantiality with the subject represented. The premium that the Farnese placed on skillfully wrought portraits has been noted by a number of historians, who usually cite Giovio's comments that Vasari's portraits in the Sala dei Cento Giorni had left something to be desired and the production of an astounding series of family portraits by Titian in the mid-1540s [**Figs. 3.1 and 3.2**], whose enthusiastic reception by the patrons is testified to by both contemporary correspondence and Vasari's *vita*.[39] It is thus worth reconsidering one aspect of the Farnese's

39 Robertson, "Il Gran Cardinale." pp. 58-59, citing Giovio *Lettere*, ed. G.G. Ferrero (Rome, 1956-58), 2:61, no. 243. This chapter (pp. 69-72) also discusses Titian's portraits for the Farnese, mostly citing Zapperi.

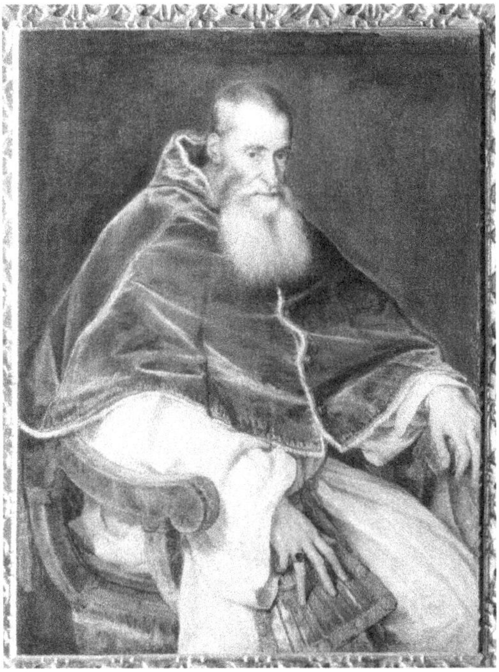

FIGURE 3.1
*Titian,* Paul III, *oil on canvas, 1545-46, Naples, Museo di Capodimonte* (PHOTO: ©SCALA, FLORENCE, 2015).

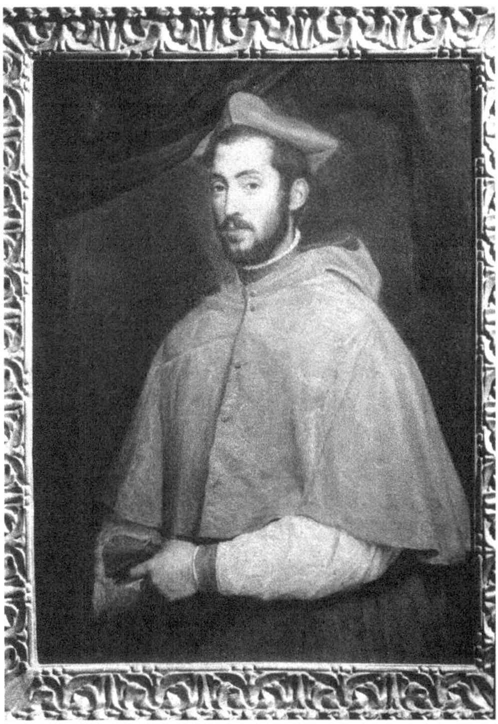

FIGURE 3.2
*Titian,* Alessandro Farnese, *oil on canvas, 1545-46, Naples, Museo di Capodimonte* (PHOTO: ©SCALA, FLORENCE, 2015).

patronage of Titian in light of Sebastiano's ability to tie his papal office to his role as a portraitist: their proposal to make the Venetian painter a *piombatore* during this period.[40] Most scholars have attributed Titian's polite refusal of the lucrative sinecure to his personal and professional ambitions, both for his son Pomponio and for himself as an Imperial artist, and his attachment to Venice. In the context of the present argument, it seems more interesting to ask *why* the Farnese offered Titian the position of *piombatore* and to note that Titian only painted on a stone support after his Roman sojourn of 1545-46, the well-known *Ecce Homo* [**Plate 5**] for Charles V now in the Museo del Prado.[41]

According to Vasari, Titian had already painted this subject as a gift for Paul III while working in his Belvedere studio, where the biographer saw it with other works including the Capodimonte portrait of Paul III, Cardinal Farnese and Ottavio Farnese [**Fig. 3.3**], and the *Danaë*. His passing mention of an *Ecce Homo* made for the pope actually functions as kind of rhetorical and theoretical transition from the praise acknowledging the excellence of Titian's portraiture, as demonstrated by the triple portrait of the Farnese, to criticism of the Venetian painter's one weakness according to the Tuscan artist, his *disegno*. Indeed, Vasari's comment about the *Ecce Homo* that he saw in Titian's Roman studio precedes the well-known passage in which Michelangelo praises the *Danaë* while lamenting the Venetian's (and his fellow compatriots') lack of training in *disegno*.[42] Vasari does not indicate the material support of this *Ecce Homo*, only commenting that it was weaker than his other paintings, perhaps losing some of its force in comparison to the works of Michelangelo, Raphael and Polidoro. While Vasari slights, by omission, Sebastiano, whose paintings on stone would have been Titian's principal point of comparison when he painted the later version for Charles V, Aretino does not fail to recommend Sebastiano in a letter of October 1545. Listing the most important works to see

40 See Amadio Ronchini, "Relazioni di Tiziano coi Farnesi," *Atti e Memorie delle R. Deputazioni di Storia Patria per le provincie modenesi et parmensi* 2 (1864): pp. 129-46, esp. 136-37.

41 Both Baker-Bates and Nygren have discussed this painting in the context of Titian's practice of painting on stone and his emulation, certainly, of Sebastiano's technique. See Baker-Bates, "The `Cloister Life' of Emperor Charles V: Art and Ideology at Juste," *Hispanic Research Journal* 14 (2013), 427-45; and Nygren, "Vibrant Icons," as in note 20 above, and idem "Titian's *Ecce Homo* on Slate."

42 Vasari, *Vite*, 6:164 (Giuntina); "A persuasione de' quali fece, per donare al Papa, un Cristo dal mezzo in su, in forma di Ecce Homo; la quale opera, o fosse che le cose di Michelagnolo, di Raffaello, di Pulidoro e d'altri l'avessono fatto perdere, o qualche altra cagione, non parve ai pittori, tuttoché fusse buon'oper, di quell'eccellenza che motre altre sue, e particolarmente i rittratti."

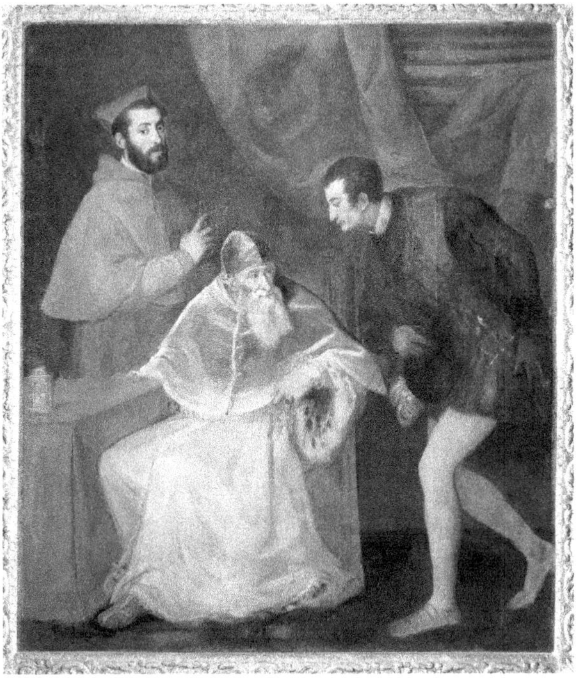

FIGURE 3.3
*Titian,* Paul III, Cardinal Alessandro and Ottavio Farnese, *oil on canvas, 1545-46, Naples, Museo di Capodimonte* (PHOTO: ©SCALA, FLORENCE, 2015).

while the painter was working in Rome, Aretino advises him to keep in mind the methods of certain famous painters, "and of our Fra Bastiano in particular."[43]

It seems telling that it was a subject normally associated with these paintings on stone that introduced Vasari's critique of Titian. Certainly, he would have seen examples of Sebastiano's invention while in Rome. As Nygren has made clear, Titian clearly understood the metaphoric and onto-theological possibilities of Sebastiano's use of stone when he painted the *Ecce Homo*, which surely meant to equal or surpass Sebastiano's accomplishment in the Úbeda *Pietà* [**Plate 1**]. To what extent the metaphor of the touchstone and the technique remained tied to the papal office and papal portraiture by Titian is uncertain. The timing of the papal reward, however, was clearly related to both Titian's trip and his recent work for the Farnese; a letter from Titian to Cardinal Farnese dated 18 June 1547 notes that he had been offered the position three times, presumably the last time when Sebastiano's recent death left the office open.[44] Given the Farnese patrons' satisfaction with the series of family por-

---

43 See Titian, *L'Epistolario*, ed. Lionello Puppi (Florence, 2012), doc. 78, pp. 120-21; "Tenete a mente il far di ciascun pittore famoso, e del nostro fra' Bastiano in spezie."

44 *L'Epistolario*, doc. 98, p. 135: "se ben la Signoria Vostra m'imponesse questa terza volta che accettasse il cappuccio del già Fra Bastiano (ASPr., Epistolario scelto, b. 19, fasc. 16).

traits that Titian produced between 1542 and 1546, their desire to reward the painter and tie him to the papal household is not surprising, especially as the office would require his presence in Rome. It is also worth noting that since the time of Bramante one of the two *piombatore* positions had been occupied by an artist. Still, the excellence of both Sebastiano and Titian in portraiture suggests that the former artist had successfully tied this practice to the office of the *piombatore*.[45] If the portraits of these artists were rendered in stone, moreover, the connection was made not simply through skill in naturalistic representation but the through the cohesion of the likeness and the papal seal, as suggested above in section two.

That Sebastiano's eventual successor in the Piombo was another favored portraitist of the Farnese, Guglielmo della Porta, who worked in stone supports this connection. When Sebastiano died in June of 1547, the sculptor had already completed a portrait bust of the Paul III now in Naples [**Fig. 3.4**], and would eventually receive the commission for the pope's tomb in St. Peter's Basilica.[46] Vasari, in fact, closely ties the commission for the tomb to della Porta's award of the office, which he received through the favor and help of Michelangelo.[47]

---

Ronchini, "Relazioni di Tiziano," pp. 136-37; Robertson, "Il Gran Cardinale," p. 70; see also Charles Hope, *Titian*, pp. 99-102.

45 The inventory of the Farnese advisor, Fulvio Orsini, demonstrates the interest of the Farnese and their advisors in these portraits on stone. According to this document, Orsini owned a *Visitation* in chiaroscuro, a portrait of Giulia Gonzaga, a portrait of Clement VII without a beard, another of the Medici pope with a beard, another of Clement on *pietra di Genova* (slate), a portrait of Paul III and Duke Ottavio on *pietra di Genova*, a small portrait (*quadretto*) of Cardinal Ippolito de' Medici in secular dress, sketched by Sebastiano, and a small, framed work of a head. For Orsini's inventory, see Pierre de Nolhac, "Une Galerie de peinture au XVI^e^ siècle. Les collections de Fulvio Orsini," *Gazette des Beaux-Arts* 29, no. 2 (1884), 427-36; and Hochmann, "Les Dessins et les Peintures". Michael Hirst, *Sebastiano del Piombo* (Oxford, 1981), pp. 156-57, includes these pieces from Orsini's inventory.

46 In addition to Cardinal Farnese's mediation in the Úbeda commission, Hirst, *Sebastiano del Piombo*, pp. 115-16, 116, n. 110, 152, n. 1, has shown the extent to which the Cardinal oversaw the management of Sebastiano's estate after the artist's death in June of 1547. The inventory of Sebastiano's property, taken on 25 June 1547, indicates that there were many unfinished works on stone supports: ibid., pp. 155-56 (Appendix B).

47 Vasari, *Vite*, 6:205 (in the Life of Leone Leoni), "mentre dunque Guglielmo andava restaurando le statue che sono oggi nel palazzo de' Farnese...morì, nell'anno 1547, fra Bastiano Veneziano, che lavorava, come s'è detto, l'uffizio del Piombo; onde tanto operò Guglielmo col favore di Michelangelo e d'altri col Papa, che ebbe il detto ufficio del Piombo, con carico di fare la sepoltura di esso Papa Paulo Terzo da porsi in San Piero"; and at the end of his description of della Porta's works (6:207), "l'uffizio del Piombo è di tanto gran

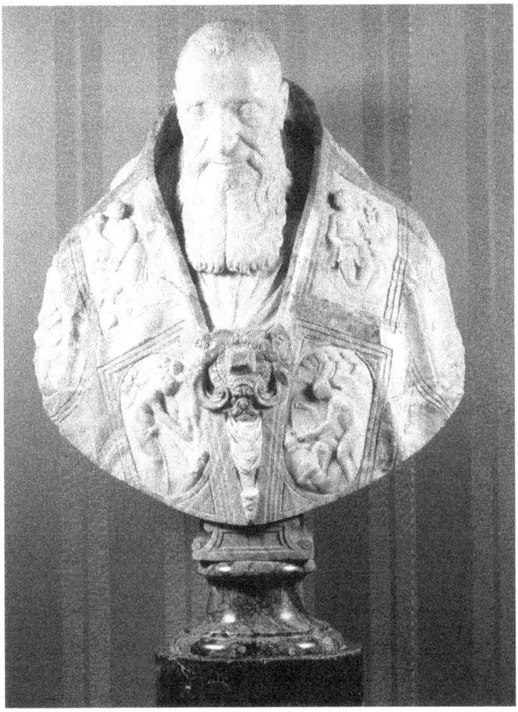

FIGURE 3.4
*Guglielmo della Porta,* Paul III, *marble, ca. 1547, Naples, Museo di Capodimonte.*

In the passage that follows, Vasari's comments further relate della Porta to Sebastiano by noting that, like the painter, the sculptor never finished very much after receiving the generous revenues of the *piombatore* because the office caused those who occupied it to become lazy. That the medium for some of the first portraits remained the same, stone, is significant; both the rhetoric of an object 'poco meno che eterna' and the reference to St. Peter and foundation of Church were preserved in della Porta's works.[48]

rendita, che si può studiare et affaticarsi per la gloria: il che non può fare chi non ha tante comodità. E nondimeno non ha condotto fra' Guglielmo di opere finite dal 1547 infino al questo anno 1567; ma è proprietà di chi ha quell'uffizio impigrire e diventare infingardo." Dorigen Caldwell has recently tied Guglielmo della Porta's marble portrait of Paul III to ideas circulating at the Farnese court; see her essay, "A Knowing Likeness": Artists and *Letterati* at the Farnese Court in mid-16th century Rome," in *Et Amicorum*: *Studies in the History of Humanism and Philosophy in Honour of Jill Kraye*, ed. M. Osserve and A. Ossa Richardson (Leiden, 2017), pp. 159-76.

48 That Guglielmo also made portraits of Paul III in bronze, especially that of the tomb in St. Peter's Basilica, does not necessarily undermine the connection between the *piombatore* and the production of naturalistic representation that absorb the function or correct the deficiencies of the more general, metal, seal.

Whatever the degree to which the Farnese associated Sebastiano's innovation of painting on stone to the office of the *piombatore*, they continued to commission or acquire works in the technique after Sebastiano's death, as the altarpiece in San Lorenzo in Damaso, their collections in Parma and Naples, and the inventory of their influential secretary Fulvio Orsini document.[49] With the exception of Sebastiano's unfinished portrait of Paul III and Ottavio Farnese, the work most indicative of the medium's links to authoritative papal portraiture at the Farnese court is the altarpiece completed by Francesco Salviati for the private chapel of Cardinal Alessandro Farnese, the Capella del Pallio in the Palazzo della Cancelleria, on local peperino [**Plate 15**]. Using a stone support for the altarpiece surely recalled both the Úbeda and Chigi altarpieces by Sebastiano, the latter possibly completed by Salviati. The inclusion of the portraits of Cardinal Farnese and Paul III around 1549, however, suggests a more specific, Farnese interest in maintaining the metaphoric potential of the stone for their own likenesses. As Patricia Rubin noted in her influential study of the chapel, the most recent conservation of the altarpiece revealed that the portraits had been painted over an initial layer of varnish.[50] The late addition of the portraits to the composition explains the somewhat awkward, cramped sense of the space in the work, but despite this less than felicitous outcome, the altarpiece is striking in its attempt to marry Titian's recent portraits of the Farnese with Sebastiano's technique in the Chigi Chapel, presumably at the patron's request.

According to Vasari, Annibale Caro and Giulio Clovio advised the Cardinal to hire Salviati for the work in his private chapel. Both men, like their patron, would have known of and appreciated Sebastiano's invention of a pictorial touchstone and embodiment of Clement VII's *impresa*, as would the ever-present Paolo Giovio. Indeed, Giovio documented a similar device of Fabrizio Colonna in his dialogue on *imprese*, in which he describes the image of a touchstone with the motto "Fides Hoc Uno, Virtusque Probantur" (Virtue and Faith are tested by this one/alone).[51] Clovio, as Vasari wrote at the end of Sebastiano's *Vita* in 1550, was one of the men in Rome with whom Sebastiano frequently conversed, we can presume at least partly about the theory and

49 See Pierre Nolhac, "Une Galerie," pp. 427-36; Hochmann, "Les Dessins et les Peintures"; Hirst, *Sebastiano del Piombo*, pp. 155-57; as in n. 45.

50 Patricia Rubin, "The Private Chapel of Cardinal Alessandro Farnese in the Cancelleria, Rome," *Journal of the Warburg and Courtauld Institutes* (1987), 82-112, 83-87 for the altarpiece; also see Robertson, "Il Gran Cardinale," pp. 151-56.

51 *Ragionamento di Monsignor Paolo Giovio sopra i motti ed i disegni d'arme e d'amore comunemente chiamano imprese* (Venice, 1556) f. 44.

practice of their art.[52] Also, available to Salviati at the court of Cardinal Farnese was Claudio Tolomei, whose 1543 letter to Sebastiano is frequently cited. In it Tolomei praises Sebastiano's portraits, which "seduce the eyes, delight the soul, and nurture the intellect; marveling the learned while stupefying the common man."[53] His desire to be portrayed by Sebastiano, he writes, is like that "severe judgment" of Alexander the Great, who only allowed Apelles to paint his portrait.[54] After this commonplace, Tolomei informs Sebastiano that he will think of his portrait of him as a divine mirror, in which he will see both himself and the artist.[55] Recognizing in the portrait Sebastiano's "singular *virtù*" and "marvelous artifice," Tolomei explains, will compel him to look at his own image with a desire to purge his soul of its shortcomings and flaws.[56] This language and the dynamic that Tolomei suggests between the painter's virtue and the sitter's contemplation of his image approximates the conceit of Clement VII's portrait, whose likeness made by the virtuosic painter's hand appears true and

52 Vasari, *Le Vite*, 5:102-3, "Vero è che per la conversazione sua dolse a molti amici et ad alcuni artefici ancora; come particularmente a don Giulio Corvatto miniature, che appresso il reverendissimo Farnese…"

53 *Delle Lettere di M. Claudio Tolomei Libri Sette* (Venice, 1589) f. 96v, "Di cui (from Sebastiano's hands) escono opere che invaghiscono gli occhi dilettano l'anima, nutriscono l'intelletto; le quali con maraviglia son considerate da dotti, con istupor mirate dal volgo." See also Hirst, *Sebastiano del Piombo*, pp. 115, 119, n. 124.

54 *Lettere di M. Claudio Tolomei*, f. 96v. "Ne so già qui io di quel severo giudizio che fu Alessandro magno, il quale non voleva ch'altri lo dipingesse, se non Apelle, anzi per lo contrario, per che voi mi dipingeste, non mi curarei che mille altri, men che mezzani dipintori, mi dipingesseno."

55 In an essay addressing the theory of portraiture, "Inventive Translation, Portraiture and Spanish Habsburg Taste in the Sixteenth Century," in *The Spanish Presence in Sixteenth-Century Italy: Images of Iberia*, eds. Piers Baker-Bates and Miles Pattenden (Farnham, 2015), pp. 175-97, I discuss this letter and the significance of the mirror metaphor for the artist and patron and Sebastiano's invention for Clement VII (pp. 181-83).

56 *Lettere di M. Claudio Tolomei*, f. 97. Tolomei ends citing Socrates' prescription for youths to look at themselves in mirrors, "Solo vi dirò che quando da voi mi venga tal grazia (come spero) allora mi parerà haver guadagnato uno specchio, il quale io sempre chiamarò specchio divino, percio che in quello vedrò voi, e me stesso insieme. Voi, vedendo ne l'imagine mia la vostra singolar virtu, e'l vostro maraviglioso artifizio. Me, vedendo ne l'arte vostra espressa vivamente la mia imagine, la quale mi sarà continuo stimolo a purgare l'anima di molti suoi mancamenti; non solo per quel rispetto, per lo qual Socrate voleva che i gioveni si guardasseno ne lo specchio; ma molto piu, perche vedendovi dentro molti luminosi raggi de le vostre virtu, mi s'accenderà l'anima a bel disiderio d'honore, e di gloria." Carl Brandon Strehlke has related this last passage to a drawing by Pontormo of two men examining themselves in a mirror; see the exhibition catalogue, *Pontormo, Bronzino, and the Medici*, ed. Carl Brandon Strehlke (Philadelphia, 2004), cat. no. 1, p. 56 and n. 1.

pure, like gold upon a touchstone. While few would argue that Salviati's portraits in the altarpiece for the Capella del Pallio rise to the expectations implied by Clement's *impresa* or the divine mirror found in Sebastiano's portraits, Tolomei's comments in 1543 do indicate the extent to which the Farnese and their court admired artists whose invention and technical prowess might match the virtue that they hoped to communicate in their painted likenesses.

Sebastiano del Piombo's critical fortune suffered seemingly from the moment he accepted the office of the *piombatore*. Certainly, Vasari attributed the material comforts gained through the office as source for both Sebastiano's and, as we have seen, Guglielmo della Porta's subsequent lack of productivity. Professional lassitude, however, was only one criticism; Ludovico Dolce accused Sebastiano of having succeeded through the intervention of Michelangelo's designs. Vasari's criticism of Titian's *Ecce Homo* imputed failure in comparison to the models of Michelangelo, Raphael and Polidoro da Caravaggio available in Rome, completely neglecting Sebastiano, even though the *Ecce Homo* that Titian next produced was surely conceived in competition with Sebastiano's Úbeda *Pietà*. Titian, in any case, did not look to Sebastiano for the figure of Christ, which derived from Michelangelo's design, but rather for the technical innovation that had proved so popular in Rome and in Spain.[57] Dolce's comments represent a constant refrain in the scholarship that has judged Sebastiano's painting as virtuosic but lacking in invention.[58] The area in which he clearly excelled, portraiture, is rarely discussed in terms of theoretical ambition or invention but rather judged mostly according to technical mastery. Yet Tolomei's comments, and I would add Francisco de Holanda's treatise on portraiture, suggest that there was a possibility of artistic invention outside of the *istoria*.[59] Sebastiano's inventive prowess is evidenced by his material experimentation and interest in the discourse of art, by his ability to produce meaning through the material supports of his works, which

57 As Nygren argues, Titian's looks to another model, the micromosaic of Santa Croce in Gerusalemme for his figure of Christ; see his "Titian's *Ecce Homo* on Slate," pp. 46-49.

58 For Dolce's criticism of Sebastiano, see Roskill, *L'Aretino*, pp. 94-95; and Morten Steen Hansen's comments in "Disegno and Authorship in Daniele da Volterra's Late Church Commissions," *Notizie da Palazzo Albani* 36-37 (2007-08), 57-78, 77.

59 Renaissance portraiture remains under-theorized, though Belting's discussion in *The Anthropology of Images* provides an important foundation. Francisco de Holanda's treatise *Do tirar polo natural* (1549) has received substantial consideration from Joanna Woodall in her monograph *Anthonis Mor: Art and Authority* (Zwolle, 2007). Joanna Woods-Marsden's appraisal of the state of the question in *Renaissance Theory*, eds. James Elkins and Robert Williams (New York and London, 2008), pp. 260-66, mentions it in passing (p. 264) as following Alberti's principal point of art improving upon nature.

might reference both antique practices described in Pliny and the personal *impresa* of his patron, and, in the case of papal portraiture, the seal of the ecclesiastical office. As Ana González Mozo notes in the previous essay we might even defend Sebastiano by considering some of Sebastiano's works on stone purposefully unfinished;[60] the signifying aspect of the stone matter in the papal portraits and in the *Madonna del Velo* is, after all, only amplified by its visual foregrounding. Sebastiano's accomplishment in portraiture and his will to invent in a mode of painting not concerned with *istoria* demand that we reconsider his skills apart from those of Michelangelo, whose friendship with the artist and provision of drawings might be the single most damaging factor in his later critical reception, beyond Vasari's fixation with Sebastiano's indolence.

In terms of Sebastiano's legacy, his technique of painting in oils on stone was much vaster than most realize, considering that almost every significant church in Rome had such a work by the first decade of 17th century, as Anne Laure Collomb's recent survey makes clear. Several of the artists who worked in this technique were also closely associated with Farnese patronage: such as Taddeo Zuccaro (*Conversion of Saul* in San Marcello al Corso), his brother Federico, whose massive *Coronation of the Virgin* altarpiece was made for the church adjoining Cardinal Alessandro's Chancery palace, San Lorenzo in Damaso, Girolamo Muziano (*Deposition,* Sta. Caterina dei Funari), or Marcello Venusti (*Saint Catherine* and *Saint John the Baptist,* Sta. Caterina dei Funari).[61] Whether their patronage and collecting of works on stone were strictly tied to Sebastiano's initial invention of a pictorial touchstone and material embodiment of the papal office, the Farnese and their *familiari* nevertheless demonstrated an immediate appreciation of Sebastiano's invention, both while he remained as one of the *piombatore* at Paul III's court and in the succeeding decades during which the proliferation of works in this technique indicate its profound influence as something almost eternal.

60 González Mozo's discussion of the non-finito and also the way in which Sebastiano's figures are painted in a way that uses the surface of the stone to suggest a kind of emergence from the material support represents the rich possibilities that further technical analysis can provide to the interpretation of these works; see her essay in this volume, pp. 91-92.

61 See Anne-Laure Collomb's *Splendeurs d'Italie; la Peinture sur Pierre à la Rénaissance* (Rennes, 2012), chapter two, "Les Peintures d'Autel," pp. 63-96, for a survey and discussion of these mid-century altarpieces in Rome.

## Bibliography

Baker-Bates, Piers. "Beyond Rome: Sebastiano del Piombo as a Painter of Diplomatic Gifts." *Renaissance Studies* 27, no. 1 (2013), 51-72.

Baker-Bates, "The `Cloister Life' of Emperor Charles V: Art and Ideology at Juste." *Hispanic Research Journal* 14 (2013), 427-45;

Baker-Bates, Piers. "Sebastianello nostro Venetiano: The Career of Sebastiano del Piombo and the Concept of a 'Court Artist' in Renaissance Rome." *The Court Historian* 16, no. 1 (2011), 7-22.

Barbieri, Costanza. "Fishing for Offices: Sebastiano del Piombo as Piombatore." *Burlington Magazine* 150, no. 1258 (2008), 35-36.

Barbieri, Costanza. "Sebastiano's Portraits in *Paragone*: Sculpted Paintings, Praise through Images." In *Sebastiano del Piombo, 1485-1547*, edited by Carlo Strinati, 53-57. Rome, 2008.

Barbieri, Costanza. "'Tu, che lo stile con mirabil cura pareggi col martello.' Fortune e sfortune di Sebastiano." In *La* Pietà *di Sebastiano a Viterbo. Storia e tecniche a confronti*, edited by Costanza Barbieri, Enrico Parlato, and Simona Rinaldi, pp. 50-65. Rome, 2009.

Barbieri, Costanza. *I Ritratti: Committenti, artisti e letterati nella Roma del Cinquecento*. Rome, 2011.

Barocchi, Paola, and R. Ristori. *Il Carteggio di Michelangelo*. 5 vols. Florence, 1965-73.

Bedos-Rezak, Brigitte. "Signes et insignes du pouvoir royal et seigneurial au Moyen Age: Le témoignage des sceaux." In *Form and Order in Medieval France: Studies in Social and Quantitative Sigillography*, pp. 47-62. Aldershot, 1993.

Bedos-Rezak, Brigitte. "Medieval Identity: a Sign and a Concept." *American Historical Review* 105, no. 5 (2000), 1488-1533.

Belting, Hans. *An Anthropology of Images: Picture, Medium, Body*. Translated by Thomas Dunlap. Princeton and Oxford, 2011.

*Bronzino. Pittore e poeta alla corte dei Medici*. Edited by Carlo Falciani and Antonio Natali, Exhibition catalogue, Palazzo Strozzi, Florence. Florence, 2010.

Butters, Suzanne. *The Triumph of Vulcan: Sculptors' Tools, Porphyry, and the Prince in Ducal Florence*. 2 vols. Florence, 1996.

Caldwell, Dorigen. "A Knowing Likeness": Artists and *Letterati* at the Farnese Court in mid-16th century Rome." In *Studies in the History of Humanism and Philosophy*, edited by M. Osserve and A. Ossa Richardson, pp. 159-76. Leiden, 2017.

Calvillo, Elena M. "Authoritative Copies and Divine Originals: Lucretian Metaphor, Painting on Stone, and the Problem of Originality in Michelangelo's Rome," *Renaissance Quarterly* 66, no. 2 (2013), 453-508.

Calvillo, Elena M. "'*Il Gran Miniatore*' at the Court of Cardinal Alessandro Farnese." In *Artists at Court: Image-Making and Identity 1350-1600*, ed. Stephen J. Campbell, pp. 163-75, 238-44. Boston, 2004.

Calvillo, Elena M. "Inventive Translation, Portraiture and Spanish Habsburg Taste in the Sixteenth Century." In *The Spanish Presence in Sixteenth-Century Italy: Images of Iberia*, edited by Piers Baker-Bates and Miles Pattenden, pp. 175-97. Farnham, 2015.

Calvillo, Elena M. "Reading Pliny in Francisco de Holanda's Roman Dialogues." In *Gifts in Exchange: Essays in Honor of Charles Dempsey*, edited Melinda Schlitt, pp. 263-96. Toronto, 2012.

Casaburo, Mario. "La Pittura su Pietra in Italia. Primi Scenari di Ricerca: Appunti sulle techniche, i materiali e le Opere," *Per la Conoscenza dei Beni Culturali* III (2010), 239-57.

Cerasuolo, Angela. "I dipinti di Sebastiano del Piombo del Museo di Capodimonte. Note sulla tecnica." In *La* Pietà *di Sebastiano a Viterbo. Storia e tecniche a confronti*, edited by Costanza Barbieri, Enrico Parlato, and Simona Rinaldi, pp. 128-47. Rome, 2009.

Collomb, Anne-Laure. "Les traits artistiques et la peinture sur pierre, XVI^e^–XVII^e^ siècle," *Histoire de l'art* 52 (2003), 111-20.

Collomb, Anne-Laure. *Splendeurs d'Italie; la Peinture sur Pierre à la Rénaissance*. Rennes, 2012.

*Delle Lettere a diversi, re et principi et cardinali et altri homini dotti a mons. Pietro Bembo*. Venice, 1560.

Falomir Faus, Miguel. "Sebastiano and 'Spanish Taste.'" In *Sebastiano del Piombo, 1485-1547*, edited by Carlo Strinati, pp. 67-71. Rome, 2008.

Gáldy, Andrea. "Identity and Likeness: Bronzino's Medici Portratis," *Agnolo Bronzino: Medici Court Artist in Context*, edited by Andrea Gáldy, 31-50. Newcastle on Tyne, 2013.

Giovio, Paolo. *Notable Men and Women of our Time*. Edited and translated by Kenneth Gouwens. Cambridge, MA, 2013.

Giovio, Paolo. *Ragionamento di Monsignor Paolo Giovo sopra ì motti e disegni d'arme e d'amore comunemente chiamano imprese*. Venice, 1556.

Groebner, Valentin. *Who are You? Identifiation, Deception, and Surveillance in Early Modern Europe*. New York, 2007.

Hansen, Morten Steen. "*Disegno* and Authorship in Daniele da Volterra's Late Church Commissions." *Notizie da Palazzo Albani* 36-37 (2007-08), 57-78.

Hessler, Christianne. "The Man on Slate: Sebastiano del Piombo's Portrait of Baccio Valori and Valori the Younger's Speech in Borghini's *Il Riposo*." *Source* (2006), 18-22.

Hirst, Michael. "Sebastiano's Pietà for the *Comendador Mayor*." *The Burlington Magazine* 114, no. 834 (1972), 585-93, 595.

Hirst, Michael. *Sebastiano del Piombo*. Oxford, 1981.

Hochmann, Michel. "Les Dessins et les peintures de Fulvio Orsini et la collection Farnèse," *Mélanges de l'École Française de Rome. Italie et Méditerrané* 105, no. 1 (1993), 49-91.

Hollanda, Francisco de. *On Antique Painting*. Trans. Alice Sedgwick Wohl, University Park, 2013.

Langedijk, Karla. *The Portraits of the Medici. 15th–18th Centuries.* 3 vols. Florence, 1981-83.

Mendelsohn, Leatrice. *Paragoni: Benedetto Varchi's* Due Lezzioni *and Cinquecento Art Theory*. Edited by Donald B. Kuspit. Ann Arbor, 1982.

Nagel, Alexander, and Christopher S. Wood. *Anachronic Renaissance*. New York, 2010.

Nolhac, Pierre de. "Une galerie de peinture au XVIe siècle. Les collections de Fulvio Orsini." *Gazette des Beaux-Arts* 29, no. 2 (1884), 427-36.

Nygren, Christopher J. "Vibrant Icons: Titian's Art and the Tradition of Christian Image-making." PhD diss., Johns Hopkins University, 2011.

Nygren, Christopher J. "Titian's *Ecce Homo* on Slate: Stone, Oil and the Transubstantiation of Paint." *Art Bulletin* 99, no. 1 (2017), 36-66.

Perkinson, Stephen. *The Likeness of the King: A Prehistory of Portraiture in Late Medieval France*. Chicago, 2009.

Pliny the Elder. *Natural History*. Vol. 9, *Books 33-35*. Edited by G.P. Goold and translated by H. Rackham. London and Cambridge, MA, 1952; reprint edition, Cambridge, MA, 1999.

*Pontormo, Bronzino, and the Medici: the Transformation of the Renaissance Portrait in Florence*. Edited by Carl Brandon Strehlke. Philadelphia, 2004.

Preimesberger, Rudolph. "Ein 'Prüfstein der Malerei,' bei Jan van Eyck?" In *Der Künstler über sich in seinem Werk*, ed. Matthias Winner, pp. 85-100. Weinheim, 1992.

Reiss, Sheryl. "Cardinal Giulio de' Medici as a Patron of Art, 1513-1523." PhD. diss., Princeton University, 1992.

Robertson, Clare. "Il Gran Cardinale"*: Alessandro Farnese Patron of the Arts*. New Haven, 1992.

Ronchini, Amadeo. `Delle relazioni di Tiziano coi Farnese.' *Atti e Memorie delle R.R. Deputazioni di Storia Patria per le Provincie Modenesi e Parmensi* (1864), 129-46.

Rubin, Patricia. "The Private Chapel of Cardinal Alessandro Farnese in the Cancelleria, Rome." *Journal of the Warburg and Courtauld Institutes* 50 (1987), 82-112.

*Sebastiano del Piombo. 1485-1547*. Edited by Claudio Strinati. Rome, 2008.

Seifertova, Hana. *Malby na kameni: umělecký experiment v 16. A na začátku 17. Století / Painting on Stone: An Artistic Experiment in the 16th and Early 17th Centuries*. Prague, 2007.

Titian. *L'Epistolario*. Edited by Lionello Puppi. Florence, 2012.

Tolomei, Claudio. *Delle Lettere di M. Claudio Tolomei Libri Sette*. Venice, 1589.

Vasari, Giorgio. *Lo Zibaldone*. Edited by Alessandro del Vita. Arezzo, 1938.

Vasari, Giorgio. *Le Vite de' più eccellenti architetti, pittori, ed scultori nelle redazioni del 1550 e 1568*. Edited by Rosanna Bettarini and Paola Barocchi, 8 vols. Florence, 1966-87.

*Vittoria Colonna: Dichterin und Muse Michelangelos*. Edited by Sylvia Ferino-Pagden, Exhibition catalogue. Vienna, 1997.

Weil-Garris Brandt, Kathleen. *Leonardo and Central Italian Art, 1515-1550*. New York, 1974.

Woodall, Joanna. *Anthonis Mor: Art and Authority*. Zwolle, 2007.
Woods-Marsden, Joanna. "Thinking Renaissance Portraiture." In *Renaissance Theory*, edited by James Elkins and Robert Williams, pp. 260-66. New York and London, 2008.
Zeri, Federico. "*Me pinxit*. Salviati e Jacopino del Conte." *Proporzioni* 2 (1948), 180-83.

# PART 2

# Ars et Natura: *The Poetics and Collecting of Paintings on Stone*

CHAPTER 4

# The Matter of Similitude: Stone Paintings and the Limits of Representation in Cavaliere d'Arpino's *Perseus and Andromeda* and Jacques Stella's *Jacob's Dream*

*Christopher J. Nygren*

> … In that Empire, the Art of Cartography attained such Perfection that the map of a single Province occupied the entirety of a City, and the map of the Empire, the entirety of a Province. In time, those Unconscionable Maps no longer satisfied, and the Cartographers Guilds struck a Map of the Empire whose size was that of the Empire, and which coincided point for point with it. The following Generations, who were not so fond of the Study of Cartography as their Forebears had been, saw that that vast map was Useless…
>
> JORGE LUIS BORGES, *"On Exactitude in Science."*[1]

Borges' paragraph-long story of an empire committed to creating a map in a one-to-one relation to the kingdom's actual territorial boundaries reveals the tensions underpinning much of the West's understanding of how images (and maps) operate. In the pre-modern world images were representational, of necessity something different than the objects they claimed to represent. Yet Borges' story underlines that countries, individuals, and religions are committed to the proposition that it need not be so: images can circumvent the slippage endemic to the mimetic process by making present the target of representation.[2] Miraculous images instantiate the deity, portraits of emperors

1 Jorge Luis Borges, *Collected Fictions*, trans. Andrew Hurley (New York, 1998), p. 325.

2 Bruno Latour, "Visualisation and Social Reproduction: Opening One Eye While Closing the Other…A Note on Some Religious Paintings," in *Picturing Power: Visual Depictions and Social Relations*, ed. Gorden Fyfe and John Law (London, 1988), pp. 15-38.

 | DOI 10.1163/9789004361492_007

and presidents make present the rule of law, and pictures alleviate the painful absence of the missing beloved.[3] Borges' map does something similar by offering a perfectly accurate topographical model of the Empire. Yet the map is useless. Rather than represent the distance between two points using a reduced scale, Borges' map embodies that essential information. This achievement, though perhaps admirable, is misplaced. In order to discharge its function as a useful representation of territory, a map must renounce its presentational ambitions and accept its status as a scaled and schematic likeness of an expanse of territory that ought not be re-embodied, but instead represented at reduced scale.[4] The representational mandate of Borges' map is undercut by the drive to attain a state of perfect similitude.

Borges' short story is not an intuitive place to begin an essay dealing with paintings on stone; however, the thrust of this intervention is to demonstrate that the issues of similitude, re-presentation, and substitution raised in that story unexpectedly align with the ambitions underwriting the emergence of stone painting in Europe in the 16th century. This essay will suggest that Renaissance and Baroque painters often appealed to stone as an artistic support in an attempt to break down the barrier between presentation and representation in ways that are analogous to the Cartographers Guild in Borges' story. Rather than simply be concerned with the commodification of painting or the so-called *paragone* between painting and sculpture, paintings on stone sought to challenge the implicit distinction between essence and representation at play in Borges' short story. While Borges' map does not resemble the kingdom, it does convey the essential information about topography and distance through a relationship of perfect correspondence. This article will suggest that a notion of deep similitude in excess of visual likeness often underwrote painters' adaptation of a stone support. This will be demonstrated through an analysis of two Baroque paintings on stone, *Perseus Liberating Andromeda* [**Plate 16**] by Cavaliere d'Arpino and *Jacob's Dream* [**Plate 17**] by Jacques Stella. Both painters used their lithic support as an integral part of the fictive ensemble. While it may be tempting to dismiss their deployment of stone as merely a form of visual 'cleverness,' akin to the *concettismo* seen in poets like Giambattista Marino, I think there is something more artistically

3 Horst Bredekamp, *Theorie des Bildakts* (Berlin: Suhrkamp, 2010) and Hans Belting, *An Anthropology of Images: Picture, Medium, Body*, trans. Thomas Dunlap (Princeton, 2011).

4 Bruno Latour, "Visualization and Cognition: Thinking with Eyes and Hands," *Knowledge and Society* 6 (1986), 1-40.

and intellectually ambitious underpinning this painting on stone.[5] By drawing attention to the problematic relationship between figure and ground as well as the interaction between the pictures and their source texts, this essay will suggest that these paintings are strangely resonant with Borges' map and notions of similitude, resemblance, and equality that it raises. Comparing these two pictures to earlier stone paintings by Sebastiano del Piombo will highlight how the concerns motivating artists' deployment of stone shifted from the Renaissance into the Baroque period.

Cavaliere d'Arpino painted *Perseus Liberating Andromeda* around 1595. The picture presents the well-known narrative found in Ovid, which tells how the poor damsel was chained to a rock as a sacrifice to satiate a sea monster that had been ravaging her village. Il Cavaliere depicts the moment at which Perseus arrives on the scene to liberate the maiden. The picture was painted on a small piece of precious lapis lazuli. However, the artist left a large swath of the blue stone completely untouched. In Il Cavaliere's painting, we see him cunningly deploy the rich blue stone substrate as the raw material for his airy skyscape; gently covered with floating algae and cresting waves, the lapis is transmuted into the liquid expanse of the Mediterranean Sea. Moreover, Il Cavaliere cunningly incorporated a crack running through the lower-right corner of the blue stone to inject greater drama into the story: Andromeda's toes nestle up against this crack, and her foot and knee seems to instinctively pull away. This motif suggests that the stone outcropping upon which the maiden has been placed is collapsing under her feet. She fears not only the sea monster, but also plummeting off the rock precipice into the blue-green sea below.

Stella's painting is from slightly later, around 1650. It relates the story of Jacob's vision found in the book of Genesis. The details of that story will be discussed later. For the moment it is important to draw attention to certain qualities of the picture. Painted on a piece of blue onyx, Stella cunningly enlisted the striations of the lithic material to provide the basic spatial coordinates of the composition. Steaks of a yellow-orange mineral deposit are visible throughout the composition, particularly in the sky at the upper-right. Once the eye has been calibrated to this warm hue, it becomes legible throughout the composition. The orange tone is seen peeking through the thin layer of paint used for the sheep at the bottom edge; serving as the hillside support for the town perched in the background; and giving a discordant radiance to the otherwise marmoreal body of Jacob, particularly his torso and legs. In the upper-left hand corner of the picture, a yellow knot in the stone is redeployed

5 Giorgio Baiardi, "Il dialogo *Del concetto poetico* di Camillo Pellegrino," *La Rassegna della Letteratura Italiana* 7 (1958), 370-74.

as the sun, marking out the heavenly realm inhabited by God the Father and angels. From this heavenly perch descends a blazing yellow ladder, once again painted over yellow-orange striations that descend toward the prophet Jacob, who sleeps with his head conspicuously resting upon a rock.

Il Cavaliere and Stella both conscripted stone in the project of figuration in ways that draw on incipient 'images' already present in the stone itself. Stone's natural striations provide the scaffolding for the pictorial structure erected by each artist. Leonardo had described how artists could extract pictorial inventions from the figural formations in stone. As Leonardo observes, the figurative striations of stone are suggestive: "… if you look at a wall spotted with stains, or with a mixture of stone, if you have to devise some scene, you may discover a resemblance to various landscapes, beautified with mountains, rivers, rocks, trees, plains, wide valleys and hills in varied arrangement; or again you may see battles and figures in action; or strange faces and costumes, and an endless variety of objects, which you could reduce to complete and well-drawn forms. And these appear on such walls confusedly, like the sound of bells in whose jangle you may find any name or word you choose to imagine."[6] Leonardo suggests that these incipient images spur the artist's creative capacity; like a poet stumbling upon what will become a new trope, recognizing images in stone becomes a mode of pictorial invention. Figuratively suggestive stones impelled the imagination in new directions and provided an engine of invention to the artist whose fantasy had stalled by pointing to modes of figuration that are not predicated upon artistic naturalism, but rather suggest forms of inchoate similitude.

In pre-modernity, similitude was a powerful concept not limited to the realm of visual resemblance. For the Latin West, this division was already embedded in the story of mankind's creation, where the deity specified that humans should be made "in our image and likeness" (*ad imaginem et similitudinem nostram*).[7] As Augustine made clear, to be fashioned in God's image invokes a relationship of physical resemblance. The relationship of similitude

---

6 "… se tu riguarderai in alcuni mvri inbrattati di uarie machie o pietre di uari misti, se avrai a inventionare qualche sito potrai lì vedere similitudine di diuersi paesi, ornati di montagnie, fiumi, sassi, albori, pianvre, grandi valli e colli in diuersi modi ancora potrai vedere diverse battaglie e atti pronti di figure, strane arie di uolti e abiti e infinte cose, le quali potrai ridurre in integra e bona forma, e interviene in simili mvri e misti come del suono di campane che ne'loro tochi vi troverai ogni nome e vocabolo che tu immaginerai." Leonardo da Vinci, *The Literary Works of Leonardo da Vinci*, ed. Jean Paul Richter, 2 vols. (London, 1883), v. 1, p. 2, pg. 254 (paragraph 508). See also Jurgis Baltrušaitis, *Aberrations: An Essay on the Legend of Forms* (Cambridge and London, 1989), p. 61.

7 Gen. 1:26.

is more complicated. Similitude speaks of "a certain proximity between things distant from each other, not proximity of place but of a sort of imitation."[8] Similitude, then, is relational. An object can express an ontological affinity to some other thing without necessarily being in a mimetic relationship to that thing. The inscription accompanying an 11th-century relief sculpture depicting *Christ in Glory* [**Fig. 4.1**] from the Abbey of St. Emmeram at Regensberg reveals that such notions of material similitude have long been operative in the realm of Christian art; "Since Christ is called a rock on account of his firm majesty, it is fitting enough that His image be in stone."[9] This inscription clearly articulates a notion of similitude that goes beyond visual resemblance: as Christ is stone-like, so too should this image be made out of stone. The image-text relationship reveals that artistic material was understood as a malleable and charged vehicle for expressing an essential similitude between an image and its target of representation. This modality of similitude was not exhausted by the image's visible resemblance to Christ, but instead depended upon the similitude between Christ and the stone out of which the image was fashioned. A similar logic can be found at play in many later paintings on stone.

Baroque paintings on stone often employed precious or semi-precious stones; as described above, the artists relied upon the incipient figuration of mineral deposits in the stone to provide the basic elements of the pictorial narrative. Thus the relationship between figure and ground is unstable. The beholder struggles to identify where the color of the stone ends and the brushstrokes of the artist begin.[10] This sort of game made these paintings particularly apt for the *Kunstkammer*, where they were appreciated as collector's pieces.[11] Within that rarefied context, these paintings troubled the distinction between *naturalia* and *artificialia*. Indeed, inventories from the period testify that beholders enjoyed the struggle of a playful game in which they were tasked with distinguishing between those parts of the painting created by nature and those elements that had been painted by an artist. 'Art' was taken as one cate-

8 Augustine, *The Trinity* (*De trinitate*), trans. by Edmund Hill (New York, 1991), 7.12 (p. 235).

9 Calvin Kendall, *The Allegory of the Church: Romanesque Portals and Their Verse Inscriptions* (Toronto, Buffalo, and London, 1998), pp. 260-61.

10 Baltrušaitis, *Aberrations*, pp. 60-105, esp. 70.

11 Roger Callois, *The Writing of Stones*, trans. Barbara Bray (Charlottesville, 1985), pp. 32-6; Lorraine Daston, "Nature by Design," in *Picturing Science, Producing Art*, ed. Caroline A. Jones and Peter Galison (New York and London, 1998), pp. 232-53; and Claudia Blümle, "Mineralischer Sturm. Steinbilder und Landschaftsmalerei," in *Vermessen: Landschaft und Ungegenständlichkeit*, ed. Werner Busch and Oliver Jehle (Zürich, 2007), pp. 73-95. See also *Pietra dipinta: tesori nascoti del '500 e del '600 da una collezione privata Milanese*, ed. Marco Bona Castelloti (Milan, 2000), pp. 183-227.

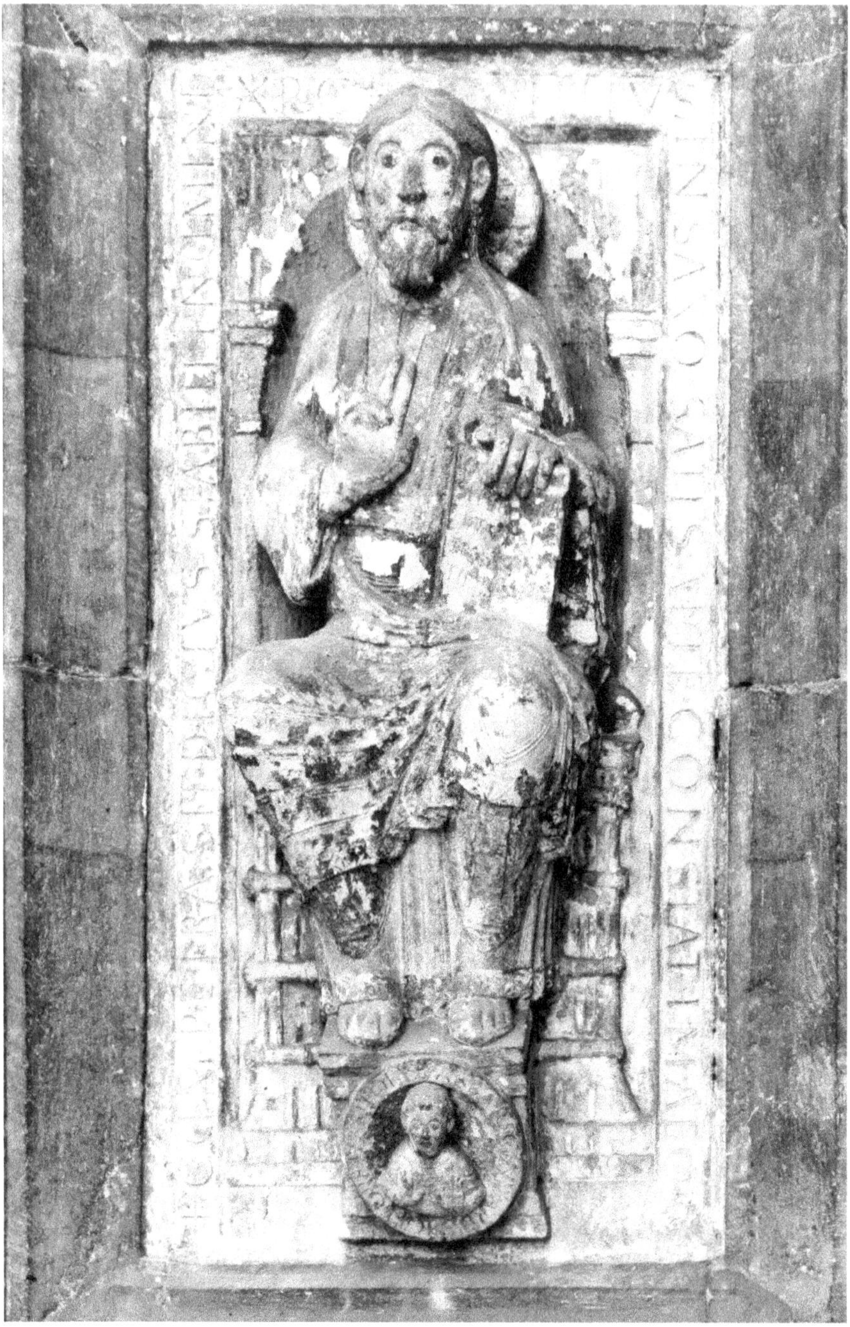

FIGURE 4.1 Christ in Glory, *11th century, stone, Abbey of St Emmeram, Regensburg* (ARTWORK IN THE PUBLIC DOMAIN; PHOTOGRAPH PROVIDED BY HIRMER FOTOARCHIVE MÜNCHEN).

gory of creation that depended upon a human intervention of painting (*ars pinxit*) while 'Natura' was taken as a separate source of imagistic generation that was devoid of human agency and intervention (*a natura sine omnis artis ministerio* or more simply *a natura depicti*).[12]

Localizing Baroque stone paintings within the emergent art collections of the *Kunstkammer* is undoubtedly a historically valid approach to understanding many of the extant exemplars. However, emphasizing the commodification of painting sidesteps the questions of similitude and resemblance that animate paintings like Il Cavaliere's *Perseus* and Stella's *Jacob*. Instead, this essay puts pressure on what Thomas Raff has called the 'material iconography' of images by reading stone pictures against the growing literature in numerous fields that challenges the hard ontological distinction between intentional beings and inert matter, which has been central to post-Enlightenment Western epistemology.[13] It will be argued here that by enlisting stone as an actant in their economy of re-presentation, these pictures strove to attain an essential shared identity with their target of representation.[14] Like Borges' imperial map, these paintings come tantalizingly close to capturing the most salient features of the subjects they purport to represent. In many Baroque stone paintings like those under examination here, the pictorial surface is no longer neutral. When a sky is painted over canvas, it is the work of the artist that is foregrounded. Instead, when Il Cavaliere 'painted' the sky in his *Perseus and Andromeda*, nature herself was enlisted to make present her own essence. The wager of this essay is that only by recuperating a pre-modern understanding of the vibrancy of matter can scholars fully appreciate the presentational ambition of paintings on stone.[15] Painting on stone allowed artist to realign resemblance and similitude in a manner that is both unexpected and visually seductive.

Stone paintings call attention to their material aspect in ways that paintings on canvas or wood panel do not. Their lithic surface overturns the dominant paradigm used to describe most Renaissance paintings, which is Alberti's metaphor of the picture as a window onto an adjacent world of illusionistic

---

12 Paolo Maria Terzago, *Musaevm Septalianvm Manfredi Septalae Patritii Mediolanensis Indvstrioso Labore Conscrvctvm* (Tortona, 1664), pp. 41 and 142.

13 Thomas Raff, *Die Sprache der Materialien: Anleitung zu einer Ikonologie der Werkstoffe* (Münster, 2008).

14 On representational economies, see Webb Keane, *Christian Moderns: Freedom and Fetish in the Mission Encounter* (Berkeley, Los Angeles, and London, 2007).

15 For the term 'vibrant matter,' see Jane Bennett, *Vibrant Matter: A political Ecology of Things* (Durham and London, 2010)

artifice.[16] Stone paintings resist the logic of the window by calling attention to their material aspect in ways that paintings on canvas or wood panel do not. Focusing on stone paintings calls attention to the 'canvas essentialism' that currently dominates early modern art history. Here I adapt the term 'screen essentialism,' which has been developed by Robert Markley and Nick Montfort to describe the elision of the material qualities of information technology, a pernicious tendency that they identify as pervading contemporary media studies.[17] Behind the computer screen there is an agglomeration of computer chips, microprocessors, screws, and human labor that are routinely effaced in accounts of cyberspace, globalization, and the spread of technology. What I am calling 'canvas essentialism' points toward a similar effacement of materiality that is present in the study of early modern European art, which is dominated by an Albertian paradigm of picture-making that privileges the metaphor of the picture plane as a window. The significance of stone paintings resides in the ways in which their obdurate materiality resists the inclination to see 'through' these pictures.

It seems axiomatic that throughout the early modern period, as images became increasingly convincing in their representation of the human figure, the notion of similitude began to shift from something flexible to something more rigid that regarded exclusively an image's outward appearance, its *optical resemblance* to the target of representation. The inscription found inside the cavity of a late medieval crucifix carved by the Sienese artist Lando di Pietro makes this point:

> The Lord God made it possible for Lando di Pietro of Siena to carve this crucifix in this wood in the similitude of the real Jesus to remind people of the passion of Jesus Chris Son of God, and the Virgin Mary, therefore

16 Samuel Edgerton, *The Mirror, The Window, and the Telescope: How Renaissance Linear Perspective Changed Our Vision of the Universe* (Ithaca, 2009), pp. 117-74. For a critique of how Alberti's metaphor has been leveraged by modern scholars, see Joseph Masheck, "Alberti's 'Window': Art-Historical Notes on an Antimodernist Misprision," *Art Journal*, 50, no. 1 (1991), 35-41.

17 See Robert Markley, "Boundaries: Mathematics, Alienation, and the Metaphysics of Cyberspace," in *Virtual Reality and its Discontents*, ed. Robert Markley (Baltimore, 1996), pp. 55-77 and Nick Montfort, "Continuous Paper: the Early Materiality and Workings of Electronic Literature," Paper presented at the Modern Language Association Convention, Philadelphia, 28 December 2004. <http://nickm.com/writing/essays/continuous_paper_mla.html>.

> you true and holy cross of Jesus Christ Son of God, render the said Lando to God.[18]

Life-size polychrome crucifixes were widely diffused in the middle of the 14th century, and they were also becoming increasing lifelike with articulated limbs giving the impression of an animated body.[19] At first, the inscription might be taken to suggest that the Crucifix's power rests in its lifelikeness. However, Christina Neilson has recently argued that the appeal of wooden crucifixes did not rely simply on the artist's ability to carve wood into an image that visibly resembled the suffering body of Jesus Christ; instead, artists and theologians understood the material of the wooden crucifix to serve as a substitute for the body of the suffering savior. The analogy between wood and the human body is deep-seated; it maps the structure of a tree onto the human body through the presence of things like veins, blood, and humors.[20] As a theologian might say, wood is similar to the human body, even if in its raw state it does not outwardly resemble the human figure. By making a crucifix in wood, artists not only activated a series of theological tropes (the Tree of Life, etc.) but also produced a three-dimensional body whose structure—veins, blood, humors, etc.—is similar to a human body. While it might afford other things that wood does not, a crucifix made out of terra cotta offers none of material similarities of wood. Artistic materials mattered greatly in the logic of image-making.

The theologically inflected concept of similitude and analogy, as opposed to physical resemblance or representation, underwrote much of the theoretical discourses surrounding the ambitions and limitations of the mimetic arts in the Renaissance. In the Renaissance, all acts of artistic creation were adumbrated in God's sculpting of the human figure from mud (*de limo terrae*. Gen. 2:7). This point is made most forcefully by Cennino Cennini,[21] but it was a commonplace shared by essentially all Renaissance theorists of art,

---

18 Donal Cooper "Projecting presence: the monumental cross in the Italian Church interior," in *Presence: The Inherence of the Prototype within Images and Other Objects*, ed. Robert Maniura and Rupert Shepherd (Burlington and Aldershot, 2005), pp. 47-69. See also Catherine King, "Effigies: Human and Divine," in *Siena, Florence, and Padua: Art, Society, and Religion, 1280-1400*, ed. Diana Norman (New Haven and London, 1995), pp. 124-27.

19 Kamil Kopania, *Animated Sculptures of the Crucified Christ in the Religious Culture of the Latin Middle Ages* (Warsaw, 2010).

20 Christina Neilson, "Carving Life: the Meaning of Wood in Early Modern European Sculpture," in *The Matter of Art: Materials, Practices, Cultural Logics, c. 1250-1750*, edited by Christy Anderson, Anne Dunlop, and Pamela H. Smith (Manchester, 2014), pp. 223-39.

21 Cennino d'Andrea Cennini, *Il libro dell'arte, The Craftsman's Handbook*, trans. Daniel v. Thompson, Jr. (New Haven and London, 1936), pp. 1-2.

even those who wished to expose the limits of this analogy.[22] Beyond this general analogy, however, Renaissance critics turned to the language of similitude and analogy when discursive language seemed to fail them. Lodovico Dolce described Titian's deployment of paint on rough canvas through an analogy with pliant, living flesh.[23] Facing the same difficulty of describing the life-like quality of Titian's pictures, Sperone Speroni deployed a radical analogy with the Eucharistic theology of transubstantiation to describe Titian's ability to turn paint and canvas into living bodies.[24] Speroni describes an artistic miracle by which Titian's paintings *become* the bodies that they substitute.

The notion of ontological similitude was also central to the Renaissance debate over the relative merits of painting and sculpture, now commonly called the *paragone*.[25] While modern scholarship has focused on claims regarding the nobility or difficulty of each medium, during the Renaissance the stakes of that debate were much higher: at issue was the referential capacity of art itself, art's ability to produce objects that exceeded mere representation. Authors on opposite sides of the debate, like Benedetto Varchi and Speroni Sperone, agreed that, as three-dimensional objects, sculptures shared a deep similitude, an 'essential identity,' with the bodies they represent. Paintings, they also agreed, lack this 'essential identity' with the target of representation.[26] Use of the term 'essential' here is quite charged, and invokes the long-standing distinction in Western philosophy between substance and accidents. At the center of the debate between painting and sculpture was the question of which medium could better convey the 'essential' qualities of objects. Varchi, the most influential writer on this topic, clearly believed that sculpture created

22 Stephen J. Campbell, "*Fare una Cosa Morta Parer Viva*: Michelangelo, Rosso, and the (Un) Divinity of Art," *Art Bulletin* 84 (2002), 596-620.

23 Mark W. Roskill, *Dolce's Aretino and Venetian Art Theory of the Cinquecento* (Toronto, 2000), pp. 213-17.

24 Sperone Speroni, "Dialogo d'amore," in *Trattatisti del cinquecento*, ed. Mario Pozzi (Milan and Naples, 1968), 1: 547-48. See also Christopher J. Nygren, "Titian's Miracles: Artistry and Efficacy Between the San Rocco *Christ* and the Accademia *Pietà*," *Mitteilungen des Kunsthistorischen Institutes in Florenz* 57, no. 3 (2015), 320-49, esp. 346-48.

25 See Claire J. Farago, *Leonardo da Vinci's* Paragone*: A Critical Interpretation with a New Edition of the Text in the* Codex Urbinas (Leiden, New York, Copenhagen, and Cologne, 1992) and Leatrice Mendelsohn, *Paragoni: Benedetto Varchi's* Due Lezzioni *and Cinquecento Art Theory* (Ann Arbor, 1982).

26 Benedetto Varchi, "Della maggioranza delle arti," in *Trattati d'arte del Cinquecento*, edited by Paola Barocchi (Bari, 1960), 1:42 and 47. See also Sperone Speroni, "Discorso in lode della pittura," in *Opere di M. Sperone Speroni degli Alvarotti* (Venice, 1740), 3:441-446, esp. 445.

essential similitude better than painting: "... and thus being able to imitate the two things that are found in all bodies, that is to say substance and accidents, we would have to say that [sculptors] imitate more the substance than the accidents and painters imitate more the accidents than the substance."[27] Painting, he suggests, cannot rival sculpture when it comes to creating a relationship of deep similitude with the object being represented.

It is often suggested that painting on stone emerged out of a concern for this debate: paintings on stone, so the thinking goes, last as long as sculpture and is therefore 'essentially' the same as a work in three-dimensions. This point was made in Vittorio Soranzo's famous letter that has been discussed elsewhere in this volume. Indeed, other Renaissance authors more sympathetic to the art of painting made this argument even more explicitly, noting that painting on stone allowed pictures to "equal the hammer, and the grandeur that previously only sculpture had possessed."[28] These words, written by Francesco Maria Molza in response to a portrait of Giulia Gonzaga painted on slate, leave little doubt that Molza understood the painting to challenge the mandate of sculpture.[29] Beyond the question of durability, though, Varchi clearly understood Molza to endorse the notion that painting on stone created an essential similitude between the image and the target of representation. Varchi took Molza to suggest that Sebastiano's painting on stone instantiates Giulia's essential (and eternal) beauty better than any artistic medium (aside from Molza's own poetry). Varchi reads Molza this way in order to dismiss this line of reasoning entirely. Discussing the portrait of Giulia Gonzaga painted on stone, Varchi explicitly rejected the suggestion that executing a painting on stone abrogates the gap between the painted image and its worldly referent. In doing so, Varchi specifically cited Molza's poem about Giulia Gonzaga's portrait in order to undercut the equivalence between painting and sculpture that Molza had proposed.[30]

---

27 Varchi, "Della maggioranza delle arti," p. 47.

28 Francesco Maria Molza, *Delle poesie volgari e Latine di Francesco Maria Molza corrette, illuftrate, e accreiciute colla vita dell'autore scritta da Pierantonio Serassi* (Bergamo, 1747), p. 253.

29 See also Costanza Barbieri, "I ritratti di Sebastiano a *paragone*: pitture scolpite, elogi per immagini," in *Sebastiano del Piombo, 1485-1547* (Milan, 2008), pp. 53-7, and *Idem*, "'Tu, che lo stile con mirabil cura pareggi col martello.' Fortunae e sfortune di Sebastiano," in *La Pietà di Sebastiano a Viterbo. Storia e tecniche a confronto*, ed, Costanza Barbieri, Ernrico Parlato, and Simona Rinaldi (Rome, 2009), pp. 51-65. For a more detailed discussion of this exchange, see Christopher J. Nygren, "Titian's *Ecce Homo* on Slate: Stone, Oil, and the Transubstantiation of Painting," *Art Bulletin* 99, no. 1 (2017), 36-66.

30 Varchi, "Della maggioranza delle arti," pp. 41-42.

The disagreement between Varchi and Molza underlines an important point about the rivalry between painting and sculpture in the Renaissance. Although tallying points about the nobility of painting versus sculpture was certainly part of the strategy for asserting the primacy of one medium over the other, at stake for authors like Varchi and Molza was not so much a zero-sum competition between painting and sculpture as much as a multi-faceted consideration of the referential capacity of each medium regarding the intended object of representation.[31] Within this context, the shortcomings of one medium could be used to underline the strengths of the other. These authors were concerned with interrogating how paintings and sculptures resemble the target of representation, but both Varchi and Molza seem to insist that resemblance is not merely a visual phenomenon. When understood within the context of the philosophical debate between Varchi and Molza, stone paintings divert attention away from visual resemblance and direct it toward modes of similitude that are predicated upon the image's materiality. These deeper forms of similitude respond to Varchi's critique that painting does not engage the 'essential' quality of the target of representation.

Varchi's argument turned on the relationship between the human body and three-dimensional sculptures made in stone; he understood paintings on stone to encroach upon this unique form of mimesis by creating painted stones that substitute human bodies. Varchi's mode of reasoning is tightly focused, though a broader application of his logic can be understood to underwrite many of the Renaissance experiments with painting on stone. Renaissance and Baroque paintings on stone do not simply argue for their own 'eternal' durability, rather they deploy stone in such a way as to draw on stone's manifold connotations within western philosophy, theology, and literature. By calling attention to their materiality, paintings on stone underline rock's status as a privileged vehicle for considering the relationship between essence and appearance.[32]

Unlike paintings on canvas or wood panel, stone painting invariably calls attention to the material that lies beneath the picture surface. Stone paintings could therefore evoke a series of metaphorical and theological associations without calling undue attention to the materiality of the lithic surface, as is the case with Sebastiano del Piombo's *Portrait of Clement VII* [**Plate 4**], which is

31 Recently, scholars have begun to think about the rivalry between painting and sculpture as a form of 'comradeship' rather than as a direct, head-to-head competition. Joris van Gastel, Yannis Hadjinicolau, and Markus Rath, "Paragone as Comradeship," in *Paragone als Mitstreit*, ed. Joris van Gastel, Yannis Hadjinicolau, and Markus Rath (Berlin, 2014), pp. 33-47.

32 Jeffery Jerome Cohen, *Stone: An Ecology of the Inhuman* (Minneapolis and London, 2015), pp. 195-252.

discussed elsewhere in this volume. In cases like this, the stone matrix remains palpable, but it does not participate in the visual economy of the picture. Figure and ground are distinct from one another; they are not coordinated in the manufacture of a unified fictive ensemble. Il Cavaliere's painting of *Perseus* is of a different order. In this case, stone is invoked as part of the visual structure of the painting itself. Figure and ground unite in the creation of an image-object that is simultaneously fictive painting and intractable object. However, Il Cavaliere's appeal to stone was anything but naive, as it invoked a material laden with figurative potential that became an actor in the process of creating meaning. The 'non-mimetic' stone simultaneously *is* and *represents* the natural world.[33] Yet following the dictum establish by Ovid's description of the Palace of the Sun, Il Cavaliere made sure that his craftsmanship augmented this material. As Ovid notes, *Materiam superabat opus*.[34] In other words, lithic matter functions not only as a precious object but also operates as part of an aesthetic and conceptual ensemble. Stone serves at least a dual function.

My thinking about stone's pluripotency owes much to the writings of Jane Bennett and Bruno Latour, both of whom have approached the agency of the material world in novel ways. Developing ideas present in Latour's Actor-Network Theory, Bennett is concerned with breaking down the onto-theological binary between living beings and inert matter. For Bennett, conceiving of matter as 'vibrant' helps to dissipate that distinction by underlining the vitality of the material world: matter "at one moment discloses [itself] as dead stuff and at the next as living presence."[35] Bennett's insistence on the ecology of agency points toward a modern theoretical concept—vibrant matter—that is useful in articulating a pre-modern disposition toward the material world and might help us address this fundamental question: Why did artists unforeseeably turn to stone as an artistic support in the middle of the 16th century? Artists, I maintain, did not understand stone as an inert material; rather, for them it possessed a 'thing-power' or latent force that exceeded its material presence.[36]

The reading developed here aligns with—and is perhaps reinforced by—observations that Herbert Kessler and Caroline Walker Bynum (among others)

33 On non-semiotic images see James Elkins, "Marks, Traces, 'Traits,' Contours, 'Orli,' and 'Splendores': Nonsemiotic Elements in Pictures," *Critical Inquiry* 21 (1995), 822-60.

34 Ovid, *Metamorphoses*, 2.5. The idea was also present in Alberti, who states, "you can conceive of almost nothing so precious which is not made far richer and much more beautiful by association with painting. Ivory, gems and similar expensive things become more precious when worked by the hand of the painter." Leon Battista Alberti, *On Painting and On Sculpture*, ed. Cecil Grayson (London, 1972), pp. 60-61.

35 Bennett, *Vibrant Matter*, p. 5.

36 See also Cohen, *Stone: An Ecology of the Inhuman*.

have recently made regarding matter (*materia*) in pre-modern Christianity. As they insist, during the late medieval and early modern period the godhead made itself manifest through the material world.[37] In late medieval aesthetics, the road to transcendence inevitably traveled the path of materiality. Objects transcended their own materiality precisely by foregrounding that materiality. And while Bynum insists on a categorical distinction between medieval aesthetics and the emergence of art in the Renaissance, this article demonstrates that the early modern period shared with the Middle Ages a fundamental disposition that foregrounded materials in an artistic context.[38] It is only by foregrounding the material of artistic creation that artists can hope to overcome the natural limits of their medium.

Bynum's basic observation about similitude is animated by the conviction that in the Middle Ages likeness was a malleable category that depended on the presence of material objects, and those objects most often did not look like the things they purported to represent.[39] I suggest that the emergence of paintings on stone in the 16th century was part of the 'long tail' of this medieval interrogation of the material world aimed at revealing unexpected forms of resemblance and similitude. Scholars are quite comfortable asserting that a medieval reliquary cannot be fully understood in the absence of a synthetic understanding of how the materials deployed were understood at the time.[40] Anna Kim's discussion of the micro-mosaic icon in Santa Croce in Gerusalemme in this volume is an excellent example of how the materiality of reliquaries is integral to their mode of producing meaning. The same ought to be true of stone paintings, which are objects that similarly foreground their materiality. In early modern paintings on stone, the material substrate inflects the 'meaning' of the image.

To get a better handle on how this occurs I would like to introduce the notion of 'affordances,' which was developed by James Gibson. Gibson was primarily concerned with psychological aspects of visual perception, uncovering how agents *perceive* the environment.[41] Affordance is a fundamentally ecologi-

37 Caroline Walker Bynum, *Christian Materiality: An Essay on Religion in Late Medieval Europe* (New York, 2011) and Herbert L. Kessler, *Seeing Medieval Art* (Orchard Park, 2004), pp. 19-42.

38 Bynum, *Christian Materiality*, p. 28.

39 Bynum, *Christian Materiality*, p. 126.

40 Martina Bagnoli, "The Stuff of Heaven: Materials and Craftsmanship in Medieval Reliquaries," in *Treasures of Heaven: Saints, Relics, and Devotion in Medieval Europe*, edited by Martina Bagnoli et al. (New Haven and London, 2010), pp. 137-47.

41 James J. Gibson, *The Ecological Approach to Visual Perception* (London and Hillsdale, 1986), p. 127.

cal concept. For Gibson, it is fundamental to note that affordances emerge from the object itself: "An affordance is not bestowed upon an object by a need of an observer and his act of perceiving it. The object offers what it does because it is what it is."[42] While water *affords* human beings a number of things (hydration, sanitation, immersion, etc.), water bugs find that water *affords* support. I introduce the notion of affordances because it helps formulate the question that animates my interrogation of early modern paintings on stone: what did the stone pictorial support *afford* artists that canvas and wooden panel could not? The deployment of stone, I suggest, created a systematic homology between signs and things.

Despite the semi-precious nature of the stones often used as a painterly supports, within early modern ontologies stone remained one of the most humble of all materials, lacking all of the Aristotelian functions and potentials.[43] On a basic level, by using stone to depict living figures—figures so real they seem to lack only breath, as Renaissance authors would have said—the painter demonstrated his ability to enliven inert material, thus becoming co-extant with *Natura*, the first artist who made living images in stone. At multiple points, Pliny described *Natura* as an artist who made paintings in stone.[44] He reports many 'natural images' made out of stone, such as a likeness of Silenus uncovered by the stone-breakers at Carrara, or an even more marvelous agate gemstone "on which could be seen Nine Muses with Apollo holding his lyre."[45] Moving from Antiquity into the middle ages, there is also a long tradition in Christian sources of finding naturally produced images of the Holy Face—the *acheiropita*—in book-matched marble slabs, which decorated the interiors of many churches.[46] Together, these ancient and medieval sources testify to the chthonic origins of images in stone: these are images not-made-by-human-hands. Baroque paintings on stone fed into this tradition by making the beholder hesitate, momentarily falling into the belief that the picture had been created by Nature (*a natura depicti*) rather than a human artificer (*ars*

42 Gibson, *The Ecological Approach*, p. 139.

43 On the Aristotelian categories of animate versus inanimate (i.e. nutrition, locomotion, intellect), see Aristotle *De Anima*, 413a. For its continuation into the Renaissance, see Roskill, *Dolce's Arentino and Venetian Art Theory of the Cinquecento* (Toronto, 2000), p. 117.

44 In the *Natural History* Pliny describes *Natura* as the "supreme artificer and mother of all." Pliny, *Natural History*. 21.56.117.

45 Pliny, 37.3.5 and 36.4.14.

46 See John Onians, "Abstraction and Imagination in Late Antiquity," *Art History* 3 (1980), 1-24, and James Trilling, "The Image Not Made by Hands and the Byzantine Way of Seeing," in *The Holy Face and the Paradox of Representation*, ed. Herbert L. Kessler and Gerhard Wolf (Bologna, 1998), pp. 109-27.

*pinxit*). By obfuscating the distinction between *naturalia* and *artificialia*, stone pictures like Il Cavaliere's and Stella's arrogate to themselves a potency that was analogous to Pliny's stone pictures and the miraculous *acheiropita*.

Like similitude, analogy is a concept that cuts to the heart of the mimetic arts.[47] Raymond Bellour has argued that figurative images are always in an analogous relationship to their target of representation. In Bellour's formulation, 'analogy' is a variable quantity; images are constantly attempting to increase the 'amount of analogy' that they convey. The 'quantitative' approach to analogy dates back at least to Aristotle, who presented analogy as the paradigmatic mathematical equation *a:b :: c:d*, which describes how the relation of a to b is quantitatively similar to the relationship between c and d.[48] Bellour notes that this potential for resemblance is historically contingent; 'similitude' and 'analogy' are notions that must continuously be redefined in light of historical circumstances.[49] What counted as a compelling similitude in one epoch or culture may be seen as inadequate in another time or place.

The zero point of analogical predication—the point at which it ceases to be sensible to describe something as proportionally similar to something else—is the state of equality. This is seen in the map Borges describes in the epigraph to this essay, where the scale of the map reaches 1:1, thus making it spatially equivalent to the territory it purports to describe. However, when something attains the state of sameness with its predicate, the parallax effect that allows for meaningful difference to emerge no longer obtains. This was the unattainable goal of most Renaissance images, which took as their model a Eucharistic mode of mimesis that turns on the notion of perfect re-presentation.[50] This is how Gabriele Paleotti shifted his discussion of pictures from the language of analogy to the concept of equality. Paleotti was a Bolognese prelate who wrote an important treatise on the use of sacred images after the Council of Trent. Paleotti was concerned with describing how an image and its referent could attain a resemblance that was perfectly analogical, a relationship of equality. "The true is taken various ways, as the authors write, but we will leave the other

47 Barbara Maria Stafford, *Visual Analogy: Consciousness as the Art of Connecting* (Cambridge and London, 1999).

48 Aristotle, *Topics*, 108a.13 (Bk. I, ch. 17).

49 Raymond Bellour, "The Double Helix," in *Electronic Culture: Technology and Visual Representation*, ed. Timothy Druckrey (Denville, 1996), pp. 173-99.

50 Alexander Nagel, *The Controversy of Renaissance Art* (Chicago and London, 2011), p. 204, and Stephen J. Campbell, "Renaissance Naturalism and the Jewish Bible: Ferrara, Brescia, Bergamo, 1520-1540," in *Judaism and Christian Art: Aesthetic Anxieties from the Catacombs to Colonialism*, ed. Herbert L. Kessler and David Nirenberg (Philadelphia, 2011), pp. 291-327.

definition aside and take the true 'to mean the equality of the sign and signified' (*prout est aequalitas signi ad rem significatam*)."[51] Here Paleotti leans on a long tradition present in scholastic theology that discusses analogy not only as a proportional equation, but also in terms of 'participation.'[52] But while the theology and philosophy underwriting analogy have been dealt with, the question remains: how is this process of analogical participation made legible? Painting on stone, I suggest, was particularly adept at discharging this function.

Baroque paintings on stone tended to illustrate stories or figures that capitalize on stone's materiality as a means of troubling the representational relationship between the image and its subject. In stone paintings, the process of re-presentation is not an articulation of difference between signifier and signified, but instead must be reconceptualized as an attempt to find commonality. For early modern artists who painted on stone, the articulation of sameness was just as important as a semiotics predicated on slippage from signifier to signified. The stone substrate functioned as a mediating substance that allowed for analogical relationships to become visible. What we see in the emergence of stone painting is an attempt to attain analogy or equality through new means not predicated on illusionism but rather material affinity.

The famous letter from Vittore Soranzo documenting Sebastiano del Piombo's discovery of the technique of painting on stone reveals that his first stone painting was a picture of Christ (*una imagine di Christo*).[53] Recent work by Elena Calvillo, as well as numerous contributions to this volume, has demonstrated that Sebastiano was well aware of the theological resonances that were activated when the artist elected to overlay a painting of Christ onto a lithic substrate.[54] Indeed, most (though not all) of Sebastiano's paintings on stone take Christ as their subject matter. Sebastiano's appeal to stone for paintings of Christ reinforces the claim that stone afforded a higher quantity of analogy between signifier and signified than the traditional media of Renaissance painting would have allowed. Rather than rely exclusively on his powers of mimesis to create the fictive presence of the prototype (i.e. Christ), Sebastiano appealed to a painterly technique that allowed for the induction of

51 Gabriele Paleotti, *Discourse on Sacred and Profane Images*, trans. by William McCuaig (Los Angeles, 2012), p. 222.

52 David C. Burrell, *Analogy and Philosophical Language* (New Haven and London, 1973), p. 84.

53 Cited in Michael Hirst, *Sebastiano del Piombo* (Oxford, 1981), p. 124.

54 Elena Calvillo, "Authoritative Copies and Divine Originals: Lucretian Metaphors, Painting on Stone, and the Problem of Originality in Michelangelo's Rome," *Renaissance Quarterly* 66, no. 2 (2013), 453-508.

the prototype into the material object. Sebastiano's process of invention mirrors the stone relief from Regensburg, discussed above. Christ, described as stone in scripture, is painted on stone.[55]

One notable exception to Sebastiano's use of stone for images of Christ is the *Portrait of Pope Clement VII* [**Plate 4**], also painted on slate. Though not a picture of Christ, this image is nevertheless highly invested in the theologically charged concept of stone: just as Christ was equated with stone, so too was the papal office equated with the material solidity of stone. As Christ said to Peter, "I say to thee: That thou art Peter (*Petrus*); and upon this rock (*petram*) I will build my church, and the gates of hell shall not prevail against it."[56] Here we see that the subject of the picture—the Pope, successor of St. Peter, the rock upon whom the church is constructed—is imbricated in the manifold figural significance of stone, the material support of the picture itself. This isn't simply clever; it is meant to probe the limits of what portraiture can accomplish. The picture reveals something important about the Petrine office. What we see here is the creation of a visual analogy that underlines a deep-seated, ontological similitude. Peter was a stone, and the Pope is the successor of St. Peter. Thus it is fitting that the Pope be painted on stone, because as the holder of the Petrine office he too shares in the rupestrian stability that Christ imputed to St. Peter. Sperone described Titian's portraits as 'miracles' because they were 'not painted' but instead seemed to be the very bodies of the people they represented.[57] Sebastiano attained a similar effect through different means; he capitalized on the material affinity between the pictorial substrate and his Petrine subject.

Artists first recognized the figurative potency of painting on stone within the realm of religious imagery. However, the conceptual game that I see at work in these pictures did not hold only within the realm of sacred painting. By the late 16th century it had migrated into the realm of secular painting, as is illustrated in Cavaliere d'Arpino's painting of *Perseus Liberating Andromeda*. Il Cavaliere's engagement with the material qualities of the stone support expands beyond the aesthetic entanglement of stone and paint (*ars pinxit* vs. *a natura depicti*) to encompass the broader import of the Ovidian myth that it illustrates. Through a visual analogy that mobilized the material qualities of the stone support, Il Cavaliere created a dynamic image that serves not only as

55 For references to Christ as stone, see Mark 12:10; Matt. 21:42; Luke 20:17; Acts 4:11; Rom. 9:32ff; 1 Cor. 3:10ff; Eph. 2:20; and 1 Pet. 2:4.

56 Matt. 16:18. Unless otherwise noted, all biblical translations are taken from the Douay-Rheims edition.

57 Speroni, 1:547-8.

a reflection on the conceit of petrification, but also as an embodiment of a form of material analogy that sheds new light on Early Modern notions of similitude.

Il Cavaliere's St. Louis painting is part of a series of paintings of *Perseus and Andromeda* that the artist created around 1595. In total, he painted about a dozen versions of this story. Aside from the St. Louis painting on lapis, Il Cavaliere painted numerous other versions of *Perseus and Andromeda* on various stone supports, including slate and agate.[58] However, among the surviving paintings on stone the St. Louis painting stands out as unique, since it is the only picture in which the material qualities of the stone are used as a part of the fictive ensemble, and it is for this reason that our analysis will focus on the St. Louis painting. Il Cavaliere also painted at least six versions on canvas and panel.[59]

Most of the versions on stone follow the same model used in the St. Louis picture, and these paintings can be described as following the 'St. Louis-type.'[60] The only exception is the painting in Rhode Island, which departs from this model in a number of important ways [**Plate 18**]. First, in the majority of the stone paintings, Perseus arrives from the beholder's left; in the Rhode Island painting, by contrast, Perseus arrives from the beholder's right, where he emerges from behind the stone outcropping to which Andromeda has been shackled. Moreover, Andromeda herself is also quite different in the Rhode Island painting. Whereas in the St. Louis-type she looks down in trepidation at the monster ominously approaching her feet (and at the crack in the rock outcropping), in the Rhode Island picture her attention has been diverted upward, toward Perseus, whom she recognizes as her salvation. As a result of this transferal of attention, Andromeda's body has similarly shifted. In the St. Louis-type paintings Andromeda's weight is borne on her left leg and her right leg is relaxed, recoiling slightly from the sea monster; in the Rhode Island painting, by contrast, her weight is carried on her right leg, and her left knee has been thrust forward, tilting her entire body in the direction of Perseus.

Perhaps the most important distinction between the Rhode Island painting and the St. Louis-type is the positioning of the principle prop: the Gorgon

58 Herwarth Röttgen. *Il Cavalier Giuseppe Cesari d'Arpino: un grande pittore nello splendore della fama e nell'incostanza della fortuna* (Rome, 2002), cats. 36, 47, 57, and 98. The catalog entry on the St. Louis painting (Cat. 47) mentions at least three other versions of *Perseus and Andromeda* on stone that were recorded in early documents but that do not survive.

59 Röttgen. *Il Cavalier,* cats. 58 and 99-103.

60 Röttgen has suggested that there are three 'types' of *Perseus and Andromeda*: the St Louis type, the Rhode Island type, and the Berlin type. However, the compositional variations between the St. Louis type and the Berlin type are minimal.

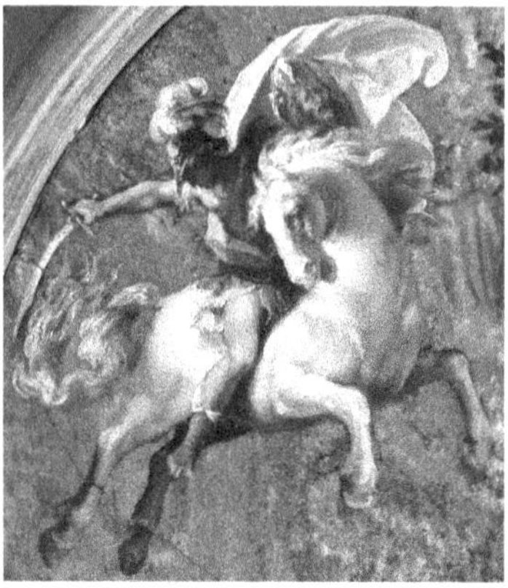

FIGURE 4.2
*Cavaliere d'Arpino,* Perseus Liberating Andromeda, *oil on lapis lazuli, ca. 1593-1594, St Louis, St Louis Art Museum. Detail* (PHOTO: ©ST LOUIS ART MUSEUM).

head. In the St. Louis painting, the Gorgon's head is mounted on Perseus' shield, which is borne on the paladin's left forearm [**Fig. 4.2**]. Thus, it is on the far side of Perseus's body, separated from the beholder by the massive presence of Pegasus, whose mane seems almost an extension of the Gorgon's distinctive snake hair. This object possessed a marvelous power of petrification. As Ovid describes, this severed head petrified all those who gazed upon it. The poet describes a desolate scene of petrified corpses left in the wake of exposure to Medusa's head: "Along the fields and paths where he had trekked, he saw the forms of men and animals who had been changed to stone because their gaze had dared to spy upon Medusa's face."[61] Ovid notes that the Gorgon head could also petrify through contact. When Perseus finally laid aside the severed head so that he might embrace his new wife, the seaweed and other plants upon which it had been laid soaked up the power of the Gorgon's blood and quickly began to stiffen into stone.[62] The quasi-miraculous power of the Gorgon head was of great interest to Baroque painters, thus it is somewhat surprising to see that in the St. Louis-type its presence is sublimated.

Scholars have drawn attention to how artists like Caravaggio and Annibale Carracci deployed the head of Medusa as synecdoche for the marvelous and stupefying power of the visual arts.[63] Caravaggio's painted shield [**Fig. 4.3**] serves as a substitute for the severed head mounted on Athena's aegis, and

61 Ovid, *The Metamorphoses of Ovid*, trans. Allen Mandelbaum (New York, 1993), 4.776-81.

62 Ovid, *Metamorphoses*, 4.740-52. See also Michael Cole, "Cellini's blood," *Art bulletin* 81 (1999), 215-35.

63 Charles Dempsey, "L'impression de merveilleux à la galerie du palais Farnèse," in *Andromede, ou, Le heros a l'epreuve de la beaute: actes du colloque international organise au musee du Louvre par l'universite de Montreal* (Paris, 1996), pp. 195-221; Elizabeth Cropper, "The Petrifying Art: Marino's Poetry and Caravaggio," *Metropolitan Museum Journal* 26

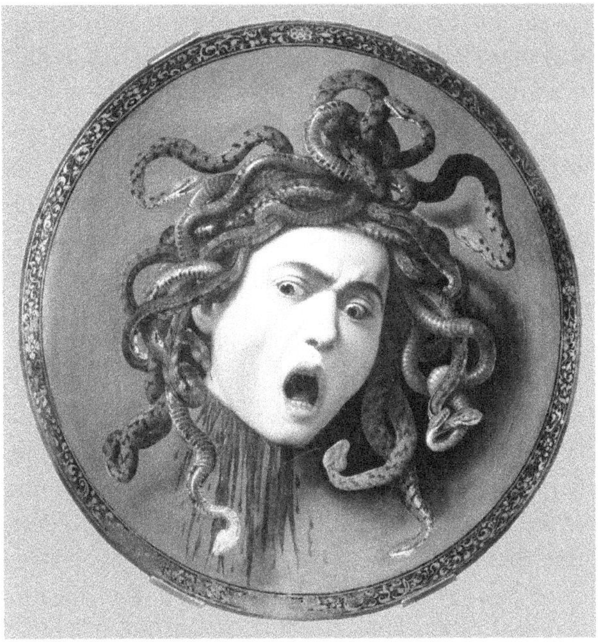

FIGURE 4.3 *Caravaggio,* Medusa, *oil on canvas glued to wood, 1590-1600, Florence, Uffizi* (PHOTO: SCALA: MINISTERO PER I BENI E LE ATTIVITÀ CULTURALI/ ART RESOURCE).

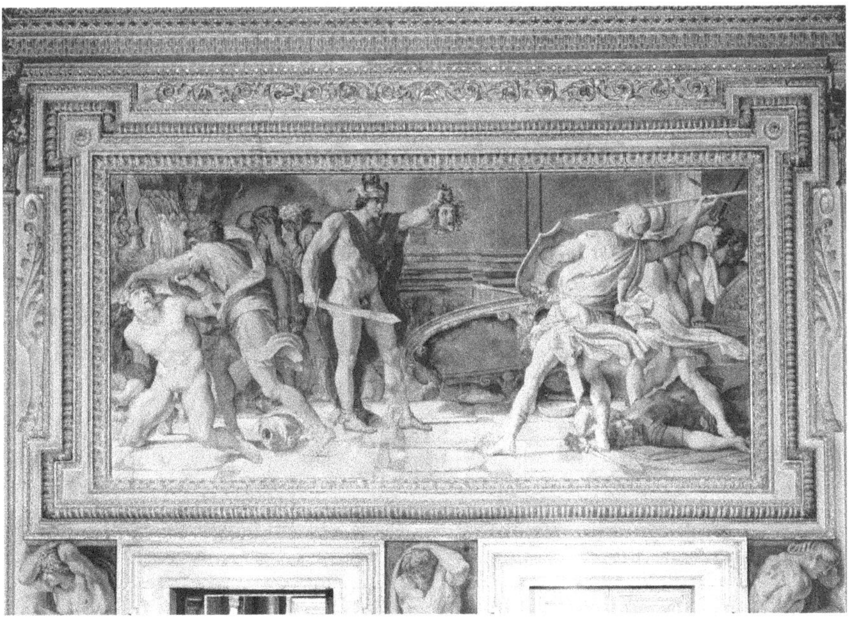

FIGURE 4.4 *Annibale Carracci,* Perseus and Andromeda, *fresco, ca. 1597-1600, Palazzo Farnese, Rome* (PHOTO: ALINARI)

Carracci illustrated [**Fig. 4.4**] for the beholder the process of petrification, showing bodies in various states of marmoreal metamorphosis. Il Cavaliere could have easily employed the Gorgon's head as short hand for the stupefying power of his art as well as a mode of commenting on his art's primacy over other mimetic modes of representation, like painting and poetry. Indeed, this is exactly what we see in the Rhode Island painting, as well as the numerous versions of this subject done on canvas, in which Il Cavaliere leaned on the trope of the Gorgon to manufacture marvel. In those pictures, the Gorgon head is foregrounded and immediately legible. However, in the St. Louis painting Il Cavaliere did not pursue this line, since the Gorgon's head is barely visible. Or at least, he did not pursue this line in a way that is as easily recognizable as the interventions by Caravaggio and Carracci. As Giovanni Pietro Bellori stated, Annibale approached the composition of this scene "in the manner of poets" as he attempted to increase the beholder's marvel and to demonstrate his superiority over Ovid's text.[64] The poetic pretentions of Il Cavaliere's St. Louis painting are somewhat harder to recognize.[65] However, there is one detail of the painting that reveals Cavaliere's poetic disposition when ideating this painting.

Il Cavaliere's paintings of *Perseus and Andromeda* all include one detail that importantly deviates from the source text: Perseus arrives on scene riding on the back of Pegasus. Ovid's text clearly states that Perseus strapped his winged sandals onto his feet and used these to cut through the air on his way to accomplish his mission.[66] Indeed, most paintings of this myth show Perseus flying through the air with the assistance of his winged sandals. This is true of Florentine painters like, Piero di Cosimo and Giorgio Vasari, as well as Venetian painters, like Titian and Veronese. While Pegasus was occasionally included in woodcut illustrations that adorned early modern editions of Ovid's text, I am not interested in a search for an iconographic source. Instead, I am concerned with underlining the importance of Il Cavaliere's decision to include Pegasus, whose story is recounted by Ovid immediately after the narrative of Perseus and Andromeda.

(1991), 193-212; Louis Marin, *To Destroy Painting*, trans. Mette Hjort (Chicago and London, 1995), pp. 97-164.

64 Giovanni Pietro Bellori, *The Lives of the Carracci*, trans. Catherine Enggass (University Park and London, 1968), p. 49.

65 In her discussion, Siefertova makes no mention of the aesthetic or poetic ambitions outlined here. Hana Seifertova, *Malby na kameni: Uměcký experiment v 16. A Na Začátku 17 století/ Painting on Stone: an Artistic Experiment in the 16th and Early 17th centuries* (Prague, 2007), pp. 92-98.

66 Ovid, *Metamorphoses*, 4.665-67.

Pegasus was born of an excretion of blood from the severed head of Medusa.[67] But Pegasus was most famous for having been the origin of the Hippocrene Spring (or, "Horse Fountain"). This fountain was found on Mt. Helicon, and had sprung forth from the ground when Pegasus' hoof struck the rock and water gushed forth. This spring provided water to Apollo and his nine muses, who abide on Mt. Helicon and protect the sacred waters of the Hippocrene spring, the source of poetic inspiration. Pegasus was credited as the origin of this font (*Pegasus huius origo fontis*), and thus he became an emblem of poetic inspiration and expression.[68] Il Cavaliere's inclusion of this proleptic detail suggests an instance of discreet poetic awareness and invention. Il Cavaliere's poetic invention mobilizes the very material of the picture, which analogizes the painted rock not only with the stony carcasses that populate the Ovidian tale, but also with the punctured stone that gave rise to the Hippocrene spring. Thus, Il Cavaliere's image, which is painted *on stone*, registers the vibrancy of the material world in which the flesh and blood of the human body is easily converted into a stone image through the power of the gaze, and it also refers to the stony brook that is the origin of poetic inspiration. As the beholder's gaze passes over the picture surface it is naturally drawn to the severed head of Medusa. In this painting the stone support afforded the artists a means of foregrounding the parallelism between the gaze of the beholder and the gaze of those forebears who had succumbed to the petrifying power of the Gorgon. This conceit opens up a momentary hesitation, a rupture, through which the aesthetic of the marvelous takes on a tangible form as the beholder imagines himself as Medusa's newest victim. Holding this small stone painting in his hand, the beholder is struck that he too will soon be become the material that he grasps. The petrifying power of the Medusa is recalled, affirmed, and made legible through the material substrate of the picture.

Importantly, Ovid's story begins with a reflection upon the deceptive materiality of female beauty. Ovid memorably describes Andromeda as an ideal figure whose statuesque beauty so confounded Perseus that he nearly mistook her for a marble statue: "when he saw her, Perseus would have thought she was a marble statue, were it not for a light breeze that stirred her hair, and warm tears trickling down her cheeks."[69] Like Sebastiano's images of Clement VII, Il Cavaliere's picture turns on a metaphor in which a person is described in rupestrian terms; the beholder is left to parse the cognitive dissonance caused by a figure that is described as a living statue and is painted on stone.

67 Ovid, *Metamorphoses*, 4.787.

68 Ovid, *Metamorphoses*, 5.262-3.

69 Ovid, *Metamorphoses*, 4.671-75.

This picture, then, is not simply a collector's piece that plays on the distinction between *naturalia* and *artificialia*. Rather, it is an ambitious work that seeks to tease out the myriad implications of painting on stone by tethering this relatively new artistic technique to a mythological story replete with metaphors that put pressure on the vibrancy of matter. Ovid's stony verses analogized lithic material, the work of art, idealized feminine beauty, and the petrifying power of the Gorgon's head. For Ovid, stone provided a vibrant mode of creating meanings. Stone does not 'mean' one thing, but is analogized for many things. Its similitude is replete and inscrutable. Il Cavaliere took up the themes present in Ovid's ancient text and sought to give them material solidity by overlaying the narrative onto a lithic ground that he made visibly present throughout his pictorial narrative.

This points to how, in painting on stone, Baroque artists sought to foreground an allegiance between the subject-matter of their images and the stony matrix they employed, just as Sebastiano had done in his pictures of Christ and the Pope. We can see this also in Jacques Stella's painting of *Jacob's Dream*. The story of Jacob's ladder constituted one of the primary *loci* for a theologically robust understanding of how stone afforded a surface suitable for theophanic revelation.[70] Stella's decision to depict the story on a lithic surface that he left visible speaks to his understanding of stone's privileged position within premodern theological thought.

Jacob's vision is recounted in the Hebrew Bible. Jacob had duped his father, Isaac, into blessing him rather than his brother, Esau. In order to escape the ire of his deceived brother, Jacob set out into Syria, and along the way had a mystical experience that would become foundational for Western spirituality. The story merits citing at length.

> And when he was come to a certain place, and would rest in it after sunset, he took of the stones that lay there, and putting under his head, slept in the same place. And he saw in his sleep a ladder standing upon the earth, and the top thereof touching heaven: the angels also of God ascending and descending by it. And the Lord leaning upon the ladder saying to him: I am the Lord God of Abraham thy father, and the God of Isaac: The land, wherein thou sleepest, I will give to thee and to thy seed ... And when Jacob awaked out of sleep, he said: Indeed the Lord is in this place, and I knew it not. And Jacob arising in the morning, took the stone which he had laid under his head, and set it up for a title, pouring oil

70 Christian Heck, *L'échelle céleste dans l'art du Moyen âge: une image de la quête du ciel* (Paris, 1997).

> upon the top of it ... And this stone, which I have set up for a title, shall be called the house of God...[71]

By painting *Jacob's Dream* on stone, Stella's picture reifies and glosses the privileged position that stone occupies in early modern technologies of theophany. In consecrating a stone with oil, Jacob created a lithic interface between God and the Hebrew nation. Stella's unctuous stone fulfills a similar function in the economy of salvation. It is a picture whose material offers commentary on its content. This isn't simply *concettismo,* though: the act of beholding becomes a form of theological commentary and the picture becomes akin to a relic of Jacob's own rock. Even if this isn't the actual stone of Jacob's vision, it shares a material affinity with that special object, and Stella relied on the beholder's visual acumen to understand this cunning maneuver. In other words, the picture's material isn't just something that appeals to the collector's eye for precious and shiny things; it also affords Stella the physical matrix to produce a picture that triggers in the beholder's mind reflections on the spiritual and devotional power of the material world. The physical object instantiates conceptions of God, mankind, and how they interact through the material world.

As noted above, Sebastiano's appeal to a stone support sought to harness a deep analogical resemblance between the lithic matter of the picture and its target of representation. This essay asserts that this sort of analogical approach carried into much later stone paintings in which artists deployed semi-precious stones like lapis and onyx. Even if these pictures were appreciated as collectors' pieces, the significance of the invocation of stone is not explained away by an acknowledgment of *collezionismo*. Rather, the stone paintings of Il Cavaliere and Jacques Stella turn on a peculiar confederation between the material substrate of the picture and the lithic resonances embedded in the stories they portray. For Il Cavaliere and Stella, stone afforded the artist something unique, and the petrified material became an actant in the visual economy of the image, helping it attain a heightened sense of analogy between subject and object. These images present rather than represent.

I would stress that this is not necessarily true of all stone paintings, and I am not advocating that we develop a 'universal grammar' of stone pictures. Rather, I suggest that we attend more closely to the entanglement of material, aesthetic, epistemological, and perhaps ontological concerns that convinced artists and patrons that stone afforded them something different than canvas or panel. Only by pushing beyond the lithic surface of early modern stone paintings can we begin to understand how artists deployed stone with the goal

71 Gen. 28:11-13, 16-18, 22.

of capturing an essential similitude of the target of representation. Like Borges' cartographers, these artists might have ultimately failed in their attempt to create a relationship of perfect similitude; however, as with Borges' cartographers, it is more important to understand *why* and *how* they failed rather than to dismiss their efforts as useless.

## Bibliography


Alberti, Leon Battista. *On Painting and On Sculpture*. Edited by Cecil Grayson. London, 1972.

Augustine. *The Trinity (De trinitate)*. Translated by Edmund Hill. New York, 1991.

Bagnoli, Martina. "The Stuff of Heaven: Materials and Craftsmanship in Medieval Reliquaries." In *Treasures of Heaven: Saints, Relics, and Devotion in Medieval Europe*, edited by Martina Bagnoli et al., pp. 137-47. New Haven and London, 2010.

Baiardi, Giorgio. "Il dialogo *Del concetto poetico* di Camillo Pellegrino." *La Rassegna della Letteratura Italiana* 7 (1958), 370-74.

Baltrušaitis, Jurgis. *Aberrations: An Essay on the Legend of Forms*. Cambridge and London, 1989.

Barbieri, Costanza. "I ritratti di Sebastiano a *paragone*: pitture scolpite, elogi per immagini." In *Sebastiano del Piombo 1485-1547*, pp. 53-7. Milan, 2008.

Barbieri, Costanza. "'Tu, che lo stile con mirabil cura pareggi col martello.' Fortunae e sfortune di Sebastiano." In *La Pietà di Sebastiano a Viterbo. Storia e tecniche a confronto*, edited by Costanza Barbieri, Enrico Parlato, and Simona Rinaldi, pp. 51-65. Rome, 2009.

Bellour, Raymond. "The Double Helix." In *Electronic Culture: Technology and Visual Representation*, edited by Timothy Druckrey, pp. 173-99. Denville, 1996.

Belting, Hans. *An Anthropology of Images: Picture, Medium, Body*. Translated by Thomas Dunlap. Princeton, 2011.

Bellori, Giovanni Pietro. *The Lives of the Carracci*. Translated by Catherine Enggass. University Park and London, 1968.

Bennett, Jane. *Vibrant Matter: A political Ecology of Things*. Durham and London, 2010.

Blümle, Claudia. "Mineralischer Sturm. Steinbilder und Landschaftsmalerei." In *Vermessen: Landschaft und Ungegenständlichkeit*, edited by Werner Busch and Oliver Jehle, pp. 73-95. Zürich, 2007.

Borges, Jorge Luis. *Collected Fictions*. Translated by Andrew Hurley. New York, 1998.

Bredekamp, Horst. *Theorie des Bildakts*. Berlin, 2010.

Burrell, David C. *Analogy and Philosophical Language*. New Haven and London, 1973.

Bynum, Caroline Walker. *Christian Materiality: An Essay on Religion in Late Medieval Europe*. New York, 2011.

Callois, Roger. *The Writing of Stones*. Translated by Barbara Bray. Charlottesville, 1985.

Calvillo, Elena. "Authoritative Copies and Divine Originals: Lucretian Metaphors, Painting on Stone, and the Problem of Originality in Michelangelo's Rome." *Renaissance Quarterly* 66, no. 2 (2013), 453-508.

Campbell, Stephen J. "*Fare una Cosa Morta Parer Viva*: Michelangelo, Rosso, and the (Un)Divinity of Art." *Art Bulletin* 84 (2002), 596-620.

Campbell, Stephen J. "Renaissance Naturalism and the Jewish Bible: Ferrara, Brescia, Bergamo, 1520-1540." In *Judaism and Christian Art: Aesthetic Anxieties from the Catacombs to Colonialism*, edited by Herbert L. Kessler and David Nirenberg, p. 291-327. Philadelphia, 2011.

Cennini, Cennino d'Andrea. *Il libro dell'arte, The Craftsman's Handbook*. Translated by Daniel V. Thompson, jr. New Haven and London, 1936.

Cohen, Jeffery Jerome. *Stone: An Ecology of the Inhuman*. Minneapolis and London, 2015.

Cole, Michael. "Cellini's blood." *Art bulletin* 81 (1999), 215-35.

Cooper, Donal. "Projecting presence: the monumental cross in the Italian Church interior." In *Presence: The Inherence of the Prototype within Images and Other Objects*, edited by Robert Maniura and Rupert Shepherd, pp. 47-69. Burlington and Aldershot, 2005.

Cropper, Elizabeth. "The Petrifying Art: Marino's Poetry and Caravaggio." *Metropolitan Museum Journal* 26 (1991), 193-212.

Dempsey, Charles. "L'impression de merveilleux à la galerie du palais Farnèse." In *Andromède, ou, Le héros à l'épreuve de la beauté: actes du colloque international organise au musée du Louvre par l'universite de Montréal* pp. 195-221. Paris, 1996.

Edgerton, Samuel. *The Mirror, the Window, and the Telescope: How Renaissance Linear Perspective Changed Our Vision of the Universe*. Ithaca, 2009.

Elkins, James. "Marks, Traces, 'Traits,' Contours, 'Orli,' and 'Splendores': Nonsemiotic Elements in Pictures." *Critical Inquiry* 21 (1995), 822-60.

Farago, Claire J. *Leonardo da Vinci's* Paragone: *A Critical Interpretation with a New Edition of the Text in the* Codex Urbinas. Leiden, New York, Copenhagen, and Cologne, 1992.

Gastel, Joris van, Yannis Hadjinicolau, and Markus Rath. "Paragone as Comradeship." In *Paragone als Mitstreit*, edited by Joris van Gastel, Yannis Hadjinicolau, and Markus Rath, pp. 33-47. Berlin, 2014.

Gibson, James J. *The Ecological Approach to Visual Perception*. London and Hillsdale, 1986.

Heck, Christian. *L'échelle céleste dans l'art du Moyen âge: une image de la quête du ciel*. Paris, 1997.

Hirst, Michael. *Sebastiano del Piombo*. Oxford, 1981.

Keane, Webb. *Christian Moderns: Freedom and Fetish in the Mission Encounter*. Berkeley, Los Angeles, and London, 2007.

Kendall, Calvin. *The Allegory of the Church: Romanesque Portals and Their Verse Inscriptions*. Toronto, Buffalo, and London, 1998.

Kessler, Herbert L. *Seeing Medieval Art*. Orchard Park, 2004.

King, Catherine. "Effigies: Human and Divine." In *Siena, Florence, and Padua: Art, Society, and Religion, 1280-1400*. Vol. 2: *Case Studies*, edited by Diana Norman, pp. 105-28. New Haven and London, 1995.

Kopania, Kamil. *Animated Sculptures of the Crucified Christ in the Religious Culture of the Latin Middle Ages*. Warsaw, 2010.

Latour, Bruno. "Visualization and Cognition: Thinking with Eyes and Hands." *Knowledge and Society* 6 (1986), 1-40.

Latour, Bruno. "Visualisation and Social Reproduction: Opening One Eye While Closing the Other ... A Note on Some Religious Paintings." In *Picturing Power: Visual Depiction and Social Relations*, edited by Gorden Fyfe and John Law, pp. 15-38. London, 1988.

Leonardo da Vinci, *The Literary Works of Leonardo da Vinci*. 2 vols. Edited by Jean Paul Richter. London, 1883.

Marin, Louis. *To Destroy Painting*. Translated by Mette Hjort. Chicago and London, 1995.

Markley, Robert. "Boundaries: Mathematics, Alienation, and the Metaphysics of Cyberspace." In *Virtual Reality and its Discontents*, pp. 55-77. Edited by Robert Markley. Baltimore, 1996.

Masheck, Joseph. "Alberti's 'Window': Art-Historical Notes on an Antimodernist Misprision." *Art Journal*, 50, no. 1 (1991), 35-41.

Mendelsohn, Leatrice. *Paragoni: Benedetto Varchi's* Due Lezzioni_*and Cinquecento Art Theory*. Ann Arbor, 1982.

Molza, Francesco Maria. *Delle poesie volgari e Latine di Francesco Maria Molza corrette, illuftrate, e accreiciute colla vita dell'autore scritta da Pierantonio Serassi*. Vol. 1. Bergamo, 1747.

Montfort, Nick. "Continuous Paper: the Early Materiality and Workings of Electronic Literature." Paper presented at the Modern Language Association Convention, Philadelphia, 28 December 2004. <http://nickm.com/writing/essays/continuous_paper_mla.html>.

Nagel, Alexander. *The Controversy of Renaissance Art*. Chicago and London, 2011.

Neilson, Christina. "Carving Life: The Meaning of Wood in Early Modern European Sculpture." In *The Matter of Art: Materials, Practices, Cultural Logics, c. 1250-1750*, edited by Christy Anderson, Anne Dunlop, and Pamela H. Smith, pp. 223-39. Manchester, 2014.

Nygren, Christopher J. "Titian's Miracles: Artistry and Efficacy Between the San Rocco *Christ* and the Accademia *Pietà*." *Mitteilungen des Kunsthistorischen Institutes in Florenz* 57, no. 3 (2015), 320-49.

Nygren, Christopher J. "Titian's *Ecce Homo* on Slate: Stone, Oil, and the Transubstantiation of Painting." *The Art Bulletin* 99, no. 1 (2017), 36-66.

Onians, John. "Abstraction and Imagination in Late Antiquity." *Art History* 3 (1980), 1-24.

Ovid. *The Metamorphoses of Ovid*. Translated by Allen Mandelbaum. New York, 1993.

Paleotti, Gabriele. *Discourse on Sacred and Profane Images*. Translated by William McCufaig. Los Angeles, 2012.

Pietra dipinta: tesori nascoti del '500 e del '600 da una collezione privata Milanese. Edited by Marco Bona Castelloti. Milan, 2000.

Pliny. *Natural History, Loeb Classical Library*, translated by H. Rackham. Cambridge, MA and London: Harvard University Press, 1995.

Raff, Thomas. *Die Sprache der Materialien: Anleitung zu einer Ikonologie der Werkstoffe*. Münster, 2008.

Röttgen, Herwarth. *Il Cavalier Giuseppe Cesari d'Arpino: un grande pittore nello splendore della fama e nell'incostanza della fortuna*. Rome, 2002.

Roskill, Mark W. *Dolce's Aretino and Venetian Art Theory of the Cinquecento*. Toronto, 2000.

Seifertova, Hana. *Malby na kameni: Umĕcký experiment v 16. A Na Začátku 17 století/ Painting on Stone: an Artistic Experiment in the 16th and Early 17th centuries*. Prague, 2007.

Speroni, Sperone. "Discorso in lode della pittura." In *Opere di M. Sperone Speroni degli Alvarotti*, vol. 3, edited by Natale dalle Laase and Marco Forcellini, pp. 441-6. Venice, 1740.

Speroni, Sperone. "Dialogo d'amore." In *Trattatisti del cinquecento*, edited by Mario Pozzi. Milan and Naples, 1978.

Stafford, Barbara Maria. *Visual Analogy: Consciousness as the Art of Connecting*. Cambridge and London, 1999.

Terzago, Paolo Maria. *Musaevm Septalianvm Manfredi Septalae Patritii Mediolanensis Indvstrioso Labore Conscrvctvm*. Tortona, 1664.

Trilling, James. "The Image Not Made by Hands and the Byzantine Way of Seeing." In *The Holy Face and the Paradox of Representation*, edited by Herbert L. Kessler and Gerhard Wolf, pp. 109-27. Bologna, 1998.

Varchi, Benedetto. "Della maggioranza delle arti." In *Trattati d'arte del Cinquecento, fra manierismo e controriforma*, vol. 1, edited by Paola Barocchi, pp. 5-58. Bari, 1960.

PLATES

*16-32*

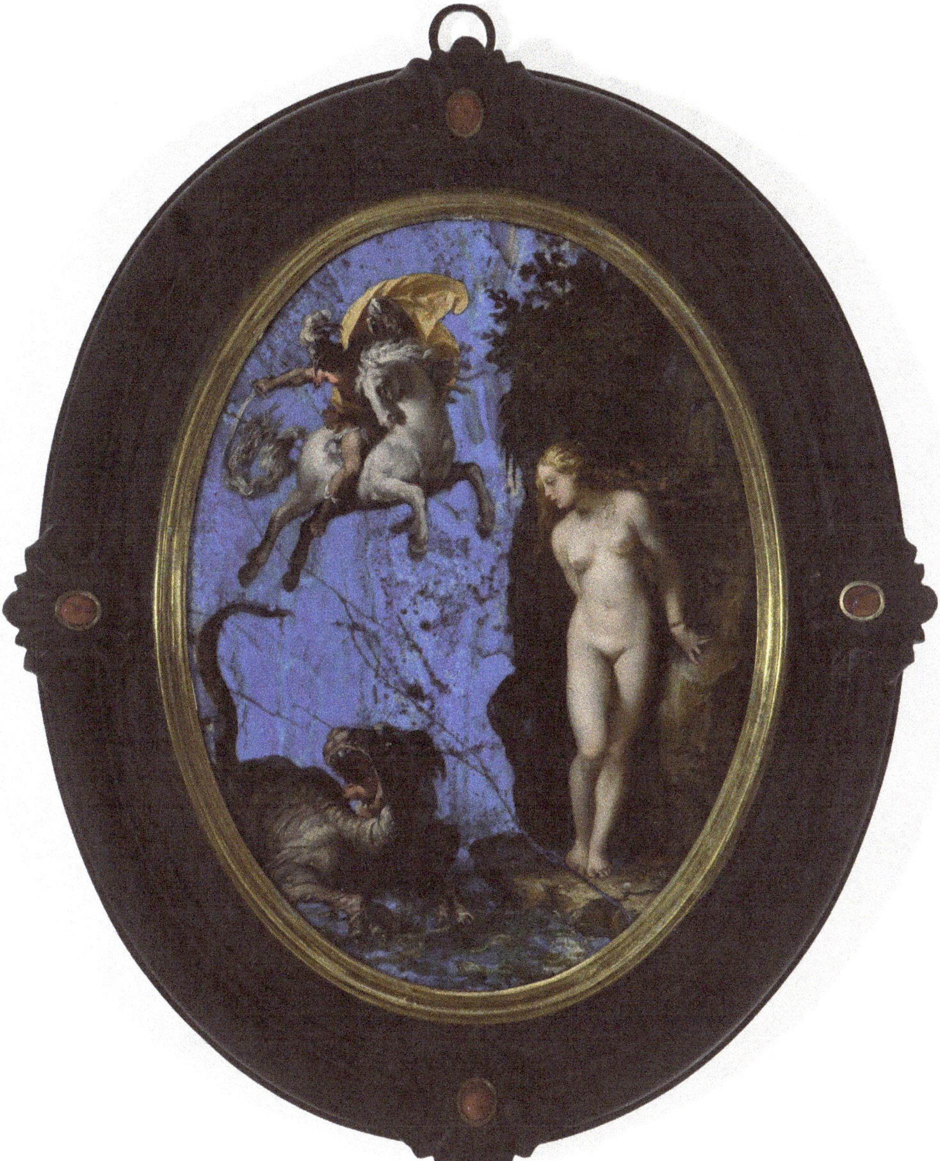

PLATE 16 *Cavaliere d'Arpino,* Perseus Liberating Andromeda, *oil on lapis lazuli, ca. 1593-1594, St Louis, St Louis Art Museum* (PHOTO: ©ST LOUIS ART MUSEUM).

 | DOI 10.1163/9789004361492_008

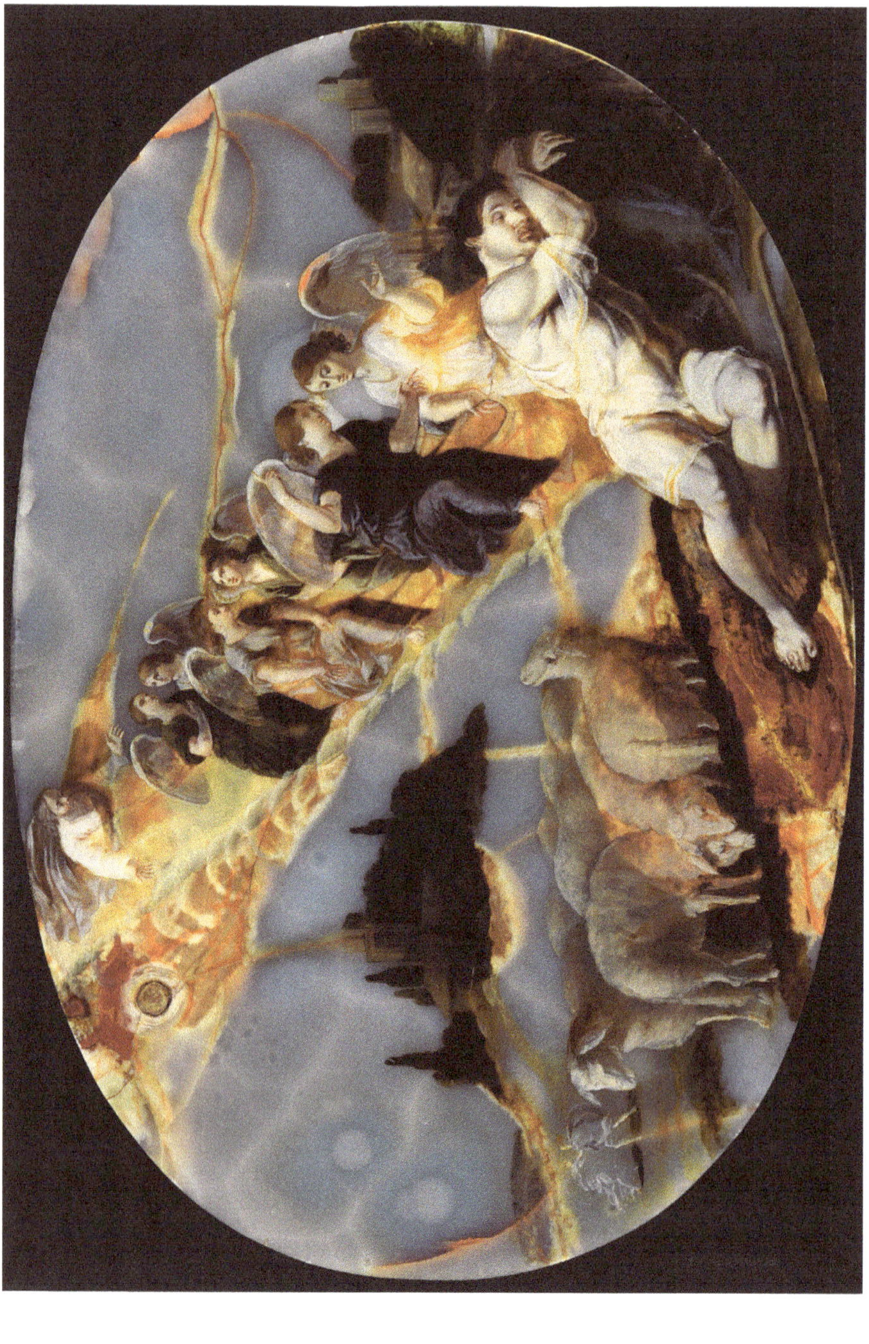

PLATE 17 *Jacques Stella,* Jacob's Dream, *ca. 1650, oil on onyx, Los Angeles, Los Angeles County Museum of Art* (PHOTO: LOS ANGELES COUNTY MUSEUM OF ART).

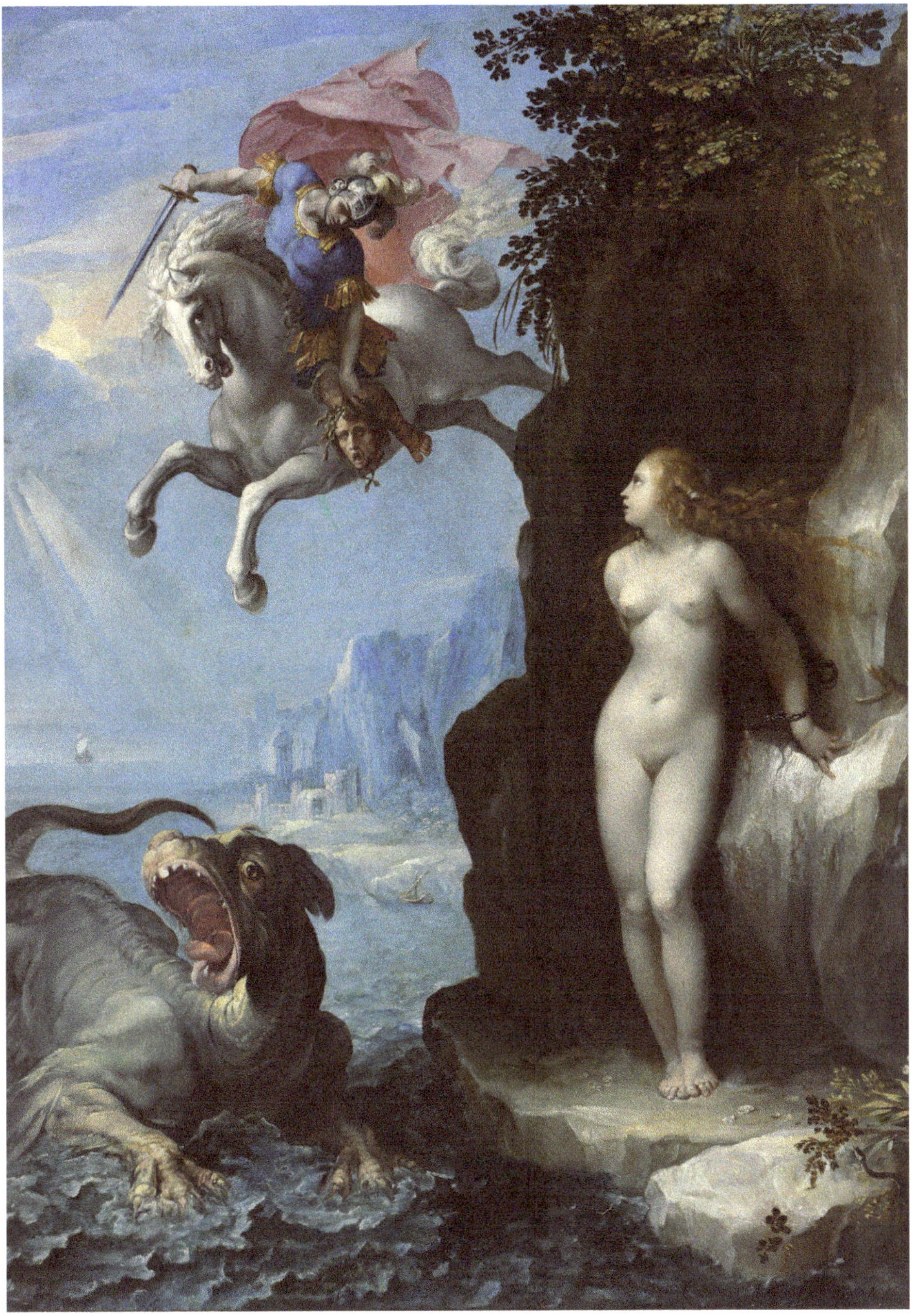

PLATE 18 *Cavaliere d'Arpino,* Perseus Liberating Andromeda, *ca. 1592, oil on slate, Providence RISD Museum* (PHOTO: MUSEUM OF ART, RHODE ISLAND SCHOOL OF DESIGN).

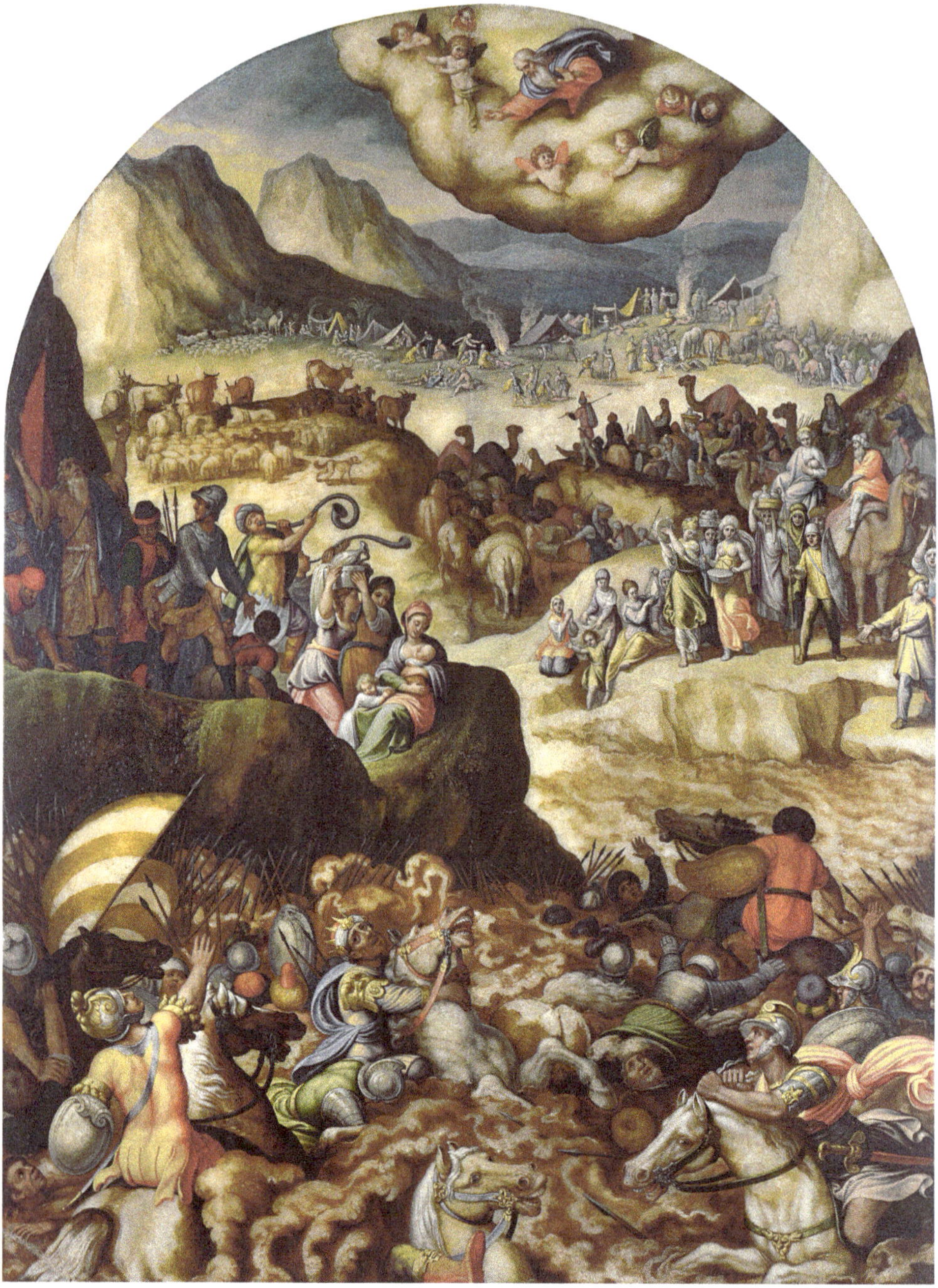

PLATE 19 *Antonio Tempesta,* Crossing of the Red Sea, *oil on alabaster, ca. 1606-1608, Milan, Giulini Collection* (PHOTO: GIULINI COLLECTION).

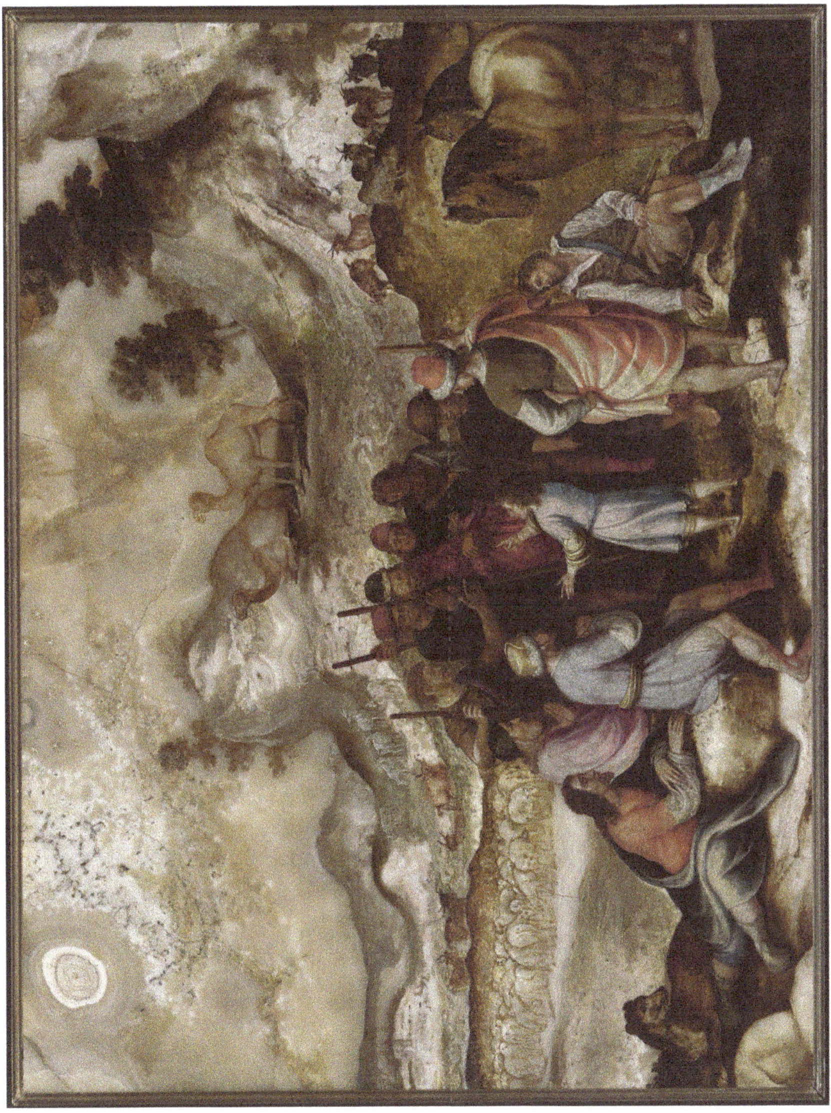

PLATE 20 *Antonio Tempesta*, Joseph explaining his dream to his brothers, *oil on alabaster*, *Boston*, *Museum of Fine Arts* (PHOTO: BOSTON, MUSEUM OF FINE ARTS).

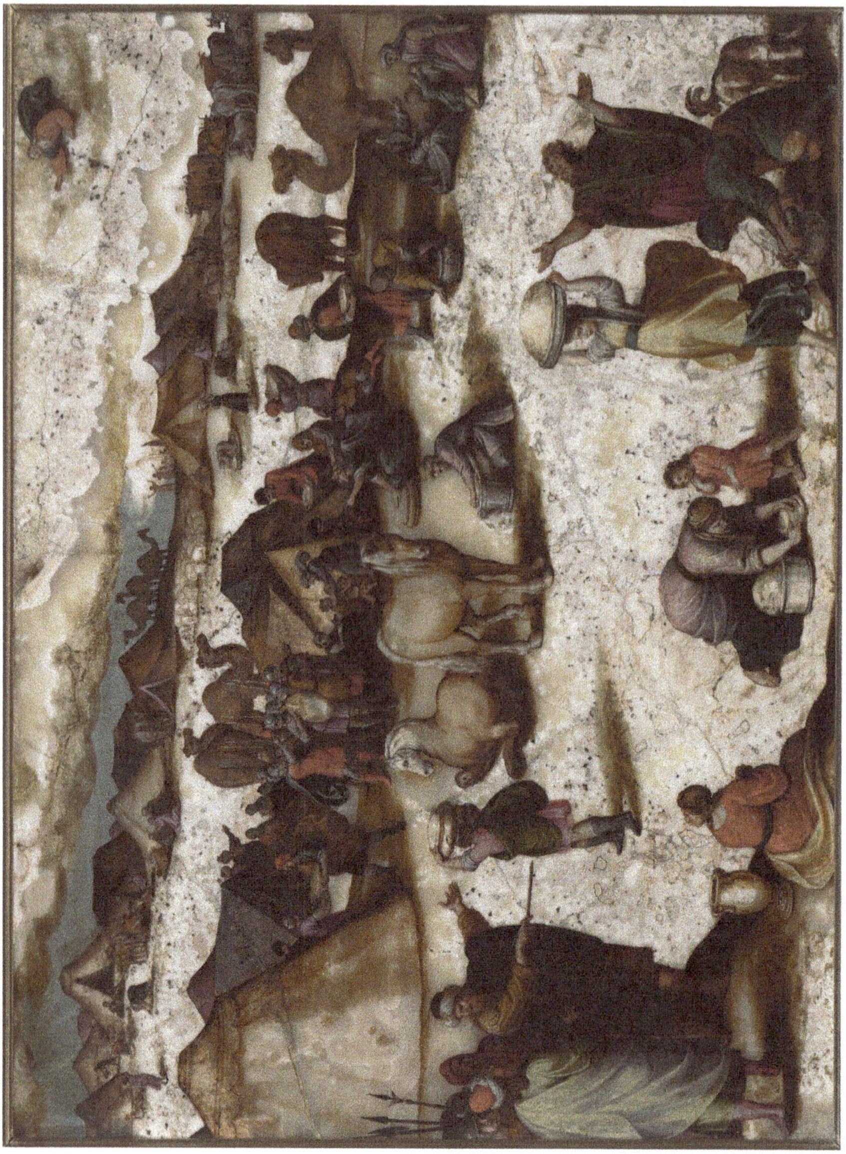

PLATE 21 *Antonio Tempesta,* Gathering of Manna, *oil on alabaster, Boston, Museum of Fine Arts* (PHOTO: BOSTON, MUSEUM OF FINE ARTS).

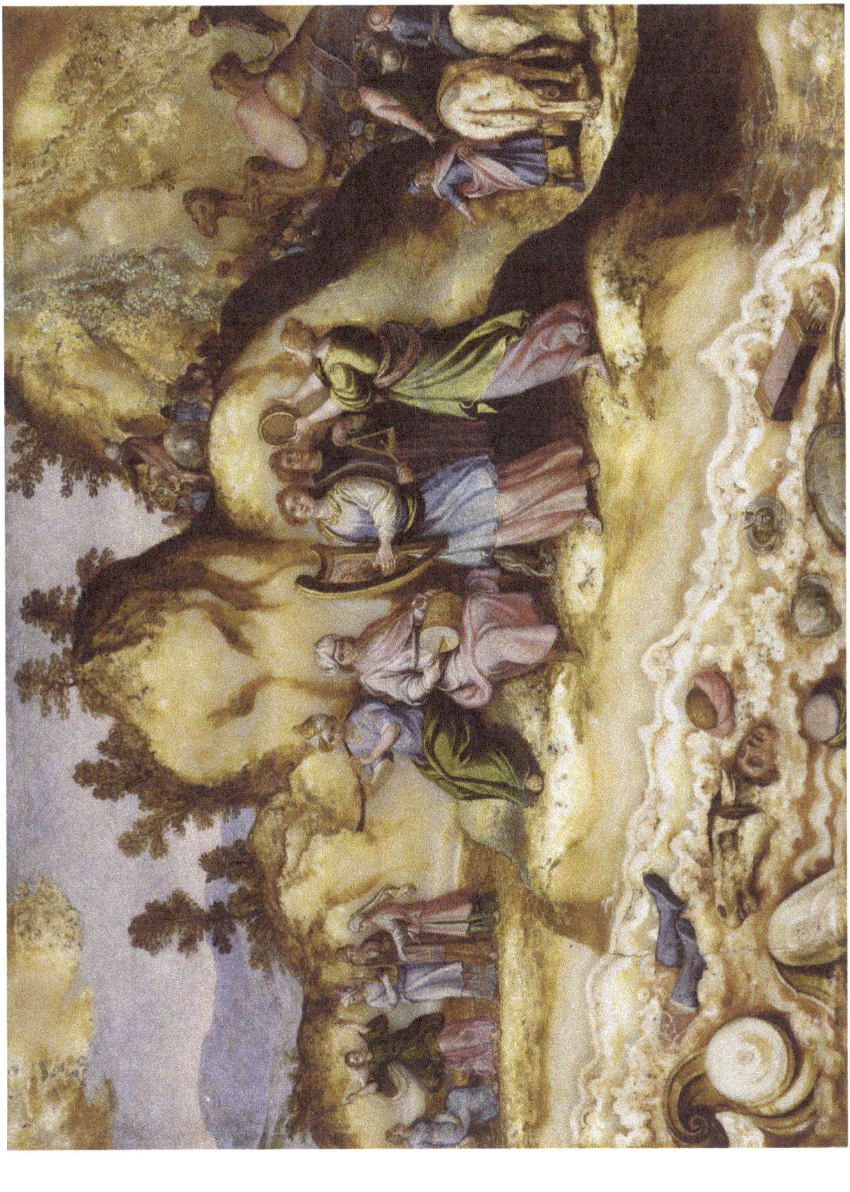

PLATE 22 *Antonio Tempesta,* Crossing of the Red Sea, *oil on alabaster, Private Collection* (PHOTO: ROBERT SIMON FINE ART, NEW YORK).

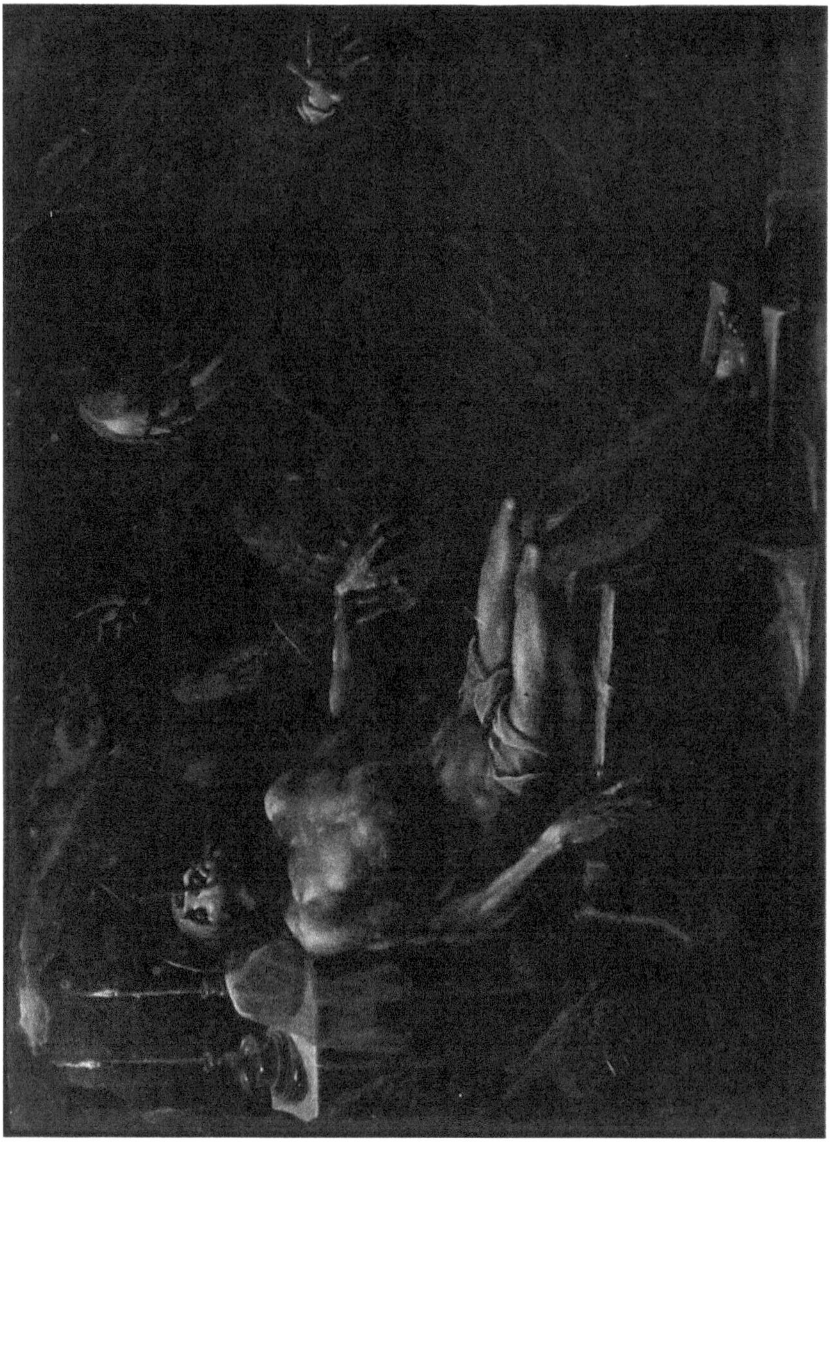

PLATE 23 *Hans von Aachen,* Pietà, *oil on black marble, 1594, Kremsmünster Abbey* (PHOTO: STIFT KREMSMÜNSTER, KUNSTSAMMLUNGEN. FOTOGRAF: JOSEF LEITHNER).

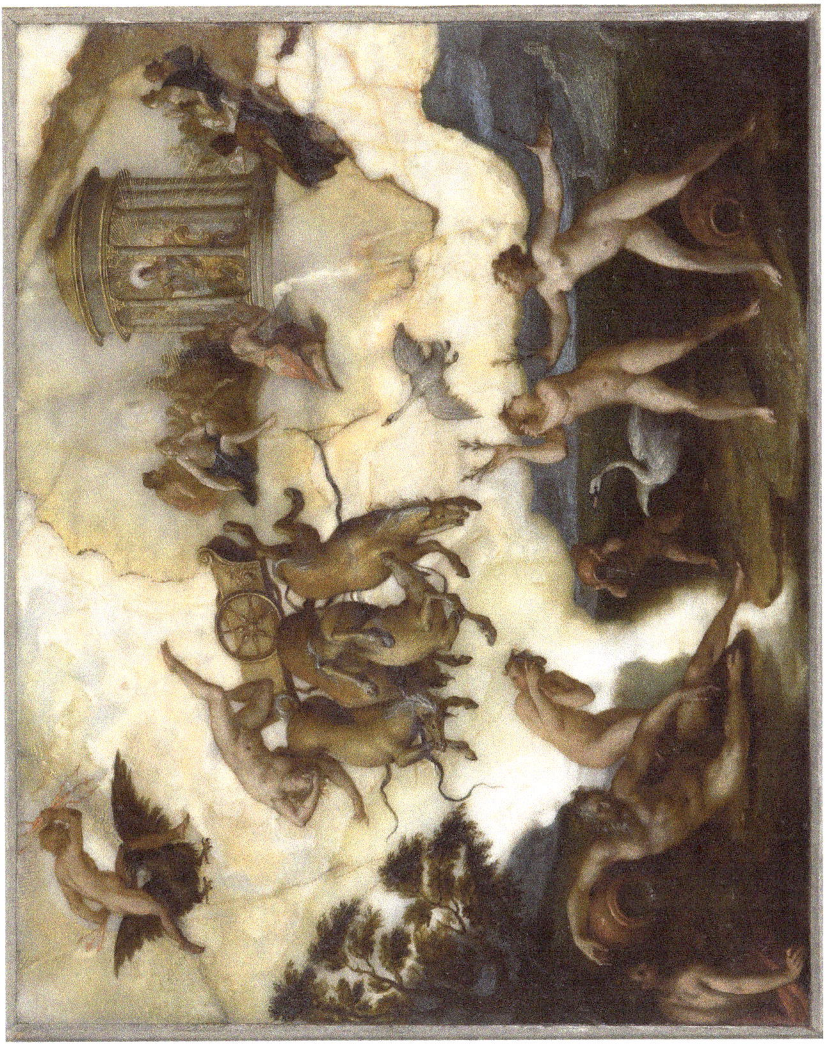

PLATE 24 *Hans von Aachen*, The Fall of Phaeton, *oil on alabaster, ca. 1600, Innsbruck, Ambras Castle* (PHOTO: KHM-MUSEUMSVERBAND).

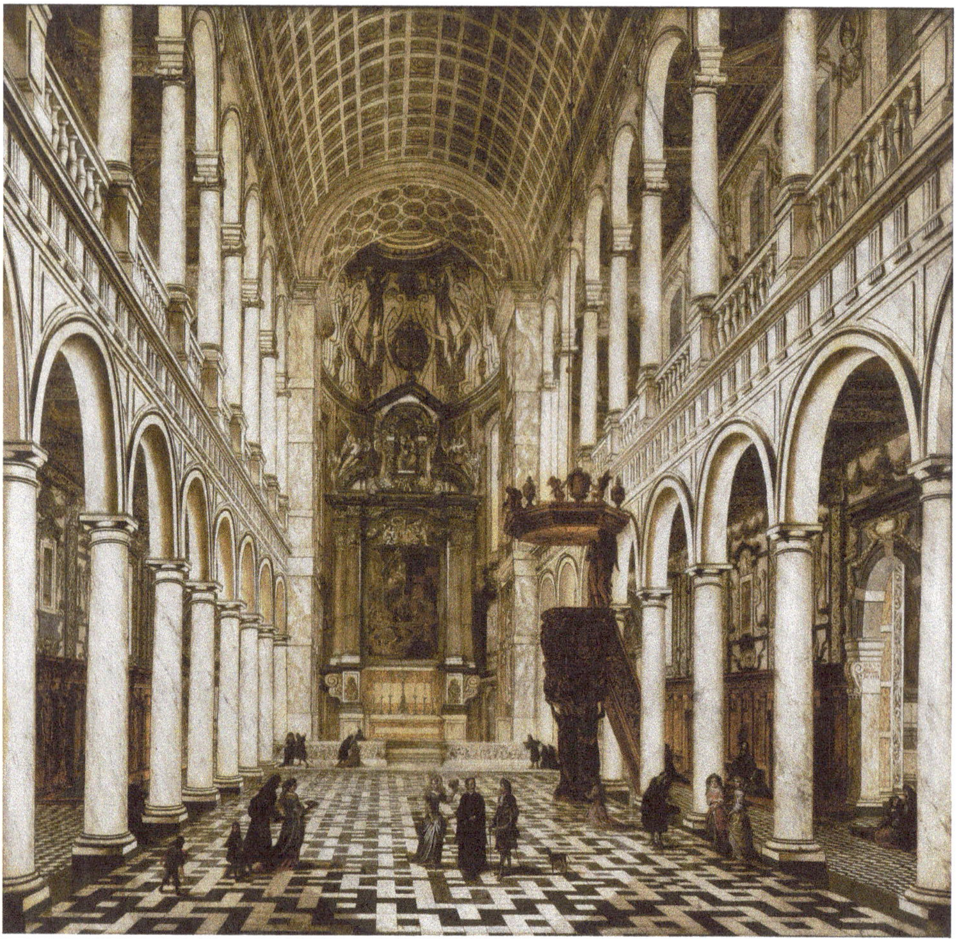

PLATE 25 *Wilhelm Schubert van Ehrenberg,* Jesuit Church in Antwerp, *1668, Antwerp, Rubenshuis* (PHOTO: RUBENSHUIS, BY MICHEL WUYTS AND LOUIS DE PEUTER).

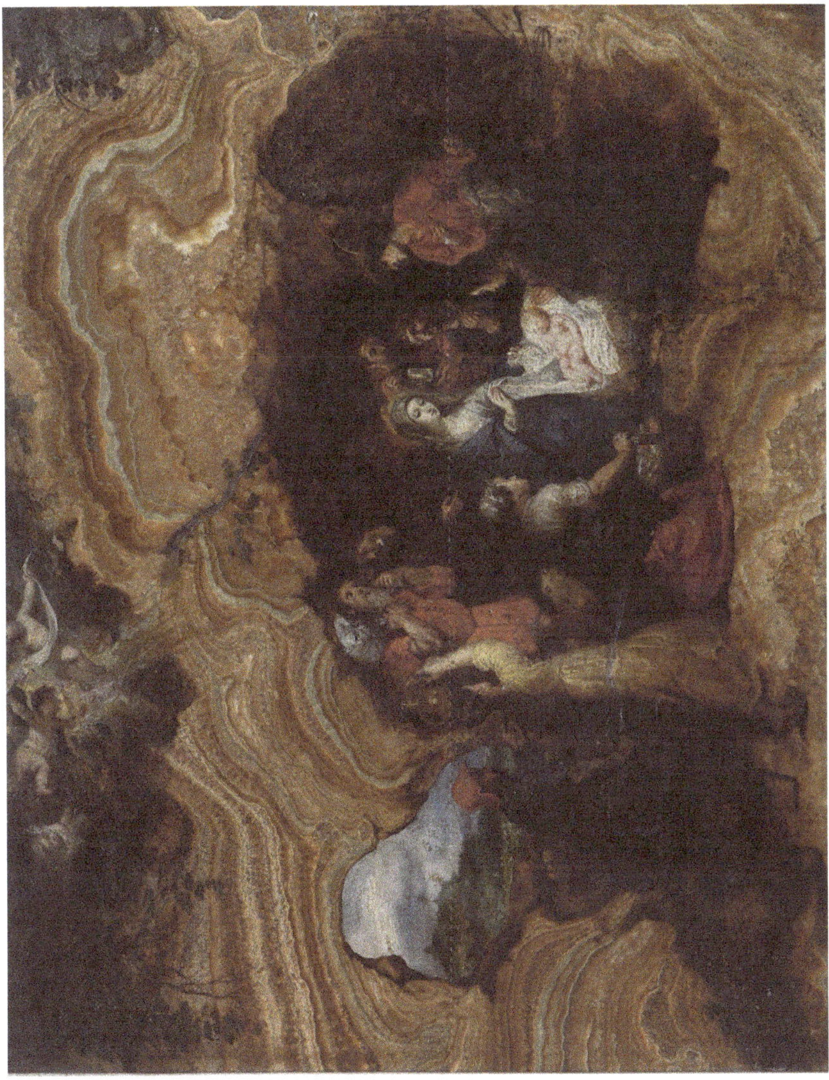

PLATE 26 *Hendrick van Balen*, Adoration of the Shepherds, *1620-1630*, *Antwerp, Sint- Carolus Borromeuskerk* (PHOTO: LUKAS–ART IN FLANDERS VZW, HUGO MAERTENS).

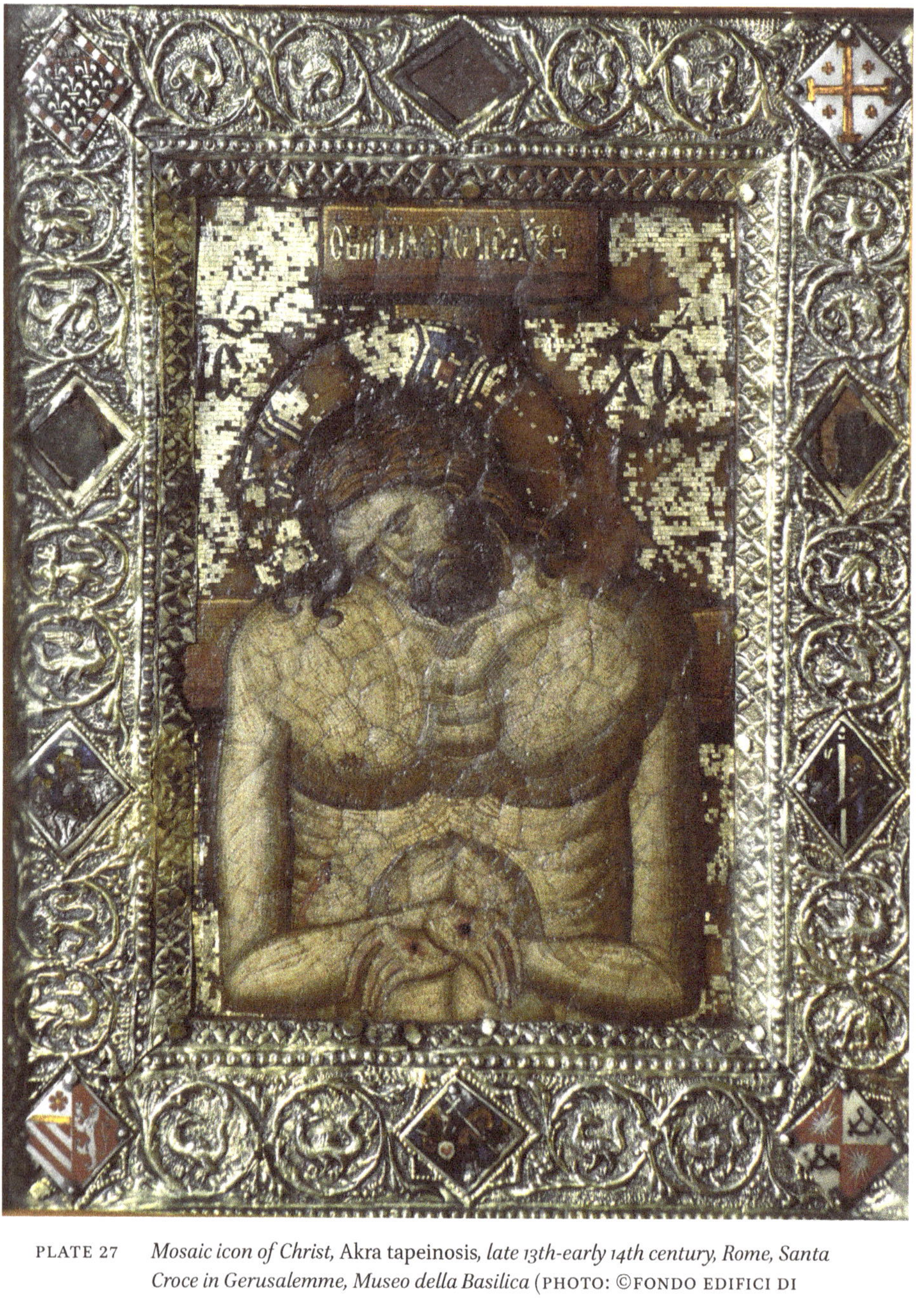

PLATE 27 *Mosaic icon of Christ,* Akra tapeinosis, *late 13th-early 14th century, Rome, Santa Croce in Gerusalemme, Museo della Basilica* (PHOTO: ©FONDO EDIFICI DI CULTO).

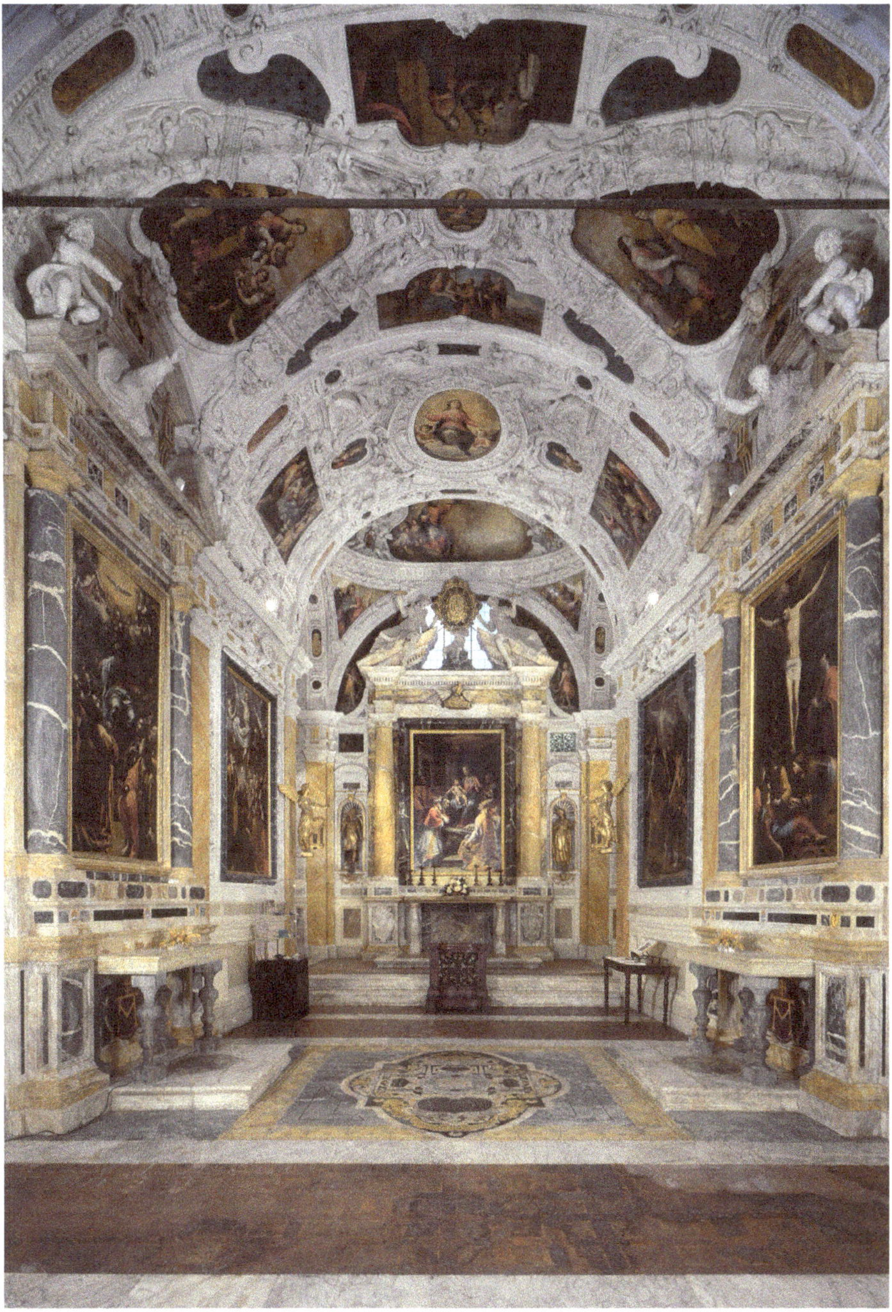

PLATE 28 *Siena, San Niccolò in Sasso, the church interior* (PHOTO: ©SCALA, FLORENCE, 2015).

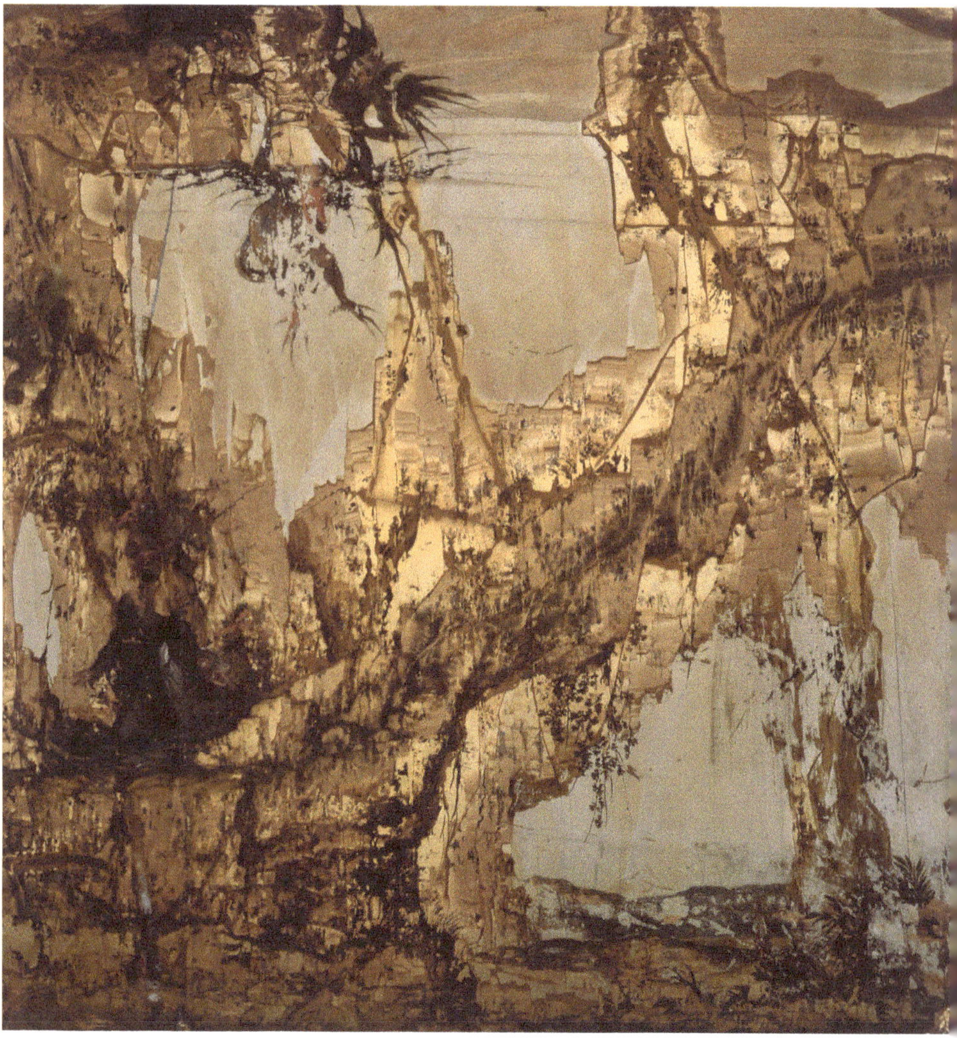

PLATE 29 *Filippo Napoletano,* Temptation of St Anthony, *pietra paesina, Florence, Palazzo Pitti* (PHOTO: FIRENZE GABINETTO DEI DISEGNI E DELLE STAMPE DEGLI UFFIZI).

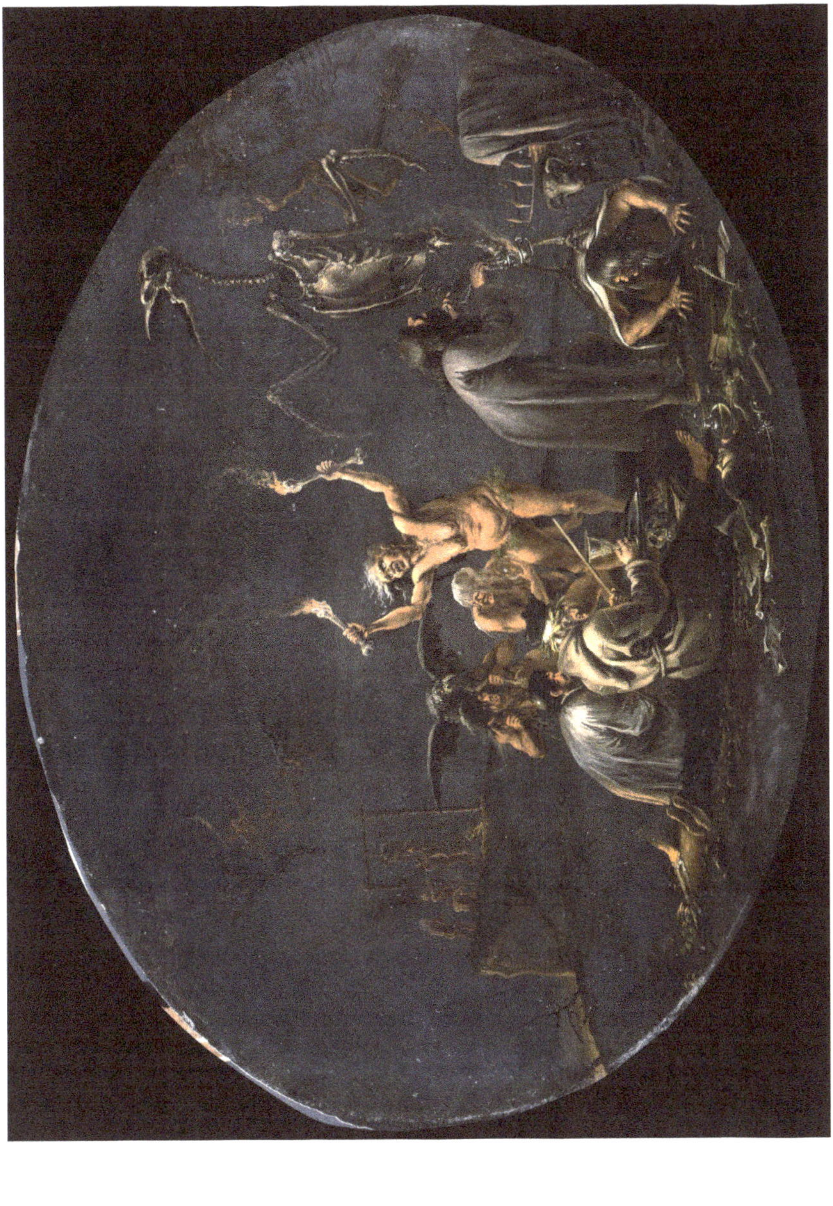

PLATE 30 *Salvator Rosa*, Scene of Witchcraft, *oil on slate, ca. 1645-46, Milan, Private Collection* (PHOTO: WITH THE MATTHIESEN GALLERY, LONDON, 1986).

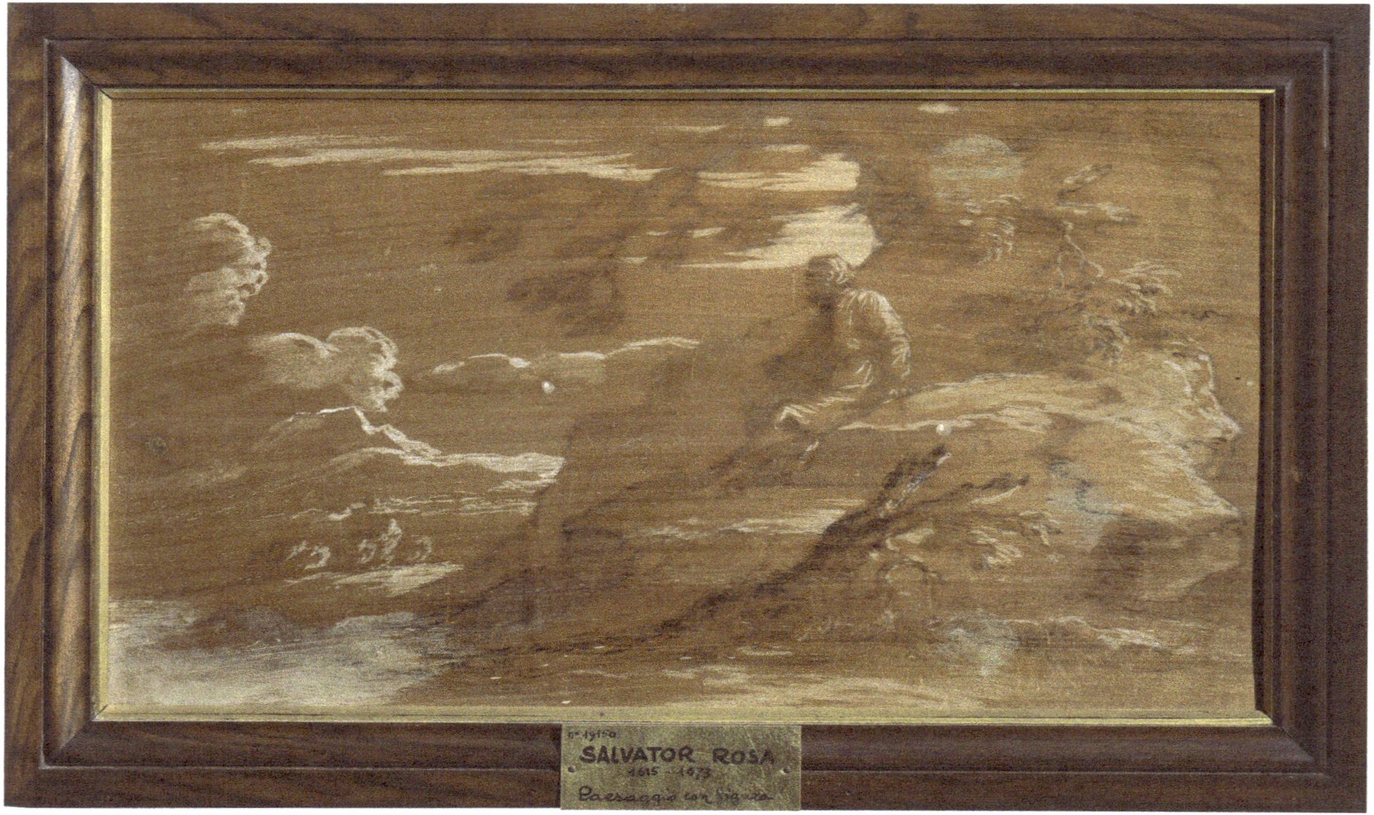


PLATE 31 *Salvator Rosa,* A man seated, facing left, in a spacious landscape, *Florence, Palazzo Pitti* (PHOTO: FIRENZE GABINETTO DEI DISEGNI E DELLE STAMPE DEGLI UFFIZI).

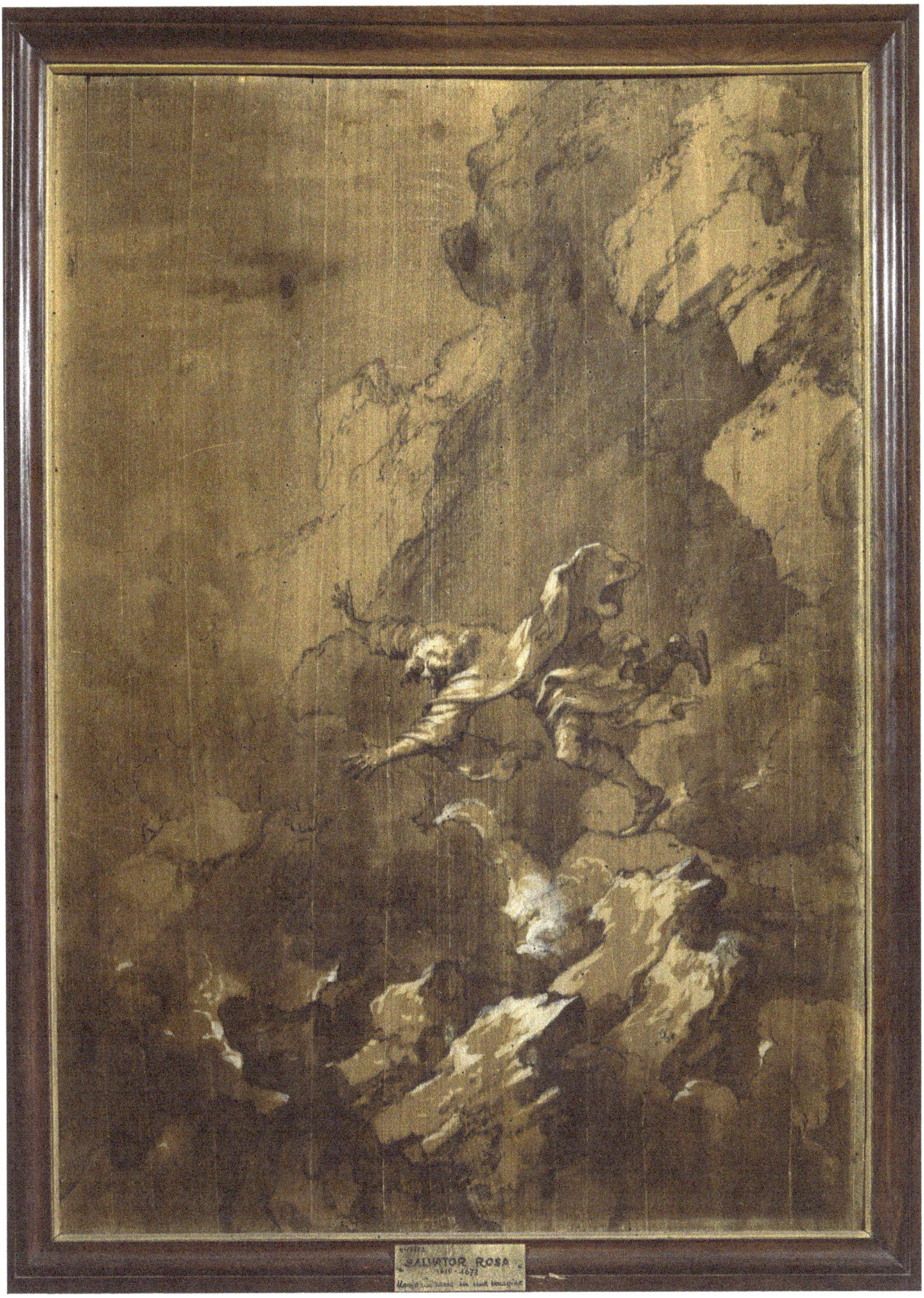


PLATE 32 *Salvator Rosa,* Death of Empedocles, *Florence, Palazzo Pitti* (PHOTO: FIRENZE GABINETTO DEI DISEGNI E DELLE STAMPE DEGLI UFFIZI).

CHAPTER 5

# Antonio Tempesta's Paintings on Stone and the Development of a Genre in 17th-Century Italy

*Johanna Beate Lohff*

In the early 17th century, small paintings on stone were highly popular with art collectors from the ecclesiastical and secular nobility.[1] Artists used different stones as supports for their paintings, leaving sections of the stone unpainted as an integral part of their compositions.[2] They had a predilection for both black stones such as slate, black marble, and *pietra paragone* as well as for patterned, colored, and semi-precious stones such as marble, alabaster, dendritic stones, ruin marble (*pietra paesina*), lapis lazuli, agate, and amethyst. The minerals used played an important role in the creative process, influencing the choice of subject matter, the composition, the color scheme, and the degree of perfection in the painting's execution. The various minerals were particularly suited for certain subjects. While black stones commonly served as a backdrop for nocturnal scenes, colored and patterned stones were mainly used in the landscape backgrounds of history paintings, which represented scenes taken from the Bible, mythology, history, or literature.

In these mostly small-scale paintings, with dimensions of less than one meter width, the painted areas and the appearance of the stone itself assume more or less equal importance. Antonio Tempesta (ca. 1555-1630) can be regarded as one of the leading artists who created paintings on stone.[3] In this paper, I will first outline the historical collecting context to which his works on stone belong. Secondly, I will examine Tempesta's individual style in painting on stones on the basis of works that have, up to now, not been discussed by

1 I would like to thank Yvonne Yiu, Basel for the translation of this essay.

2 There is a clear distinction between the mostly small format patterned stones as pictural supports discussed in this essay and the large paintings on slate like those executed by Sebastiano del Piombo or in the course of large altar pieces like Peter Paul Rubens executed for Santa Maria in Valliccella. Cf. Introduction, Baker-Bates, Calvillo, and other authors in this volume.

3 Anne-Laure Collomb and Hana Seifertová also attribute a special role to him. Seifertová, *Malby na kameni. umelecký experiment v 16. a na zacátku 17. století. Painting on Stone. An Artistic Experiment in the 16th and Early 17th Centuries* (Prague, 2007), pp. 106-26; Collomb, *Spendeurs d'Italie. La peinture sur pierre à la Renaissance* (Tours and Rennes, 2012), 197-203.

 | DOI 10.1163/9789004361492_009

scholars examining their distinctive features with respect to the artist's oeuvre on stone as a whole.

Furthermore, I will focus on the stones that served as picture supports, with the aid of mineralogical treatises of the 16th and 17th centuries, which attached a significance to these stones that improves our understanding of them. As the mineral picture support had a decisive influence on the art work as a whole, it is necessary to examine the iconography and meaning of the stone used.

## Antonio Tempesta as a Painter of Pictures on Stone

Antonio Tempesta (*c.* 1555-1630) was an exceptionally successful artist active in the Early Baroque Period who excelled both in the graphic arts as well as in painting.[45] He trained with Jan van der Straet (also called Giovanni Stradano, 1523-1605), a Netherlandish artist based in Florence, who encouraged him to explore new subjects such as hunting scenes.[6] According to Giovanni Baglione, Tempesta moved to Rome during the pontificate of Gregory XIII and stayed there until his death.[7] Initially, Tempesta worked as a fresco painter and was involved in the execution of large commissions for the Vatican and the Roman nobility.[8] When Sixtus V (Felice Peretti di Montalto) became pope in 1585, Tempesta ceased working as a fresco painter, possibly because he was excluded from the major papal projects.[9] It was probably from this time on that he focused on his talents as a graphic artist, inventing and establishing new subjects such as "hunting scenes, landscapes, battles, animals and grotesques."[10]

4 Recently, the art market and private collections have brought to light further paintings on stone attributed to Tempesta, which could not be considered in this essay: Cf. *Paintings on Stone of the 16th & 17th Centuries*, ed. Alessandra Di Castro and Alexis Kugel, exhibition catalogue, New York (Château de Saint-Rémy-en-l'Eau, 2017), no. 17.

5 Cf. the comprehensive study by Eckhard Leuschner, *Antonio Tempesta. Ein Bahnbrecher des römischen Barock und seine europäische Wirkung*, Studien zur internationalen Architektur und Kunstgeschichte, vol. 26 (Petersberg, 2005).

6 Leuschner, *Antonio Tempesta*, 24-32; *Stradanus 1523-1605. Court Artist of the Medici*, ed. Alessandra Baroni and Manfred Sellink, exhibition catalogue, London and Amsterdam (Turnhout, 2012), pp. 172-73.

7 Giovanni Baglione, *Le Vite de pittori scultori et architetti. Dal Pontificato di Gregorio XIII. del 1572 in fino a tempi di Papa Urbano Ottauo nel 1642* (Rome, 1642), 314; Leuschner, *Antonio Tempesta*, p. 39.

8 Leuschner, *Antonio Tempesta*, pp. 39-91.

9 Leuschner, *Antonio Tempesta*, p. 60.

10 Eckhard Leuschner, "Antonio Tempesta's 'New' Subjects in the Context of Roman Printmaking ca. 1600," in *The Art Market in Italy 15th– 17th Centuries*, ed. Marcello Fantoni, et al. (Modena, 2003), p. 65.

This appears to have been a marketing strategy that enabled him to approach potential patrons. Employing many different printing techniques, he not only created biblical and mythological scenes and images of saints but also illustrated books and, in 1593, made a large map of Rome. Furthermore, he executed numerous paintings on canvas, wood, copper and stone. The last feature themes from the Bible and mythology as well as hunting scenes. Tempesta chose to paint on alabaster, marble, lapis lazuli, and dendrites. Paintings on *pietra paesina* and on black stones such as slate, black marble, or *pietra paragone* may only be attributed to Tempesta, if at all, with caution.[11]

## Antonio Tempesta's Paintings on Stone as Collectors' Items

Tempesta's paintings on stone graced the aristocratic collections of the high-ranking members of the Church and nobility. His earliest documented paintings on stone can be dated to the beginning of the pontificate of Paul v (Camillo Borghese) in 1605. It is, however, possible that he already began using this technique at an earlier point in time, as he was probably acquainted with it through the work of other artists in his circle. Already in 1604 a certain Francesco Tempesta received 60 *scudi* for a "creatione del mondo" on stone.[12] Antonio had a brother called Francesco who was also a painter.[13] Nonetheless, it remains subject to debate whether the latter executed the painting, and the possibility exists that it might have been made in Antonio's workshop. The painting is mentioned in an inventory from the year 1621 listing the belongings of the Roman noble family Mattei, together with an "adoratione de maggi" on alabaster—the artist is not named—as well as a *Christ Carrying the Cross* on copper and other paintings on unspecified supports.[14]

11 Cf. my critical discussion of the current attributions of paintings on stone to this painter: Johanna Beate Lohff, *Malerei auf Stein. Antonio Tempestas Bilder auf Stein im Kontext der Kunst- und Naturtheorie seiner Zeit* (Munich, 2015), pp. 197-202.

12 "(42) Adj 24 di Aprile 1604 e pui deveno havere sc. sesanta d moneta p. tanti pagati à fran. co Tempesta p. un quadro in alabastro p. servitio mio pittura della creatione del Mondo dico 60." Francesca Cappelletti and Laura Testa, *Il Trattenimento dei virtuosi. Le collezioni seicentesche di quadri nei Palazzi Mattei di Roma* (Rome, 1994), p. 140.

13 On Francesco Tempesta cf. Leuschner, *Antonio Tempesta*, pp. 21, 36.

14 "(107) Otto quadretti di diverse figure cioè uno di Alabastro che dimostra l'adoratione de maggi, (108) Uno sim.e d'Alabastro la Creatione del mondo, (109) Uno in rame quando xpo portava la Croce, (110) Uno quando xpo fù crocifisso al Calvario, (111) uno quando andò in Gerusalemme il di delle palme, e (112—114) tre altri di diverse istorie e faccie tutte con sue cornice d'ebano." Cappelletti and Testa, *Il Trattenimento*, p. 180. As Antonio Tempesta

A *Last Judgement* on agate counts among the earliest paintings on stone that can be unequivocally attributed to Antonio Tempesta. This now lost painting is mentioned in a document pertaining to a legal dispute between Francesco Lionelli and the merchant (*rigattiere*) Domenico de Marchis.[15] On 15 April 1610 the artist had to testify that three years earlier, in 1607, he had been paid by Francesco Leonelli for a painting on agate depicting the Last Judgement. An inventory of Cinzio Passeri Aldobrandini's estate dated 1 March 1610 reveals that the painting belonged to this cardinal-nephew of Clement VIII for a certain period of time and that he kept it in a casket in his collection of paintings.[16] The inventory also informs us that Cinzio had a *studiolo* in which he dedicated himself to the study of Nature and to alchemical experiments.[17] The cardinal seems to have had a preference for minerals and especially agates that he kept in his *studiolo*. In addition to Tempesta's painting on agate he possessed "'various dried pieces of agate', a statue of Christ tied to the column made of agate (attached to the crucifix with the crystal base) and 'thirty-five pieces of glass painted to look like agate.'"[18] Painted imitations of a certain stone on glass only rarely feature as collectors' items. It is, therefore, possible that these artworks were originally created for a different context, for instance as decorative elements for a cabinet.

Following Cinzio Passeri Aldobrandini's death on 1 January 1610, Tempesta's *Last Judgement* was supposed to be returned to Francesco Lionelli. However, as de Marchis also asserted a claim on the painting, a legal dispute ensued. Francesco Lionelli, also documented as Leonelli (died *c.* 1637), was a patron of Tempesta's, and he may have introduced the artist to the Aldobrandini family for which Tempesta made several paintings.[19] On 1 March 1609, Tempesta dedicated a series of hunting scenes to Lionelli.[20]

painted several versions of the *Adoration of the Magi* on alabaster, it is possible that the painting with this subject in the Mattei collection was painted by Tempesta.

15 Cf. Lohff, *Malerei auf Stein*, pp. 82-83.

16 Karen Lloyd, "Adopted Papal Kin as Art Patrons in Early Modern Rome (1592-1676)," PhD diss., Rutgers University, 2010), pp. 120-21.

17 Lloyd,"Adopted Papal Kin," pp. 111-30.

18 Lloyd, "Adopted Papal Kin," p. 124.

19 Leuschner, *Antonio Tempesta*, p. 603, no. 57.

20 For unknown reasons the series was rededicated to Giovanni Leoncino between 1609-13, whose final letter needed to be changed to an "I" to reuse the letter at the end of Lionelli. Cf. Leuschner, *Antonio Tempesta*, pp. 405-406.

## Tempesta's *Crossing of the Red Sea* on Alabaster for Giovanni Niccolini

Roughly at the same time that Tempesta worked for Francesco Lionelli, the artist created four paintings on alabaster for Cardinal Giovanni Niccolini (1544-1611), who was Florentine ambassador in Rome from 1587-1610: The *Crossing of the Red Sea*, *Moses striking Water from the Rock*, *Jacob leaves Laban* (or *Abraham leaves Lot*), and *The Victory of the Hebrews over the Assyrians*.[21] Giovanni Cinelli mentions these paintings in his description of Florence published in 1677, which was essentially a new edition with improvements of Francesco Bocchi's *Le Bellezze della città di Fiorenza* (1591).[22] Giovanni Niccolini's accounting books reveal that, from 1606-08, Tempesta received a total of 401 *scudi* paid in irregular installments.[23] The Tuscan ambassador maintained close relations with the Florentine artists in Rome, and thus it is probable that he commissioned the paintings from Tempesta during his sojourn in Rome, taking them back to Florence in 1610, at the end of his diplomatic mission.[24]

The *Crossing of the Red Sea* now in Milan is the only painting on stone by Tempesta from Niccolini's collection that has been preserved to this day [**Plate 19**].[25] In the foreground, the swirling waves of the Red Sea engulf the Egyptian cavalry while the Israelites watch this spectacle from the safety of the rocky landscape in the middle ground. Moses stands on the left with his hands raised in thanksgiving, and, on the right, Miriam, amid her companions, beats the tambourine. In the background, the Israelites have set up camp at the foot of a mountain chain. At the apex of the painting, God the Father, surrounded by angels, emerges from a cloud to bless the scene. It initially seems to the beholder as if the veining of the stone appears in the Red Sea, he or she may also admire the artist's ability to integrate the figures of the soldiers in the waves seemingly produced by nature's hand. But, in fact, the white alabaster of the picture's support is largely masked by the painting, its pale color only shim-

---

21 Lohff, *Malerei auf Stein*, pp. 93-95.

22 Francesco Bocchi and Giovanni Cinelli, *Le Bellezze della Città di Firenze dove a pieno di pittura, di scultura, di sacri templi, di palazzi, i più notabili artifizi, e più preziosi si contengono, scritte già da M. Francesco Bocchi ed ora da M. Giovanni Cinelli ampliate ed accresciute* (Florence, 1677), p. 407.

23 Riccardo Spinelli, "Documenti artistici dall'archivio Niccolini di Camugliano—I. Marmi antichi e 'moderne pitture' di Giovanni di Agnolo (Firenze, 1544-1611)," *Paragone Arte* 61, no. 663 (2005), 86.

24 On Giovanni Niccolini cf. Spinelli, "Documenti artistici."

25 *Pietra Dipinta. Tesori nascosti del '500 e del '600 da una collezione privata milanese*, ed. Marco Bona Castellotti (Milan, 2000), pp. 92-93.

mering through in the landscape, the foam and the light-colored parts of the horses. The patterning itself is not that of the stone but of paint. Thus, the representation plays with a double illusion: the lights and light-colored areas are not painted but are an effect of the stone, while the "stone veining" is painted and not the figuration of the mineral support.

Laura Laureati was able to identify the *Crossing of the Red Sea* as part of the Niccolini collection on the basis of a print by Carlo Lasinio published in Marco Lastri's *L'Etruria Pittrice...* (1791-95).[26] It depicts Tempesta's painting of *Moses striking Water from the Rock*, which was still in the Niccolini collection in the 18th century.[27] Both artworks have an unusual portrait format with a semi-circular apex and can, therefore, be regarded as a pair.[28] On comparing the painting in Milan with Lasinio's print, it seems logical to assign the former to Niccolini's collection. These paintings are extraordinarily precious, not only because of their artistic quality but also because of the sheer amount of material used. *Moses striking Water from the Rock* as well as the other two paintings on alabaster by Tempesta, which, according to Cinelli, belonged to the collection, are lost. The information provided by Cinelli corresponds to that in Niccolini's accounting books, although *Jacob leaves Laban* is called *Abraham leaves Lot* in the archival documents. Probably, half a century later, Cinelli was not able identify the painting's iconography with absolute certainty.[29]

Like many cardinals at the time, Giovanni Niccolini was especially interested in precious stones, and decorated his family chapel in Santa Croce in Florence accordingly. Probably the alabaster panels that later served Tempesta as a particularly appreciated picture support were acquired in the course of work on the chapel. In 1579 the architect Giovanni Antonio Dosio (*c.* 1533-1610) presented the design for the chapel.[30] However, before he could actually begin building it, he had to procure the stones for its interior decoration. These mostly came from Rome and needed an export license. He devoted himself to this task in the early 1580s. Precious marbles such as *marmo africano, marmo bianco antico* and *marmo nero antico, marmo di Carrara,* and alabaster were

---

26 *Pietra Dipinta*, p. 92.

27 Marco Lastri, *L'Etruria pittrice ovvero Storia della pittura toscana dedotta dai suoi monumenti che si esibiscono in stampa dal secolo X fino al presente* (Florence, 1791-95), vol. 2, pl. 65.

28 The *Crossing of the Red Sea* measures 90 × 70 cm. In the inventories, the height of *Moses striking Water from the Rock* is given as 1 ½ braccie, which roughly corresponds to the size of the other painting.

29 Lohff, *Malerei auf Stein*, p. 94.

30 Riccardo Spinelli, "Giovanni Antonio Dosio e il progetto della cappella Niccolini in Santa Croce a Firenze. Quarantatre lettere inedite," *Rivista d'arte* 44 (1992), 199-296.

given pride of place in the chapel. Both the cardinal and the architect had a particular predilection for alabaster which was employed not only in various sections of the chapel but also in the furniture and table tops. Niccolini instructed his architect to buy every suitable piece of alabaster he chanced upon, since one could, after all, decide at a later stage for what purpose it might be used.[31]

## Tempesta's Hunting Scenes on Dendrites in the *Kunstkammer* of Emperor Rudolf II

Two hunting scenes on dendrite, a *Boar Hunt* and a *Bear Hunt* now in the Kunsthistorisches Museum, Vienna, also count among the early paintings on stone by Tempesta [**Figs. 5.1-5.2**]. Dendrites are especially reminiscent of wooded landscapes (*dendron* is classical Greek for tree). The patterning is caused by metallic deposits in the limestone that look like trees, and Tempesta used this type of stone several times for the representation of hunts.[32]

Since the veining of both pieces is similar it seems reasonable that the panels for the Vienna paintings were cut from the same block of stone. In the foreground of both paintings the hunters accompanied by their hounds pursue their prey on horseback and on foot. The hunt continues into the middle ground and mountains loom in the distance. Tall, slim trees with sparse branches contribute to the mood of the painting. The small branches and leaves are created by the veining of the stone, and Tempesta merely outlined the tree trunks, thereby both using and reinforcing the natural structure of the stone.

Both paintings originally belonged to the famous *Kunst-* und *Wunderkammer* of Rudolf II (1552-1612), Emperor of the Holy Roman Empire of the German Nation, in Prague, who was especially interested in minerals. He invited the stone cutters Ottavio Miseroni and Cosimo Castrucci to Prague, where, following Florentine models, they ran the court's own workshops.[33] In 1609, the personal physician of Rudolf II, Anselmus Boethius de Boodt, wrote the treatise *Gemmarum et lapidarum historia* in which he discussed the medicinal and magical properties of stones. The *Kunstkammer*'s holdings were rich in stones, including not only unworked stones, fossils, bezoars, which were thought to have healing properties, precious jewels, and cut stone vessels, but also stone

31 Spinelli, "Giovanni Antonio Dosio," pp. 236-37.

32 Lohff, *Malerei auf Stein*, pp. 194-97.

33 Annamaria Giusti, *Pietra Dura. Bilder aus Stein* (Munich, 2005).

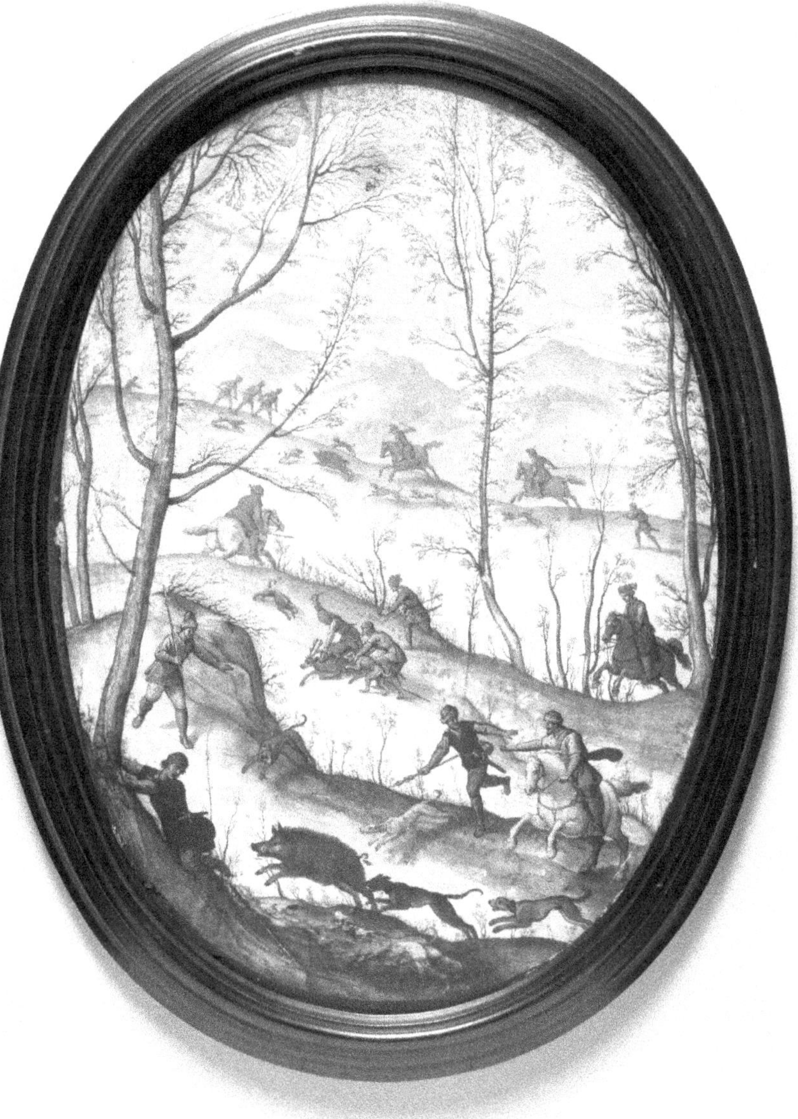

FIGURE 5.1 *Antonio Tempesta,* Landscape with Hunting on a Boar, *oil on dendrite, before 1611, Vienna, Kunsthistorisches Museum* (PHOTO: KHM-MUSEUMSVERBAND).

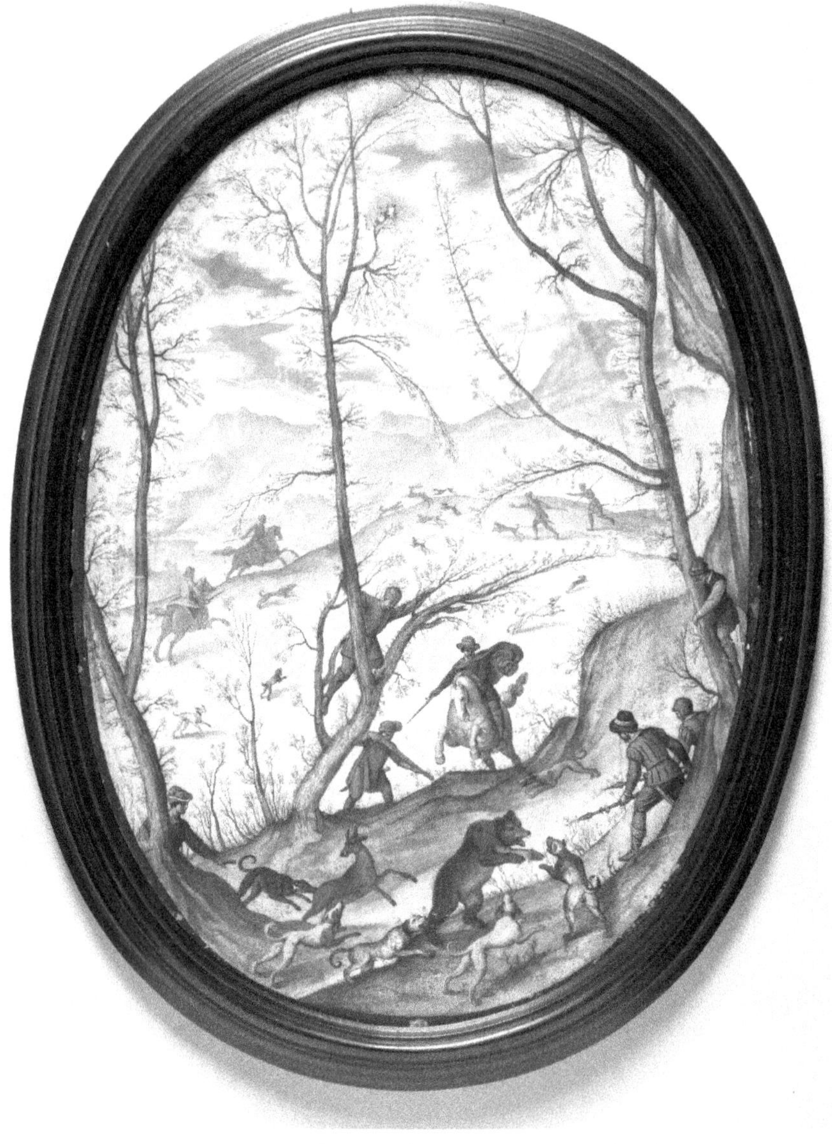

FIGURE 5.2 *Antonio Tempesta,* Landscape with Hunting on a Bear, *oil on dendrite, before 1611, Vienna, Kunsthistorisches Museum* (PHOTO: KHM-MUSEUMSVERBAND).

inlays and paintings on stone.[34] In the *Kunstkammer* inventory written by the Augsburg antiquarian and painter of miniatures, Daniel Fröschl (1563-1613), Tempesta's paintings are described as „Zwey grosse stuck stain, so von Roma kommen, darein wie bäumb oder landtschafften natürlich gewachsen und von *Anthon: Tempest* jaglandtschafften darin gemalt."[35] The inventory was begun in 1607 and updated until 1611. Thus, the paintings must have been completed before 1611 and sent to Prague soon thereafter. Rudolf Coraduz, one of the Emperor's ambassadors who, in 1595, purchased paintings for the Prague collection in Rome, may have acted as intermediary, or possibly Hans von Aachen, who also painted pictures on stone and was often in Rome, may have been instrumental in the acquisition of Tempesta's panels.[36]

In the inventory, the two paintings by Tempesta were listed in the category "Gemehld auf stain, von *alabaster, orient*: marmo und dergleichen."[37] They were stored in a chest or cupboard identified as "Truhen N° 20," together with other paintings on stone by Hans von Aachen and Joseph Heintz the Elder. The Prague court painters Hans von Aachen, Joseph Heintz the Elder, and Bartholomäus Spranger probably familiarized themselves with this technique in Italy and introduced it to the North.[38] Spranger painted *Apollo and the Nine Muses* on marble, this art work was probably also kept in the *Kunstkammer* although it is not listed among the paintings on stone in Fröschl's inventory.[39]

Paintings on stone are generally regarded as ideal *Kunst-* and *Wunderkammer* pieces because the way in which they combined *ars* and *natura* was perfectly attuned to the concept of the *Kunst-* and *Wunderkammer*. In addition to art works, natural specimens, and scientific instruments, objects that linked art and nature were especially popular *Kunst-* and *Wunderkammer* pieces. Most of Tempesta's paintings were, however, housed in Italian aristocratic collections. It follows that the categorisation of paintings on stone as typical *Kunst-* and *Wunderkammer* pieces needs to be differentiated. Natural history collections which featured a multitude of stones are often termed as *Wunderkammers*;

---

34 Rotraud Bauer and Herbert Haupt, *Das Kunstinventar Kaiser Rudolfs II. 1607-1611*, Jahrbuch der Kunsthistorischen Sammlungen in Wien, vol. 72 (Vienna, 1976).

35 Ibid., p. 140.

36 Lohff, *Malerei auf Stein*, p. 69.

37 Bauer and Haupt, *Das Kunstinventar*, pp. 3 and 139-40.

38 Eliška Fučíková, "Die Malerei am Hofe Rudolphs II," in *Prag um 1600. Kunst und Kultur am Hofe Kaiser Rudolphs II*, exhibition catalogue, Essen and Vienna (Freren, 1988), p. 188. On Hans von Aachen as a painter on stone, cf. the essay by Susanne Wegmann in this volume.

39 Yannis Hadjinicolaou, "Malerei auf Stein, steinerne Malerei," in *Paragone als Mitstreit*, ed. Joris van Gastel, et al. (Berlin, 2014), pp. 211-15.

however, they only rarely included paintings on stone. Rather, they displayed figured stones with images made by Nature.

## Tempesta's Paintings on Lapis Lazuli for Cosimo II de' Medici

In the 17th century, lapis lazuli was mined mainly in Afghanistan. It was then shipped to Venice where it was sold for prices "higher than gold."[40] The blue stone can contain either pyrite inclusions which add a golden glitter to the stone, or white or yellowish-white calcite inclusions.

An early entry for a painting on lapis lazuli by Tempesta is found in Alessandro Peretti Montalto's records: on 18 May 1615, the cardinal paid Tempesta 30 *scudi* for "una pittura fatta in doi pezzi di lapis lazzoli, cioè una Madonna assunta in cielo, et altro quando andava in Egitto."[41] Neither of the art works mentioned, an *Assumption of Mary* and a *Flight into Egypt*, can yet be identified securely.[42]

Three years later, two especially precious paintings on lapis lazuli were recorded in the collection of Cosimo II de' Medici in Florence. On 14 February 1618, Cosimo's secretary wrote to the Florentine ambassador in Rome, Piero Guicciardini, that the Florentine Piero Strozzi had presented the Grand Duke

40 Philip Poupin, "La ricerca del lapislazzuli," in *Lapislazzuli. Magia del blu*, ed. Maria Sframeli, exhibition catalogue, Florence (Florence, 2015), p. 17; on pigments made of lapis lazzuli (ultramarine) cf. Michelle O'Malley, *The Business of Art: Contracts and the Commissioning Process in Renaissance Italy* (New Haven and London, 2005), p. 66; on trading lapis lazuli cf. Anne Dunlop, "On the origin of European painting materials, real and imagined," in *The matter of art. Materials, practices, cultural logics, c. 1250-1750*, ed. Christy Anderson, et al. (Manchester, 2015), pp. 78-80.

41 Belinda Granata, "Appunti e ricerche d'archivio per il cardinal Alessandro Montalto," in *Decorazione e collezionismo a Roma nel Seicento*, ed. Francesca Cappelletti (Rome, 2003), 52; Belinda Granata, *Le passioni virtuose. Collezionismo e commitenze artistiche a Roma del cardinale Alessandro Perretto Montalto (1571-1623)* (Rome, 2012), pp. 127-28. Cardinal Peretti Montalto's collection of paintings also included a *Battle* and a painting "con faraone" by Tempesta. Granata, *Le passioni virtuose*, p. 118. Possibly the latter was a version of the *Crossing of the Red Sea* which Tempesta painted on stone several times.

42 Several paintings on lapis lazuli showing *The Flight into Egypt* are preserved until now. Cf. Lohff, *Malerei auf Stein*, p. 248; *Paintings on Stone*, no. 14. A further version which shows the Virgin with Child on the donkey held by an angel is in a private collection. The latter seems close to Tempesta's style, but non of the paintings could be attributed to an artist with certainty so far.

with a painting on lapis lazuli representing *The Pearl Divers*.[43] [**Fig. 5.3**] On 31 March 1618, the painting was located in the Guardarobe Medicea.[44] The painting depicts a rocky costal landscape peopled by a multitude of figures. The pearl fishers have rowed their boats out to sea in order to dive for the valuable pearls. In the foreground, semi-nude women remove the pearls from the shells. The huts in the rocky middle ground provide shelter and a place to trade the pearls while the background is enlivened by a plethora of miniscule figures. In this painting, Tempesta integrated the stone support into his composition. The water and sky are largely unpainted, and the white-yellow calcite inclusions in the lapis lazuli create the impression of clouds lit up by the sun. Individual details such as the dishes or the stone on the gunwale of the foremost boat are also unpainted stone.

Tempesta probably derived the exotic subject of the pearl divers from Cesare Federici's travel account *Viaggi alle Indie Orientali* which was published in 1587.[45] The Venetian jeweller describes the pearl fishery in Ceylon (modern Sri Lanka) at length, and many details from his narrative can be found again in Tempesta's painting, for instance the huts which the pearl divers erected during the harvest months, the negotiations with the traders, and the techniques of descending to the bottom of the sea with aid of a rope and of placing the shells in a basket fastened around the diver's neck.[46]

43 "Illustrissimo e Eccellentissimo mio Signore Osservandissimo, Essendo il Serenissimo padrone stato oggi fra gli altri regali presentato da monsignore Piero Strozzi d'un bel quadro di lapislazzeri figurato di pescagioni e fatto per mano del Tempesta, a Sua Altezza è piaciuto estremamente e non si è saziata di considerarlo e di lodarlo, et io ho voluto significarlo a Vostra Eccellenza, perché, intendo io ch'ella ha veduto il detto quadro, sappia in quanta stima è tenuto da Sua Altezza." Quoted after Paola Barocchi and Giovanna Gaeta Bartelà, *Collezionismo Mediceo e storia artistica*, 6 vols. (Florence, 2002), 1:154, n. 565. Cf. the important essay on this painting by Alessandro Cecchi, "Le pêche des perles aux Indes. Une peinture d'Antonio Tempesta," *La Revue du Louvre* 36, no. 1 (1986); Lohff, *Malerei auf Stein*, pp 189-90.

44 "Un quadro bello di lapislazaro dipintovi dentro a olio più figurine e paesini che pescono le perle, di mano del Tempestino, con adornamento piano e scorniciato, di piastra d'argento e dorato in parte e smaltato di più colori, con granati grossi nelli angoli delle cantonate e smeraldini e altre pietre, largo braccia 1¼ e alto braccia 7⁄8, incassato in una cassa coperta di velluto nero." Quoted after Barocchi and Gaeta Bartelà, *Collezionismo Mediceo*, 1:154, n. 565.

45 Cf. Lohff, *Malerei auf Stein*, pp. 58-59.

46 C. Federici and G. Balbi, *Viaggi di C. Federici e G. Balbi alle Indie Orientali* (Venice, 1587), ed. Olga Pinto (Rome, 1962), pp. 26-27. In this particular case, not only the subject matter of the painting but also its stone support refers to the wealth which Italy flaunted by import-

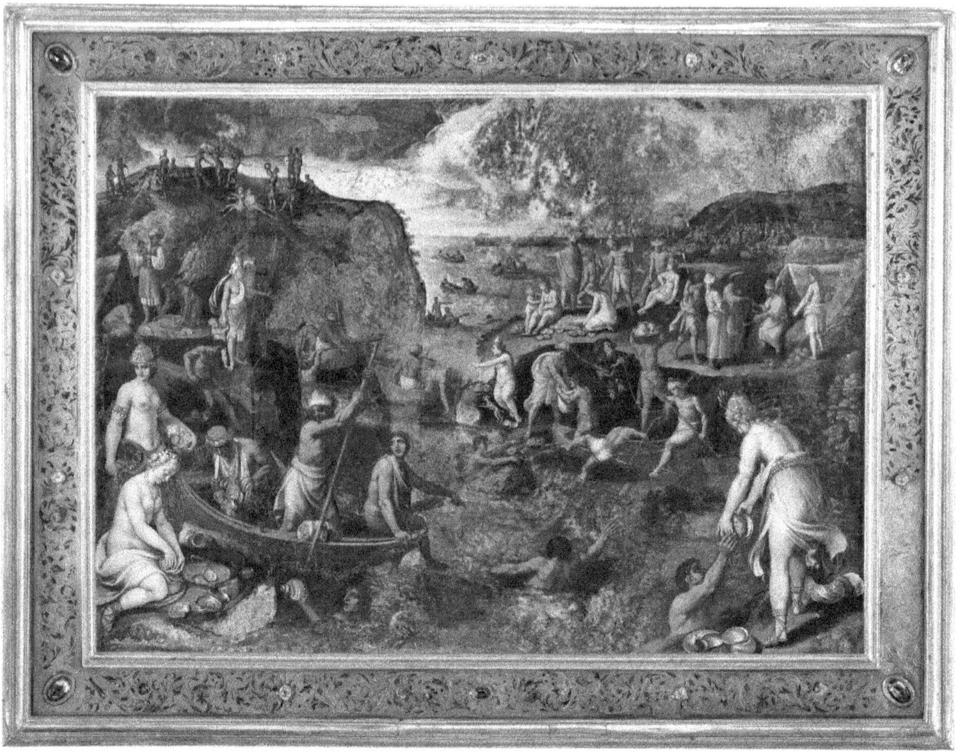

FIGURE 5.3 *Antonio Tempesta,* Pearl-fishery, *oil on lapis lazuli, before 1618, Paris, Musée du Louvre, Département des objets d'art* (PHOTO: BPK / RMN / PARIS, MUSÉE DU LOUVRE / JEAN SCHORMANS).

On 18 April 1618 another painting on lapis lazuli by Tempesta entered the Medici collection.[47] It depicts *Joseph sold by his Brothers* on one face and the *Arrest of Jesus* on the other [**Figs. 5.4-5.5**]. Again, the lapis lazuli plays an important role in the painting's composition. On both sides, the sky and the

ing precious materials from countries in the Middle East and Asia. Cf. Dunlop, "On the Origin," pp. 78-80.

47 "Un quadro di pietra in lapis-lazaro dipinto a olio da due bande, di mano del Tempestino, ch'è da una banda quando Nostro Signore fu preso ne l'orto con molte figure con li apostoli e Iuda Scariotte che lo vende e da l'altra banda è quando Ioseffe fu venduto da fratelli con più figurine e paesi, con incassatura di adornamento scorniciato da tutta dua bande d'ebano. Comesso nelli angoli e ne' mezzi di fogliami e festoni di madreperla filettati di rame dorato, con campanella e viticcio d'argento dorato." Quoted after Barocchi and Gaeta Bartelà, *Collezionismo Mediceo*, 1:154, n. 565. The painting is currently in a private collection. Cf. Lohff, *Malerei auf Stein*, p. 191, *Lapislazzuli. Magia del blu*, ed. Maria Sframeli, exhibition catalogue, Florence (Florence, 2015), pp. 344-45.

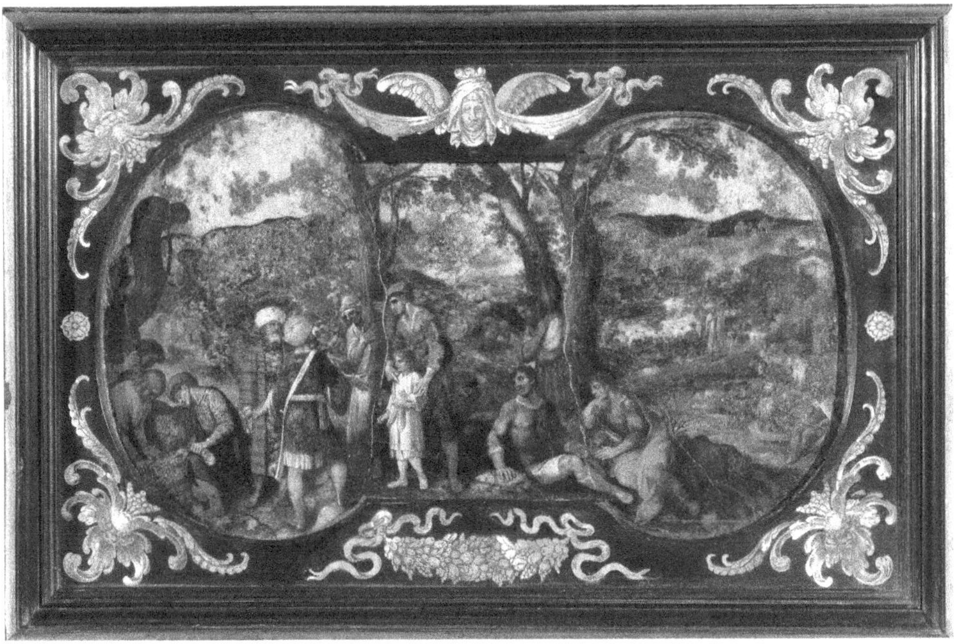

FIGURE 5.4 *Antonio Tempesta,* Joseph sold by his brothers, *oil on lapis lazuli, before 1618, private collection* (PHOTO: DI CASTRO, ROME).

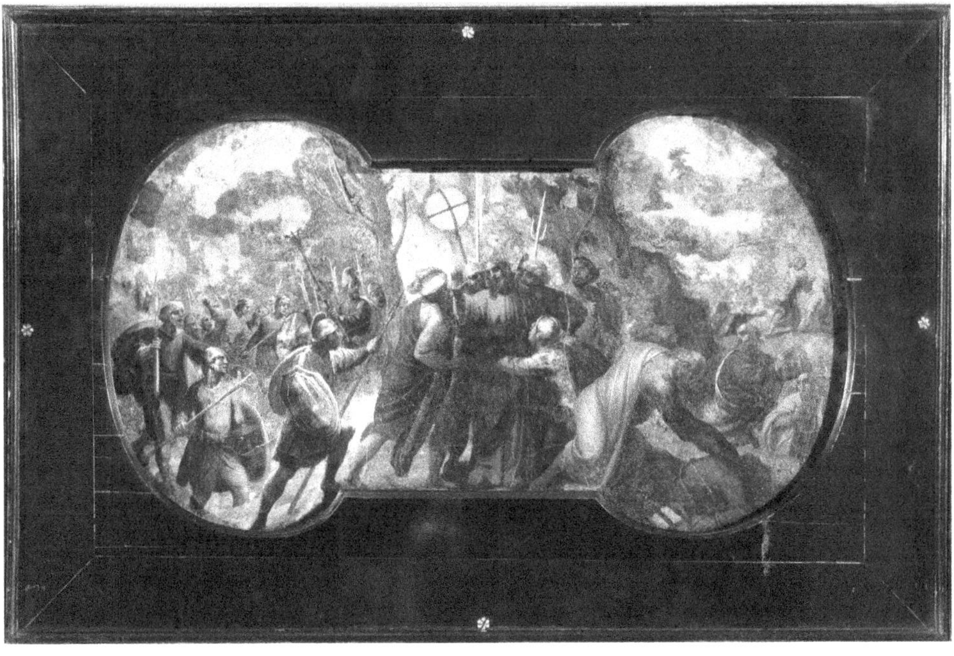

FIGURE 5.5 *Antonio Tempesta,* Capture of Christ, *oil on lapis lazuli, before 1618, private collection* (PHOTO: DI CASTRO, ROME).

background landscape is defined by the blue color of the stone. Furthermore, the lapis lazuli is integrated in the clothes of some of the figures, so that both images are characterized by the smooth transition between the painted sections and the stone.

As of 1635, both paintings are listed in the inventories of the Tribuna of the Uffizi, and they remained there until the 18th century.[48] In 1753, the panel painted on both sides with *Joseph sold by his Brothers* and the *Arrest of Jesus* is mentioned in the inventories of the Galleria, and there is evidence that it was still there in 1769. *The Pearl Divers* were still in the Tribuna in 1783; during the French occupation of Florence in 1800, the painting was probably moved to Paris, where it remains to this day.[49]

The Tribuna of the Uffizi was an ideal collection space for these two exceptional art works on lapis lazuli.[50] At Ferdinando de' Medici's behest, his architect Bernardo Buontalenti completed this octagonal room which had been begun in 1581 by the Grand Duke's brother Francesco. The architecture and decoration of the Tribuna followed a learned program dedicated to art and nature which included references to the four elements. The mother of pearl inlays in the dome symbolize water, the weather vane on top of the lantern stands for the air, and the elements of fire and water are represented by the statuettes of Apollo and Juno.[51] Apart from paintings by Raphael, Leonardo da Vinci, Titian, and many other artists, sculptures, weapons, exotica and a multitude of other precious objects, the Tribuna was also home to numerous cut stone vessels and stone inlays from the Florentine workshops. Specimens from nature could also be admired.[52] Furthermore, Cosimo II had a predilection for exotic objects and also for paintings on stone.[53] The paintings on lapis lazuli with their combination of precious material and exquisite painting were admirably suited to the context of his collection. *The Pearl Divers* were displayed

48 Cecchi, "Le pêche," p. 46; *Lapislazzuli*, pp. 344-45.

49 Antonio Tempesta, *The Pearl Divers*, before 1618, oil on lapis lazuli, 42.5 × 60 cm, Paris, Musée du Louvre, Inv. MR Suppl. 243.

50 On *The Pearl Divers* within the context of the Tribuna cf. Cecchi, "Le pêche"; Lohff, *Malerei auf Stein*, pp. 54-60.

51 Detlef Heikamp, "Zur Geschichte der Uffizien-Tribuna und die Kunstschränke in Florenz und Deutschland," *Zeitschrift für Kunstgeschichte* 26 (1963), 200-09; on the Tribuna as a collection space cf. *La tribuna del principe*. ed. Antonio Natali, et al. (Florence, 2014), pp. 218-86.

52 Anna Massinelli, "Aggiunte al tesoro della Tribuna degli Uffizi," in *La tribuna del principe*.

53 In 1618 Filippo Napoletano made several paintings on stone, mostly *pietra paesina*, for Cosimo II. Marco Chiarini, *Teodoro Filippo di Liagno detto Filippo Napoletano. 1589-1626. Vita e opere* (Florence, 2007), pp. 105-09.

between a "Madre Perla grandetta" and a *Crucifixion* painted on wood.[54] In the collection housed in the Tribuna, references to shells and precious mother of pearl played an outstanding role, all the more so as pearls were a recurrent symbol of the Medici brides.[55] Thus, Tempesta's painting stood in a direct relationship not only to the object next to it but also to the collection space with its mother of pearl dome as a whole.

## Paintings on Stone by Antonio Tempesta in the Borghese Collection

It is probable that during the pontificate of Paul V (1605-21) Tempesta painted pictures on stone for the pope or for his cardinal nephew Scipione Borghese (1576-1633). Nonetheless, the earliest reference to paintings on stone by Tempesta in the Borghese collection dates from 1693.[56] They were located in the "Stanza passata della Galleria" in the city palace on the Campo Marzio, where approximately 30 paintings on stone, including slate, *pietra paesina*, jasper, agate, amethyst, and lapis lazuli, were on display. The person who wrote the inventory associated Tempesta with the practice so closely that he attributed to him everything painted on patterned stone in small scale. These include a *Crossing of the Red Sea* and a *Calling of Peter* on *pietra paesina*.[57] At a later point in time, two further paintings from the Borghese Collection were attributed to the artist: a *Storming of a City* (*Capture of Jerusalem*) on *pietra paesina*[58]

54 Cecchi, "Le pêche," p. 46; Massinelli, "Aggiunte al tesoro," p. 211.

55 Fabian Jonietz, "Le conchiglie della Tribuna," in *La tribuna del principe*, p. 193.

56 The inventory is published in Paola Della Pergola, "L'inventario Borghese del 1693 (III)," *Arte Antica e Moderna* 30 (1965), 202-17.

57 Antonio Tempesta?, *The Crossing of the Red Sea*, oil on *pietra paesina*, 15 × 33 cm, Rome, Galleria Borghese, inv. 501; Antonio Tempesta?, *The Calling of Peter*, oil on *pietra paesina*, 14 × 32 cm, Rome, Galleria Borghese, inv. 497. On the attribution cf. Lohff, *Malerei auf Stein*, pp. 198-99.

58 This painting was described by Giacomo Manilli: "In quello d'alabastro dove la natura hà senza aiuto dell'arte formata una Città, si vede la presa dell'istessa Città fatta à forza da un'esercito Cristiano contro i Turchi: & hà forse il Pittore voluto esprimere la presa di Gierusalemme, descritta dal Tasso." Giacomo Manilli, *Villa Borghese fuori di Porta Pinciana descritta da Iacomo Manilli Romano. Guardaroba di detta Villa* (Rome, 1650), p. 113. The painting is, as a rule, attributed to Tempesta (most recently by Hermann Fiore, cf. *L'arme e gli amori. La Poesia di Ariosto, Tasso e Guarini nell'arte Fiorentina dei Seicento*, ed. Elena Fumagalli, Massimiliano, Rossi, and Riccardo Spinelli, exhibition catalogue, Florence [Florence, 2001]). However, it is similar in style to paintings of the same subject matter on *pietra paesina* by Filippo Napoletano. On the attribution cf. Lohff, *Malerei auf Stein*, pp. 197-98.

and an *Adoration of the Magi* on alabaster.[59] Paintings on lapis lazuli attributed to Tempesta formed part of the Borghese collection, too.

Tempesta's *Hunt* on lapis lazuli mentioned in the Borghese Inventory of 1693 might have been preserved to this day.[60] [**Fig. 5.6**] The painting that can be most convincingly attributed to Tempesta shows a lakeside landscape with nobles on horseback that pursue a stag in the foreground and a bear in the middle ground, while hunters on foot armed with spears and their pack of hounds hold the animals at bay. Once again, the stone itself is used to represent the sky and water. On occasion, the stone and painted areas blend together, for instance in the small waterfalls.

In the Borghese inventory, apart from this horizontal oval panel three other paintings on lapis are mentioned, a *Baptism of Christ, Christ's Entry into Jerusalem* and *Christ with his Apostles*. These are also assigned to Tempesta.[61] The last-mentioned picture is still extant, provided that the work that has come down to us is correctly identified with the painting mentioned in the inventory.[62]

By way of a hypothesis, it can be assumed that the picture supports used by Tempesta came from the Pope's stock of lapis lazuli. Paul v had procured a vast amount of this precious stone for the decoration of the Cappella Paolina in Santa Maria Maggiore. A venerated image of the Blessed Virgin Mary called *Salus Popoli Romani* graced the chapel's altar. The painting is surrounded by a frame studded with jewels and is borne aloft by gilded bronze angels. The inlays were executed entirely in lapis lazuli symbolizing the air into which the

59 The painting is mentioned in the inventory of 1693 but the name of the artist is not given; it is attributed to Tempesta for the first time in the inventory of 1790. Cf. Paola Della Pergola, *Galleria Borghese. I dipinti*, 2 vols. (Rome, 1955-59), 2:55-56; Lohff, *Malerei auf Stein*, pp. 183-84.

60 "562) Sotto al detto un quadro di lapislazzaro ovato tre quarti di altezza in circa con una Caccia dentro del N° 208 seg.to dietro cornice negra del Tempesta [...]" Paola Della Pergola, "L'inventario Borghese del 1693 ((already mentioned p. 157, n. 56)), 208; *Lapislazzuli*, p. 346.

61 "584-85) Sotto a detto due Quadrucci ovati longhi un palmo e mezzo incirca, in uno il Battesimo al Giordano, l'altro quando Nro Sig.re entra in Gerusalemme sopra un asino depinti sopra lapislazzoli con cornice attorno di rame dorata del N° [sic] del Tempesta [...] 608) Sotto all'altro di d.ti quadrucci un quadro ovato alto mezzo palmo di lapis con Christo con gli Apostoli tre altre figure in una Barca. Cornice a frontespizio nera d'ebbano con rami dorati del N° [sic] del Tempesta." Della Pergola, "L'inventario Borghese," pp. 209-10.

62 Antonio Tempesta (?), The Calling of St. Peter, oil on lapis lazuli, oval, 16.5 × 32.7 cm, Private Collection. Lohff, *Malerei auf Stein*, p. 193.

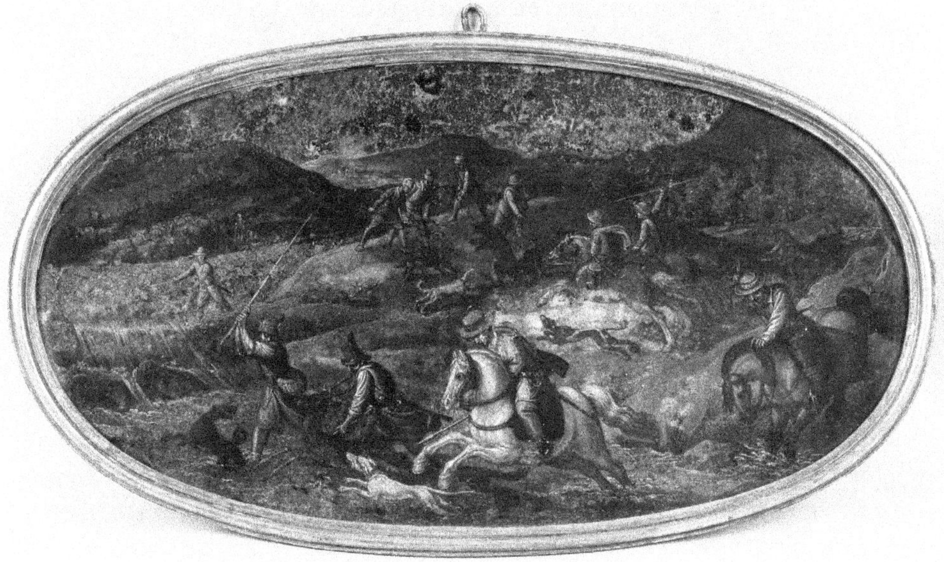

FIGURE 5.6 *Antonio Tempesta,* Hunting Scene, *oil on lapis lazuli, Rome, Collezione Di Castro* (PHOTO: DI CASTRO, ROME).

painting rises.[63] Pier Vincenzo Strozzi, who had given the *Pearl Divers* to Cosimo II as a gift in 1618, was employed as by Paul V and entrusted with the supervision of the altar's construction.[64] Circa 1617 he wrote a memorandum in which he described how the lapis lazuli had been used to decorate the chapel in order to defend himself against charges that he had misappropriated the precious stone.[65] The document reveals that a quantity of the stone was used as picture supports.[66] It is, therefore, possible that the lapis lazuli picture supports (including the relatively large panel for *The Pearl Divers*) used by Tempesta may have come from the stock of stone that was intended for the altar of the Cappella Paolina.

63 Maria Cristina Dorati, "Gli scultori della Cappella Paolina di P. Maria Maggiore," *Commentari* 18 (1967), 234.

64 Cf. Lohff, *Malerei auf Stein*, pp. 54-57.

65 Published by Isabelle Bigazzi, ed., *Un memoriale di Piero di Vincenzo Strozzi* (Florence, 1986).

66 Bigazzi, *Un memoriale*, pp. 46-47.

## Tempesta's Paintings on Stone for the Barberini Collection

By the beginning of the pontificate of Urban VIII (Maffeo Barberini) in 1623, if not earlier, Tempesta also executed paintings on stone for the Barberini family. As Taddeo and Antonio Barberini's drawing teacher, Tempesta was in fairly close contact to the papal *famiglia*.[67] In the course of Urban VIII's endeavours to negotiate peace between France and Spain, a painting on stone by Tempesta even served as a diplomatic gift. From 1624 onwards, the papal ambassador Giulio Sacchetti resided in Madrid in order to prepare the peace negotiations between the two powers.[68] In 1626 Francesco Barberini, whom Urban VIII had appointed as cardinal legate the year before, left for Spain on a diplomatic mission.[69] There is a long tradition of giving art works as diplomatic gifts, and paintings on stone were also used as such.[70] The presents which Sacchetti took with him to Madrid included a *Deluge* by Tempesta painted on *pietra alberese*. On the occasion of Francesco Barberini's visit to the convent of the church Santissima Trinidad on 21 June 1626, the Roman antiquarian Cassiano Dal Pozzo, who accompanied the legate on his journey, noted: "Vi si vidde trà altre pitture un quadro del Tempesta, che era un diluvio fatto in pietra albaresa con un'infinità di figurine, che fù donato nel principio della nunciatura del Sacchetti a S.M.tà [...]".[71] The convent was demolished in the late 17th century and Tempesta's *Deluge* on *pietra alberese* has hitherto not been traced.[72]

Works on stone by Tempesta are also found in the collections belonging to members of the Barberini family. According to Francesco Barberini's accounting books, Tempesta was paid 40 *scudi* on 7 May 1624 for a battle painting and an "adoratz.e di Magi."[73] Antonio Vannugli has linked an *Adoration of the*

67 Leuschner, *Antonio Tempesta*, p. 500.

68 Sebastian Schütze and Enrico Colle, "I Sacchetti. Forme e funzioni, strategie ed intenzioni del mecenatismo. Un'introduzione," in *Palazzo Sacchetti*, ed. Sebastian Schütze (Rome, 2003), pp. 29-30

69 Cassiano Dal Pozzo kept a diary during this journey. Published by Alessandra Anselmi, ed., *Il diario del viaggio in Spagna del cardinale Francesco Barberini scritto da Cassiano dal Pozzo* (Aranjuez, 2004).

70 Johanna Beate Lohff, "Bilder auf Stein als Diplomatengeschenke und ihre Verbreitung in Europa durch den diplomatischen Verkehr," in *Materielle Grundlagen der Diplomatie. Schenken, Sammeln und Verhandeln in Spätmittelalter und Früher Neuzeit*, ed. Mark Häberlein and Christof Jeggle, Irseer Schriften N.F., vol. 9 (Constance, 2012).

71 Anselmi, ed., *Il Diario*, p. 165.

72 Anselmi, *Il Diario*, p. 164, n. 362.

73 "330. A di detto [7 May 1624] sc. 40 m.ta in cr'o a Paoli con n'ro mandato n.o 196 al s.r. *Antonio Tempesta* per due quadretti in pietra cioè una battaglia, et una adoratz.e di

*Magi* on alabaster signed by Tempesta with the painting mentioned in the inventory.[74] Tempesta executed several versions of the *Adoration of the Magi* on alabaster;[75] in each painting a large cloud that reveals the stone support hovers above the scene of worship.

Further paintings on stone by Tempesta were to enter the Barberini collection. Francesco Barberini acquired a picture by Tempesta on *pietra alberese*, probably dendrite, from the estate of Cardinal del Monte. Probably the support was already broken at the time of the purchase.[76] On 26 October 1624, Taddeo Barberini paid a member of the Sacchetti family "100 m[one]ta in oro" for a lapis lazuli panel painted on both sides with a "creazione del mondo" and a "sommersa di faraone".[77] Francesco Barberini's inventory of 1644 mentions two paintings with a *Boar Hunt* and a "huomo ignudo giacente," and the collection built by Urban VIII still included a *Hunt* on stone by Tempesta in 1686, many years after the Pope's death.[78]

Magi—40." Marilyn Aronberg Lavin, *Seventeenth-Century Barberini Documents and Inventories of Art* (New York, 1975),p. 41.

74 Antonio Vannugli, "Antonio Tempesta. Un retablo portátil en la catedral de Segovia y otras pinturas sobre piedra," in *In Sapientia libertas. Escritos en homenaje al profesor Alfonso E. Pérez Sánchez* (Madrid, 2007), p. 233. Antonio Tempesta, *Adoration of the Magi*, oil on alabaster, 45 × 60 cm, New York, Stanley Moss & Co. Inc.

75 Lohff, *Malerei auf Stein*, pp. 182-84.

76 "Seguono gli quadri compri dall'Heredi del S.r Card.le del Monte [...] 350)—Sette paesini, Sei de quali sono in pietra di rossezzano, ch'e' macchiata a' modo di Città, e fabbriche con historiette di variate di figurini di guerra mano del med.mo con cornici di pero nero. Il 7.mo e pietra albarese Spaccata per mezzo depinta dal Tempesta–stimati, e ridotti sc. 40." Aronberg Lavin, *Seventeenth-Century Barberini Documents*, p. 90. Cf. also Aronberg Lavin, *Seventeenth-Century Barberini Documents*, p. 110.

77 "100 m[one].ta in oro a Sacchetti con mio m.to 169 al Ant Tempesta per un quadro ovale dipinto da due bande in pietra lapis lazzaris da una La creazione del Mondo, dall'altra la sommersa di faraone." Quoted after Eckhard Leuschner, "Antonio Tempesta's Drawings for a 'Gerusalemme liberata'," *Masterdrawings* 37, no. 2 (1999), 152, n. 18.

78 "614) Due Quadretti con cornice d'Ebano figure in pietra uno con una Caccia d'un Cervio e l'altro con un huomo ignudo giacente longhi l'uno poco meno d'un palmo." (Inventory of Francesco Barberini's collection, 1644) Aronberg Lavin, *Seventeenth-Century Barberini Documents*, p. 239. "515) Un quadretto di pietra, con una Caccia, lunga quasi p.i 1, e alto ¾ incirca, con cornice di pero mano del *Tempesta*," (Inventory of Urban VIII's collection, 1686) Aronberg Lavin, *Seventeenth-Century Barberini Documents*, p. 416.

## Tempesta's Paintings on Stone in Other Collections

Many of Tempesta's paintings on stone were first documented only after the artist's death: collection inventories are an important source of information for these works.[79] Presumably some of the paintings already entered these collections during the artist's lifetime, but were documented only at a later point in time. Furthermore, from the mid-17th century onwards collections were documented in greater detail.[80] Increasingly, artists were consulted to ensure that the art works were professionally appraised.

This was the case, for instance, with the Collection of the House of Savoy in Turin. From 1635 onwards, the painter Antonio della Corgna (1584-1654) was entrusted with the inventory of the Ducal collection. He assessed the paintings by Tempesta, a "caccia di leoni" and a "sommersione di faraone" as "buono."[81] Already in 1631, a "caccia," which is still in Turin today, was listed as part of the Collection.[82] In contrast to the other hunting scenes by Tempesta, the Savoy painting features a mythological scene, namely the *Caledonian Boar Hunt*.

The inventory of the estate of the Roman Cardinal Vincenzo Giustiniani dated 1638 includes a *Conversion of Saul* on stone.[83] Carlo Antonio Dal Pozzo's collection included two battle scenes, which the writer of the inventory attributed to Tempesta.[84] Carlo Antonio was the younger brother and heir of Cassiano Dal Pozzo, to whose collection of drawings of ancient art works Tempesta had contributed.[85] It is thus possible that the painting originally

79 The following study is based mainly on inventories of the 17th century.

80 Patrizia Cavazzini, *Painting as Business in Early Seventeenth-Century Rome* (University Park, 2008), pp. 81-119.

81 "411. Caccia di leoni, ovato in pietra. Del Tempesta. Buono. A.p.0.7; l.p.0,9." and "790. Sommersione di Faraone, in pietra ovata. Del Tempesta. Buono. A. p. 0.7; L.p. 0.9." Alessandro Baudi di Vesme, *La Regia Pinacoteca di Torino* (Rome, 1897), pp. 51, 65.

82 "Caccia del Tempesta in ovato sopra pietra arboraria, cornici smaltate, long.on.10, alt 9." Giuseppe Campori, *Raccolta di cataloghi ed inventari inediti di quadri* (Modena, 1870), p. 88. Antonio Tempesta, *Caledonian Boar Hunt*, oil on dendrite, 27 × 36 cm, Turin, Galleria Sabauda, inv. 601. Cf. Lohff, *Malerei auf Stein*, p. 195.

83 "244. Un quadretto picciolo di pietra con l'historia della Conversione di S. Paolo di grandezza d'un palmo in circa di mano del Gobbo Tempesta. Con cornice negra." Luigi Salerno, "The Picture Gallery of Vincenzo Giustiniani (II)," *Burlington Magazine* 102 (1960), 104

84 "[301-302] Due quadretti in pietra di battaglie si credono del Tempesta con cornici nere filettate d'oro." (Inventory of the estate of Carlo Antonio Dal Pozzo, 1689), Donatella Livia Sparti, "The dal Pozzo Collection Again. The Inventories of 1689 and 1695 and the Family Archive," *Burlington Magazine*, 132, no. 1049 (1990), 558.

85 Ingo Herklotz, *Cassiano dal Pozzo und die Archäologie des 17. Jahrhunderts* (Munich, 1999), p. 404, n. 4.

formed part of Cassiano Dal Pozzo's collection and only entered that of Carlo Antonio after the former's death. Amadeo Dal Pozzo, Cassiano's cousin in Piemonte, owned two paintings on stone by Tempesta in his considerably collection, a *St. Eustache* on lapis lazuli and a *Battle* on a fine stone.[86] The inventory of the estate of the Roman aristocrat Massimo de' Massimi dated 25 June 1644 mentions a *Landscape with Animals* by Tempesta.[87] In Naples, a "Battaglia" is documented in Partengo Partengba's collection.[88] Lastly, the collection of the Roman noble Lelio Orsini lists, in 1696, "... una caccia di Leoni e tigri..." and a *Deluge* by Tempesta, which was kept in a jewellery box.[89]

Tempesta probably also made other small-scale paintings on horizontal oval lapis lazuli panels, as is suggested by two such paintings which belonged to the collection of the Gonzaga di Novellara in the 18th century and are currently attributed to Tempesta.[90] A *Crucifixion of Peter* on lapis lazuli, which does not seem to have survived, also belonged the Novellara Collection.[91]

While inventories become more detailed in the middle of the 17th century, an increasing uncertainty on the part of their authors regarding the works of

---

86 "595. Tempesta. Un S. Eustachio, sopra un lapis lazuli, ovato opera del Tempesta largo mezzo palmo con cornice debano. D 15. [...] 507. Una battaglia del Tempesta sopra la pietra diligientissima con cornise d'ebano alto un palmo, e larga 1 ½. D 50." Arabella Cifani/ Franco Monetti, "'L'Illustrissimo cugino'. Cassiano e Amadeo dal Pozzo. Le relazioni artistiche del Marchese di Voghera e la storia della sua quadreria," in *I segreti di un collezionista*, ed. Francesco Solinas (Rome, 2001), p. 50.

87 "Il Camerino delle Colonne / li altri dui sono in pietra, uno è un paesetto con diversi animali del Tempesta e l'altro e Cristo, che aiuta S. Pietro." Quoted after Lothar Sickel, *Caravaggios Rom. Annäherungen an ein dissonantes Milieu* (Emsdetten and Berlin, 2003), pp. 245-46.

88 "Una battaglia di Tempesta sopra pietra con cornice d'oro," Gérard Labrot, *Collections of Paintings in Naples 1600-1780* (Munich, 1982). p. 60.

89 "[89] Un altro quadro ovato di pietra di grandezza due palmi incirca con grandezza d'Ebano dipintovi una Caccia di leoni e tigri con cornice d'Ebano violato [Antonio Tempesta] [...] [392] Un ottavo in pietra dentro una scattola dipintovi il diluvio portato nel Cantanaro esiste nella stanza dove dormiva il S.r.P.n.pe." Adriano Amendola, *La collezione del principe Lelio Orsini nel palazzo di piazza Navona a Roma* (Rome, 2013), pp. 109-10, 131." Both paintings were sold to the Spanish ambassador Juan Francisco Pacheco IV Duque de Uceda. Amendola, *La collezione del principe*, p. 72. Cf. Gisela Rubsamen, *The Orsini Inventories* (Malibu, 1982), p. 25.

90 "Antonio Tempesta Fiorentino. Due ovatini per illungo in lapislazzuli con cornice intagliata e dorata, una presenta il bambino Mosè ritrovato nel Nilo dalla figlia di Faraone, l'altro, un altro fatto Mosè, in tutto zecchini 22." Campori, *Raccolta*, p. 658 ; *Lapislazzuli*, p. 345.

91 "Antonio Tempesta. La crocifissione di S. Pietro dipinta sopra lapilazuli, con cornice dorata, zecchini 25," Campori, *Raccolta*, p. 665.

deceased artists can also be observed. It cannot, therefore, be excluded that already towards the end of the 17th century some paintings on stone were erroneously attributed to Tempesta. Conversely, there must also have been numerous paintings on stone by Tempesta in different collections which were not described as such, as is evinced by the paintings on stone which are nowadays attributed to Tempesta but whose provenance cannot be reconstructed with precision.

## Three Remarkable but Little-known Paintings on Stone

In the following section, I will focus on three paintings on alabaster that are generally attributed to Tempesta but have hardly been taken into consideration by modern scholarship:[92] a *Gathering of the Manna*, *Joseph Explaining his Dream to his Brothers* and *The Crossing of the Red Sea*.[93] Today, two of them are on display in the Boston Museum of Fine Arts, the third, *The Crossing of the Red Sea*, is in a private collection. Probably the Boston paintings were created as three related paintings, as is suggested by their common provenance, their almost identical dimensions, the texture of the alabaster, and the relatively large group of figures in the foreground of each painting. *The Crossing of the Red Sea* has similar dimensions and the same provenance back to 1810; however, the painting and the stone support are slightly different, e.g. in the texture of the panel in the foreground, so that it remains a matter of debate whether this painting belongs to the other two now in Boston.

92 These paintings are not included in Lohff, *Malerei auf Stein*. For the present essay I did not have the opportunity to examine the originals. Attributions cannot be made with certainty.

93 Antonio Tempesta, *Gathering of the Manna*, oil on alabaster, 43 × 57 cm, Museum of Fine Arts, Boston, inv. 2012.120; Antonio Tempesta, *Joseph Explaining his Dream to his Brothers*, oil on alabaster, 42.5 × 54.5 cm, Museum of Fine Arts, Boston, inv. 2012.121. Antonio Tempesta, *The Crossing of the Red Sea*, oil on alabaster, 45.1 × 58,.7 cm, Private Collection. Provenance of the Boston Paintings: "Martin Farqhuar Tupper (1810-1889), London. 1898, sold by J.D. Ichenhauser, London, to William Cornelius van Horne (1843-1915), Montreal; until at least 1946, with the estate of William van Horne, Montreal. November 25, 1991, anonymous sale, Waddington's, Toronto, lot 2030. 1994, sold by Stanley Moss and Co., Riverdale-on-Hudson, NY, to Joseph Koerner, London, and Cambridge, MA; 2012, gift of Joseph and Margaret Koerner to the MFA." <http://www.mfa.org/collections/object/gathering-of-the-manna-558738> (8.1.2016); <http://www.mfa.org/collections/object/joseph-explaining-his-dream-to-his-brothers-558739> (8.1.2016). Until 1994, the provenance of the *Crossing of the Red Sea* is identical to that of the two paintings in Boston. I would like to thank Stanley Moss and Robert B. Simon, New York for their kind assistance.

*Joseph Explaining his Dream to his Brothers* [**Plate 20**] is based on the Old Testament story in which Joseph dreams that the sun, moon and, eleven stars bow down to him (Gen. 37:5). The painting shows Joseph as a child surrounded by his elder brothers and pointing at the heavenly bodies in the sky. The group is framed by a dog on the left and by two horses on the right, giving the picture a genre-like character. The background landscape is largely based on the structure of the stone. Mountain crags with goats grazing on the slopes rise up on the right. The landscape opens to the left, providing a view of the brothers' herds of sheep and goats. Two dromedaries add an exotic touch to the scene. The mountains and hills are given relief by shading and are rendered more tangible by groups of trees. The color of the landscape and sky is defined by the ochre-yellow tone of the alabaster. An area of circular patterning depicts the sun and the porous structure of the alabaster represents the stars. Thus, the composition is perfectly integrated into the natural appearance of the stone, and key elements of the biblical narrative are defined by the stone, i.e. by Nature herself.

The *Gathering of the Manna* [**Plate 21**] is likewise painted on a white- to ochre-colored alabaster panel, and again the composition of the painting is adapted to the patterning of the stone.[94] Moses stands on the left and points to the miracle of the manna with his staff (Ex. 16:1-36) while turning back to face another man, probably his brother Aaron.

The panel is composed of white, seemingly porous alabaster with dark inclusions around a central patch of a beige hue. The latter shimmers through the tent on the left and the horses in the center of the composition. The white alabaster is used to represent the manna, which is described as "a fine, flake-like thing, fine as frost on the ground" (Ex. 16:15). The stone's structure is also used to depict the clouds; the only intervention by the artist is the addition of the figure of God the Father. Thus, here again, the stone participates significantly in the narrative: the manna and the cloud surrounding the Lord are represented purely by the picture support in its natural state. And thus, the most important elements of the narrative are partly constructed through the stone.

---

94 Tempesta created several prints and a fresco depicting the Gathering of the Manna. Eckhard Leuschner, ed., The Illustrated Bartsch. vol. 35, no. 1: Commentary. (New York, 2007), p. 87, no. 141; Maria Vittoria Brugnoli, "I primi affreschi nel Palazzo Bassano di Sutri," *Bollettino d'Arte* 42 (1957), p. 245, fig. 7. Furthermore, there is a drawing with this subject that is attributed to him. Antonio Tempesta (?): *The Gathering of the Manna*, pencil on blue-grey paper, 11.4 × 14.5 cm, Stuttgart, Staatsgalerie Stuttgart, inv. C 90 / 3731b. Cf. Corinna Höper, *Italienische Zeichnungen 1500-1800. Bestandskatalog*, part 2 (Stuttgart, 1992), p. 122, no. 198.

This also holds true for the *Crossing of the Red Sea* [**Plate 22**]. Tempesta painted several versions of the theme on alabaster and red marble.[95] In contrast to those in the present painting, the figures are concentrated into little groups. The drowning Egyptian soldiers are depicted in the foreground. On the shore, a craggy rock formation provides a backdrop for a group of five women making music, including Miriam who plays the tambourine. As in other versions of this theme by Tempesta, the crossing of the Red Sea (Ex: 14:1-13) and Miriam's song (Ex. 15) are shown in a single scene. However, in this painting much greater attention is lavished on the music-making women than in the comparative pieces. A further group of figures is represented on the left side of the picture and on the right a multitude of Israelites move on towards the Promised Land with their horses and camels. As in the Boston paintings, the patterning of the stone defines the landscape. This holds true for the waves in the foreground as well as for the background which is characterized by a smooth, yellow-beige surface quite different from the lively veining of the water. The rocks are subtly modeled by enhancing the stone's patterning by means of brown shadows and by painting in shrubbery; the sky and mountains are painted blue. A special feature of the stone was used in the cloud at the top left: a light-colored, circular patch is surrounded by grained stone, whereby the picture support itself creates the impression of the sun shining through clouds.

In the paintings, iconographically important motifs are represented by the stone itself and there are fluid transitions between the stone and the painted areas. In these cases the artist uses the stone meaningfully to highlight important aspects of the iconography. The fluid transitions between painting and patterning is a signature trait of this particular artist. As this handling of the mineral picture support is characteristic of art works that can be attributed to Tempesta with certainty, it seems reasonable to assume that this artist also painted these three fine panels on alabaster.

## Figured Stones as Picture Supports

The ability of Nature to create images that could be perfected by artists is first described in art-theoretical texts. At the beginning of his treatise *De Statua*, Leon Battista Alberti comments on the relationship between shapes found in natural materials and the human imagination:

95 For further versions of the *Crossing of the Red Sea* by Tempesta cf. Lohff, *Malerei auf Stein*, pp. 180-82, 186-88.

> I believe that the arts of those who attempt to create images and likenesses from bodies produced by Nature, originated in the following way: They probably occasionally observed in a tree-trunk or clod of earth and other similar inanimate objects certain outlines in which, with slight alterations, something very similar to the real faces of Nature was represented. They began, therefore, by diligently observing and studying such things, to try to see whether they could not add, take away or otherwise supply whatever seemed lacking to effect and complete the true likeness."[96]

Alberti refers to this human ability to see pictures in natural shapes as the key prerequisite for artistic activity in the broadest sense. Oskar Bätschmann argues that Alberti thought that the picture was already extant in Nature, that is was an *inventum*, a natural picture that the artist found, not an *inventio*, an invention by the artist, based on his creativity and imagination.[97] This interpretation is supported by the fact that Alberti refers to a passage from Pliny the Elder concerning the gem of Pyrrhus: "Indeed it is evident that Nature herself delights in painting, for we observe she often fashions in marble hippocentaurs and bearded faces of kings. It is also said that in a gem owned by Pyrrhus that nine Muses were clearly depicted by Nature complete with their insignia."[98]

According to Alberti, Nature possessed the ability to express herself artistically as a painter. Thus, stones in which a pictorial representation can be recognized are images made by Nature. There is a fundamental difference between these and the images made by chance which Horst W. Janson discussed in his essay *The Image made by Chance in Renaissance Thought* (1961).[99] In 1973 Janson revised his concept of the 'image made by chance' and distinguished between images that were wrought by the "divine will personified under such names as Fate, Fortune, or Nature" and "chance images ... made only in the beholder's imagination."[100]

96 Leon Battista Alberti, *On painting and On sculpture: the latin texts of* De pictura *and* De statua, ed. and trans. Cecil Grayson (London, 1972), p. 121.

97 Leon Battista Alberti *Das Standbild. Die Malkunst. Grundlagen der Malerei. De statua. De pictura. Elementa picturae*, ed. Oskar Bätschmann and Christoph Schäublin (Darmstadt: Wissenschaftliche Buchgesellschaft, 2000), pp. 31-32.

98 Alberti, *On painting*, p. 65.

99 Horst W. Janson, "The Image Made by Chance in Renaissance Thought," *De Artibus opuscula XL. Essays in Honor of Erwin Panofsky*, ed. Millard Meiss (1961).

100 Horst W. Janson, "Chance images," in *The Dictionary of the History of Ideas. Studies of Selected Pivotal Ideas*, vol. 1, ed. Philip P. Wiener (New York, 1973), p. 340.

The ability of the human imagination to recognize images in abstract shapes such as the patterning of stone is of central importance for the art of painting on stone. Leonardo da Vinci's passage on the "Method of Awakening the Mind to a Variety of Inventions" from his treatise on painting has repeatedly been discussed with regard to paintings on stone:

> I will not omit to introduce among these precepts a new kind of speculative invention, which though apparently trifling, and almost laughable, is nevertheless of great utility in assisting the genius to find variety for composition. By looking attentively at old and smeared walls, or stones and veined marble of various colors, you may fancy that you see in them several compositions, landscapes, battles, figures in quick motion, strange countenances, and dresses, with an infinity of other objects. By these confused lines the inventive genius is excited to new exertions.[101]

As relevant as this may seem to a discussion of paintings on figured stone, the passage does not shed light upon the meaning and material iconography of the mineral picture supports.[102]

A richer source of information on the meaning of figured stones, i.e. stones whose natural veining forms images, is provided by texts concerned with the theory of nature. Medieval treatises listed the different types of stones alphabetically and ascribed magical or medicinal powers to them.[103] In the 16th century, naturalists began searching for a different way of ordering stones. One of the first scientists who chose to order stones according to their visual properties was the Swiss doctor and naturalist Conrad Gesner (1516-65).[104] In his treatise *De rerum fossilium genere ...* (1565), he classified them based on their exterior shape and their patterning. He first listed stones with simple patterns such as geometrical shapes, moving on to stones containing increasingly complex pictures such as fishes and other sea creatures, which are known as fossils

101 Leonardo da Vinci, *Treatise on Painting*, rev. ed., trans. John Francis Rigaud (London: G. Bell and Sons, 1877), 62; Jurgis Baltrušaitis, *Imaginäre Realitäten. Fiktion und Illusion als produktive Kraft. Tierphysiognomik, Bilder im Stein, Waldarchitektur, Illusionsgärten* (Cologne, 1984), p. 55; Seifertová, *Malby na kameni*, p. 101-02; Collomb, *Spendeurs d'Italie*, 31.

102 Cf. Lohff, *Malerei auf Stein*, pp. 104-09.

103 Cf. Albertus Magnus, *Book of Minerals*, trans. Dorothy Wyckoff (Oxford, 1967), pp. 68-126; Camillo Leonardi, *Speculum Lapidum...* (Venice, 1502).

104 Angela Fischel, *Natur im Bild. Zeichnung und Naturerkenntnis bei Conrad Gessner und Ulisse Aldrovandi* (Berlin, 2009), pp 28-34; Lohff, *Malerei auf Stein*, pp 123-26.

today.[105] In his dedicatory letter to Andreas Schadcovius of Cracow, Gesner pointed out that figures could not only be seen in painting, which was a liberal art practiced by man and deserved admiration, but that Nature too had imprinted images and figures in various things (including stone).[106] In addition to the minerals themselves, he also described art works made of stone and discussed both objects which were created by Nature but resembled works of art and artefacts in stone which referred to things made by Nature.[107] Gesner's new method of classification was based on visual criteria; stones were ordered according to their images, and a clear distinction between art works and natural objects was not made.

One of Gesners followers was Ulisse Aldrovandi (1522-1605), professor of natural philosophy in Bologna. The fourth and last book of his treatise *De rerum metallicum* deals with stones; containing 95 chapters it is the longest section of the text.[108] Many of the stones were illustrated, and Aldrovandi described their appearance on the basis, among other things, of the pictures that one could see in the stones. It is striking that those stones that were also used as picture supports, such as jasper, agate, marble, and alabaster, were described by Aldovrandi as frequently containing images. He did not, however, refer to paintings made by artists but to images painted or drawn by Nature herself.[109] For instance, he mentions a dendrite which showed how elegantly Nature could paint trees.[110] **[Fig. 5.7]** Elsewhere, referring to the same stone, he

105 As yet, fossils were not known to be fossilized creatures. Rather, the term was used for all objects found in the ground. On the classification of fossils by Gesner cf. Fischel *Natur im Bild*, pp. 28-37; Lohff, *Malerei auf Stein*, pp. 123-26.

106 "Pictura certè ars homini propria & liberalis, admiratione simul & oblectatione plerunq; spectantes afficere solet: A natura verò rebus impressae imagines ac figurae, ceu reuera quaedam hieroglyphicae notae, veriùs quidem quàm illae Aegyptioru sacrificis celebratae, maiestatis etiam aliquid prae se ferunt, praesertim quae no fortuitae fuerint, sed planè naturales & (ut ita dicam) specificae, ac generi alicui semper ceu propriae inhaerere videtur, ut in Stirpium genere ac Foßilium." Conrad Gesner, *De omni rerum fossilium genere, gemmis, lapidibus, matellis, et huius modi, libri aliqot, plerique nunc primum editi* (Zurich, 1565), epistola dedicatoria [p. 6].

107 Cf. Chapters 5-6 in Gesner, *De omni rerum*.

108 Ulisse Aldrovandi, *Musaeum metallicum in libros IIII distributum bartholomaeum ambrosinus in patrio bono* (Bologna, 1648). The treatise was finally published in 1648, 43 years after Aldovrandi's death by his successor at the University, Bartolomeo Ambrosinus.

109 On the development of theories in palaeontological research on the basis of artistic techniques, cf. Robert Felfe, *Naturform und bildnerische Prozesse. Elemente einer Wissensgeschichte in der Kunst des 16. und 17. Jahrhunderts* (Berlin, 2015), pp. 13-162.

110 "Rursus maiori admiratione tenemur, quado iutvemur Marmora ab Opifice Dei Natura, arbusculis elegantissimè picta." Aldrovandi, *Musaeum metallicum*, p. 763.

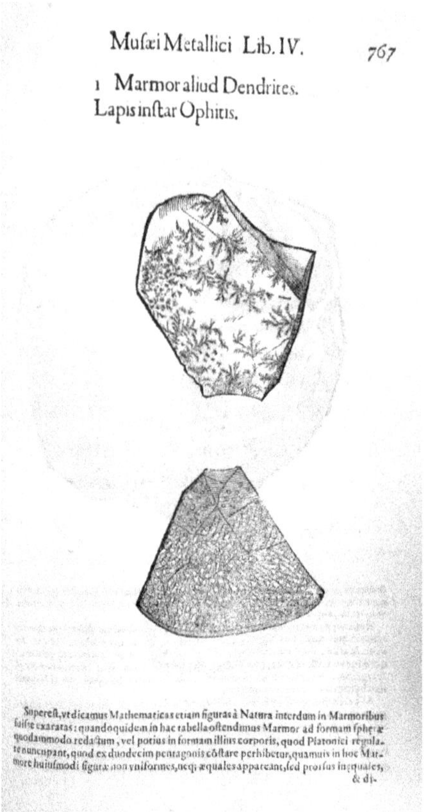


FIGURE 5.7
Marmor aliud Dendrites. Lapis instar Ophitis, *in Aldrovandi 1648, p. 767* (PHOTO: UB HEIDELBERG, T 2104 RES, P. 767).

expounds that no painter could represent a wood more convincingly with his brush than Nature.[111]

As we saw in the works in Vienna, Antonio Tempesta preferred using dendrites for his hunting scenes and used the natural forms described by Aldrovandi to enrich his paintings [**Figs. 5.1-5.2**]. The same holds true for Tempesta's treatment of the alabaster and marble panels which he employed for his versions of the *Crossing of the Red Sea*. In Aldrovandi's times, the categorization of minerals did not follow the strict criteria in force today, and so, for instance, he regarded dendrites as a form of marble. Rather, it is significant that naturalists searched for new methods of categorization to explain the creation and form of the stones on the basis of their outward appearance. For

111 "Pariter in eodem arborum rami cum folijs quasi arte depicti conspiciuntur. Est etiam alius huius Lapidis genus cinererium, in quo Syluulæ [sic] ita elegantes apparent, ut nullus pictor meliùs, & et elegantiùs eas penicillo exprimere posset." Aldrovandi, *Musaeum metallicum*, p. 443.

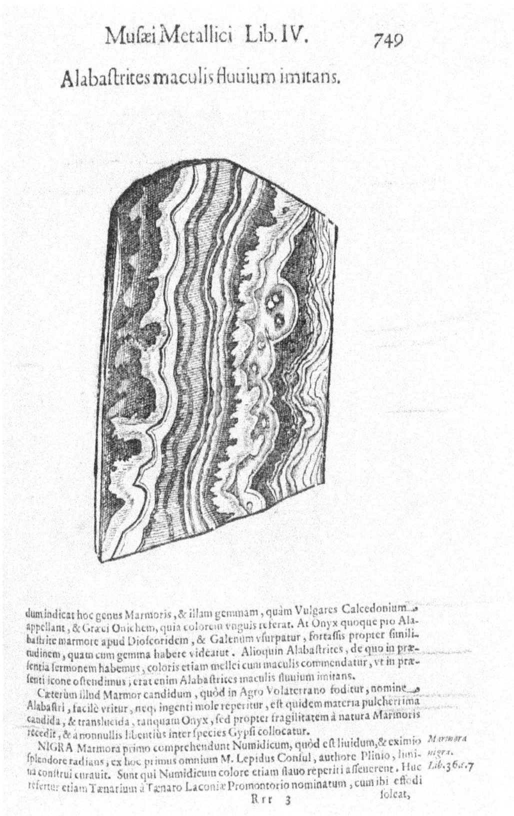

Muſæi Metallici Lib. IV. 749

Alabaſtrites maculis fluuium imitans.

dum indicat hoc genus Marmoris, & illam gemmam, quàm Vulgares Calcedonium appellant, & Græci Onichem, quia colorem vnguis referat. At Onyx quoque pro Alabaſtrite marmore apud Dioſcoridem, & Galenum vſurpatur, fortaſſis propter ſimilitudinem, quam cum gemma habere videatur. Alioquin Alabaſtrites, de quo in præſentia ſermonem habemus, coloris etiam mellei cum maculis commendatur, vt in præſenti icone oſtendimus; erat enim Alabaſtrites maculis fluuium imitans.

Cæterùm illud Marmor candidum, quòd in Agro Volaterrano foditur, nomine Alabaſtri, facilè vritur, neq; ingenti mole reperitur, eſt quidem materia pulcherrima candida, & translucida, tanquam Onyx, ſed propter fragilitatem à natura Marmoris recedit, & à nonnullis libentiùs inter ſpecies Gypſi collocatur.

NIGRA Marmora primo comprehendunt Numidicum, quòd eſt liuidum, & eximio ſplendore radians; ex hoc primus omnium M. Lepidus Conſul, authore Plinio, limina conſtrui curauit. Sunt qui Numidicum colore etiam flauo reperiri aſſeuerent. Huc refertur etiam Tænarium à Tænaro Laconiæ Promontorio nominatum, cum ibi effodi ſoleat,

*Marmora nigra. Lib. 36. c. 7*

Rrr 3

FIGURE 5.8 Alabastrites maculis fluvium imitans, *in Aldrovandi 1648, p. 749.* (PHOTO: UB HEIDELBERG, T 2104 RES, P. 749).

example, Aldrovandi illustrated an alabaster specimen whose patterning resembled a river. [**Fig. 5.8**] In general such phenomena were explained by reference to the *ludi naturae*: Nature was thought to play with her abilities and assume the role of an artist or sculptor.[112]

Tempesta appears to have been familiar with contemporary theories on the origin of stones. His illustrations for the first edition of Bernardo Amico's *Trattato delle piante et imagini de i sacri edificii di Terra Santa* (1606) include the *Plan of the Crypt in Bethlehem with the Representation of a Monk who Appears in the Marble of the Crypt.*[113] [**Fig. 5.9**] Tempesta's monk was later reused in the

112 Cf. Natascha Adamowsky, Hartmut Böhme, and Robert Felfe, eds., *Ludi Naturae. Spiele der Natur in Kunst und Wissenschaft* (Munich, 2011).

113 Bernardino Amico, *Trattato delle piante et imagini dei sacri edificii di Terra Santa disegnate in Gierusalemme secondo le regole della prospettiva, & vera misura della lor grandezza* (Rome, 1609), pl. VI. Only the first edition was illustrated with prints by Tempesta. The illustrations of the second edition (1620) were made by Jacques Callot. On Amico's book,

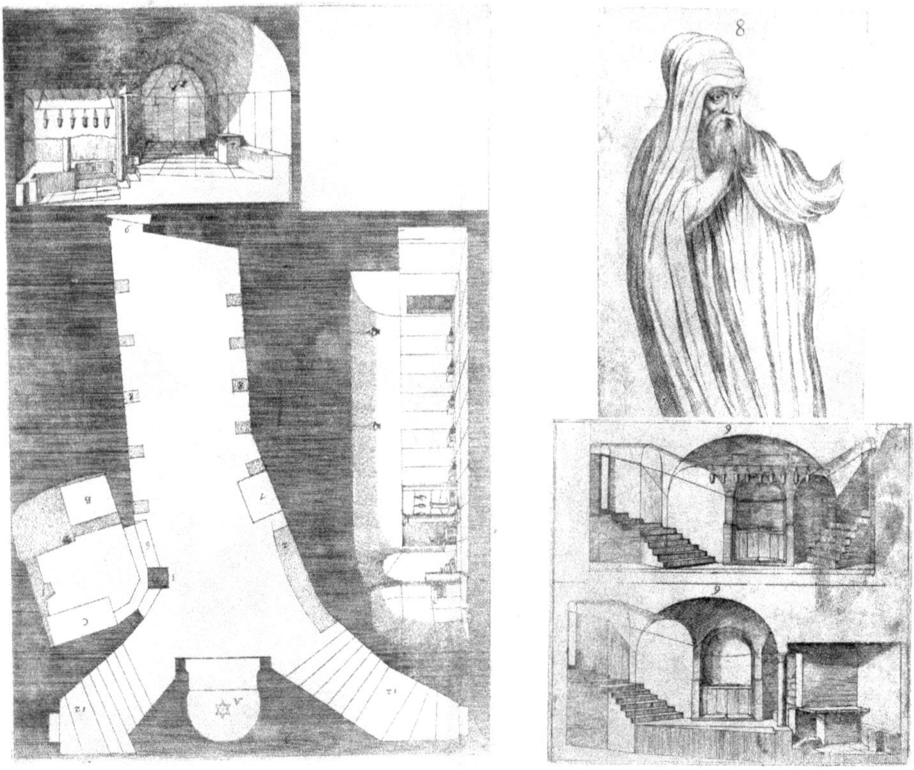

FIGURE 5.9 *Antonio Tempesta,* Plan of the crypt of Bethlehem with image of a monk appearing in the marble of the crypt, *in Amico 1609, Tav. 6* (PHOTO: BIBLIOTHECA HERTZIANA – MAX-PLANCK-INSTITUT FÜR KUNSTGESCHICHTE).

comprehensive compendium on the 'subterranean world,' *Mundus subterraneus* (1665) by the Jesuit scholar Athanasius Kircher.[114] The eighth book of the second volume is devoted to the study of stones.[115]

Like Aldrovandi, Kircher classified stones according to their images. He established a hierarchy, starting with stones with linear patterns, which, with increasing complexity, could form geometric shapes and even the letters of the

cf. Eckhard Leuschner and Yaha Kuoroshi, "Topographie und Parteilichkeit. Bernadino Amicos *Terra Santa* (Rom 1609, Florenz 1620) als Dokumentation der christlichen Stätten unter osmanischer Herrschaft," *Revista de Historiografía* 21 (2014).

114 Athanasius Kircher, *Mundus subterraneus in XII libros digestus*, ed. Gian Battista Vai, 2 vols., reprint of 1678 ed. (Bologna, 2004), 2:39.

115 Kircher, *Mundus subterraneus*, vol. 2, bk. 8 (*De lapidibus*).

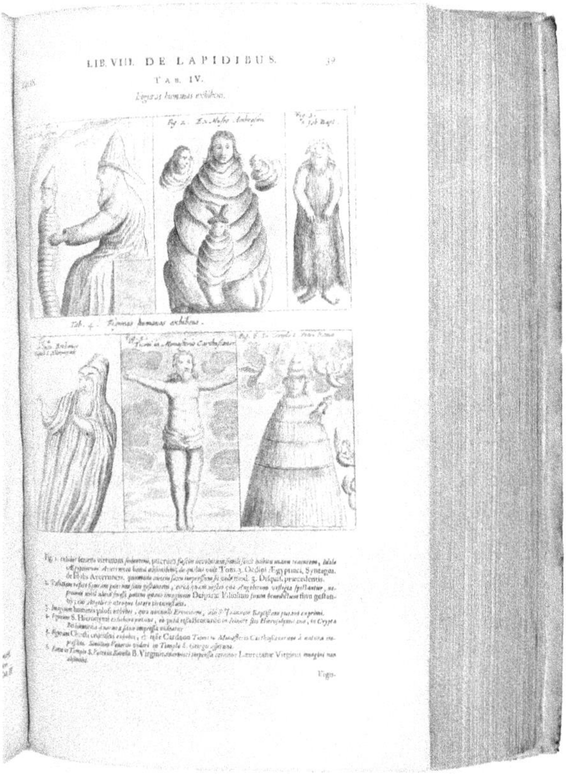

FIGURE 5.10
Figuras humanas exhibens, *in Kircher 1665, S. 39* (PHOTO: UB HEIDELBERG, O 2526 A GROSS RES ::1-2, P. 39).

alphabet.[116] Then he enumerated the images that could be found in minerals, mentioning stones in which the heavenly bodies, objects from nature and cities could be seen.[117] After mentioning stones with herbs and flowers, he proceeded to those with images of animals.[118] Finally, he discussed stones in which human beings could be seen. [**Fig. 5.10**] Tempesta's monk, who was reinterpreted as *St. Jerome*, belongs to this category.[119] Kircher assigned the

116 "Inveniuntur nonnulli lapides, qui increscentibus variæ in gemmis maculis atque verrucis linearum [...]." Kircher, *Mundus subterraneus*, 2:23.

117 "Natura quasi coelum contemplata, in iis stellas in astersismos quondam digestos, ut in Astroite apparet, vel ad pictorum invidiam depinxit [...] Natura mari, urbibus, sylvis, fluminibus, non tantum illa omnia pulchre depinxit, sed & convenientibus coloribus adumbravit ita, ut artificialibus vix cedere videantur [...]." Kircher, *Mundus Subterraneus*, 2:30.

118 "Non hîc desistere videtur, sed & Vegetabilis œconomiæ officinam ingressa, omnes generis flores [...] delineat [...]" Kircher, *Mundus Subterraneus*, 2:31.

119 Kircher, *Mundus subterraneus*, 2:39.

highest level of his hierarchy to stones depicting angels and heavenly beings.[120] Already here it can be seen that inanimate Nature was, in the realm of stones, at least given the possibility to represent animate Nature in the form of images allowing an interior order to be developed which reflected the macrocosm.[121] Kircher provided several explanations for the presence of images in stones, distinguishing between four categories. The first included stones with random images that were turned into figural representations by the imaginative powers of the beholder, just like when we see 'flying dragons' and similar apparitions while looking at the clouds.[122] Such cloud formations in which people could recognisze images were already discussed by writers in Antiquity.[123]

In the second category, Kircher listed fish which were turned into stone by a 'spiritus lapidificus,'[124] or, as he writes in a different passage, by a 'vis plastica' or a 'vis formativa."[125] Such juices or forces which could form stones were, likewise, already mentioned by ancient authors such as Theophrastus and these ideas were passed on during the Middle Ages, for instance by Avicenna and Albertus Magnus. In the 16th and 17th centuries these more or less occult forces were used not only to explain the formation of fossils but also other images in stone.[126]

---

120 "Natura ulterius progressa, & quod Naturæ terminos excedit, in nonnullis lapidibus vel ipsos Angelos Cœlitesque uti deiparam Virginem, cætersospqe Sanctos, quin & Christum Dominum crucifixum delineare attentavit, ut in sequentibus Figuris eorum apparebit." Kircher, *Mundus subterraneus*, 2:31.

121 On Kircher's systematisaion of pictorial stones, cf. Anne Eusterschulte, "Naturwunder aus dem Inneren der Erde. Athanasius Kirchers Mundus Subterraneus und die Etablierung einer geokosmischen Wissenschaft," in *Ludi Naturae. Spiele der Natur in Kunst und Wissenschaft*, ed. Natascha Adamowsky, et al. (Munich, 2011), pp. 206-16.

122 "Primo modo plerasque formas in lapidibus exortas fortuito & casu natas assero quod ut innotescat, noverit Lector, quod quemadmodum phantasia hominum, dum coelum variegate nubiumobductum videt, varia sibi, nunc dracones volantes, jam naves, modo montes, urbes & castella, nunc cruces, modo humanas figuras & similia phansmata, ex nubium incomposita repræsentatione imaginatur [...]." Kircher, *Mundus subterraneus*, 2:40.

123 For instance by Pliny the Elder and Dioscorides. Cf. Eusterschulte, "Naturwunder," p. 211.

124 "Secundus Modus est, cum terrenæ substantia materia, ex qua marmor variegatum nascitur, ita disponitur ut intra eam, tanquam intra formam quandam, quam modellam vulgo vocant, humor tinctus, tandem spiritu lapidifico in eam figuram, quam materiæ matrix referet, induratur. Estque potissimum de iis figuris intelligendum quæ perfectam aliquam figuram solidam exhibent: Cuiusmodi sunt piscium quorundam Figuræ in præcendenti Schematismo exhibitæ". Kircher, *Mundus subterraneus*, 2:41.

125 Kircher, *Mundus subterraneus*, 2:348.

126 Lohff, *Malerei auf Stein*, pp. 135-37, Felfe, *Naturform*, p. 143.

Finally, in the third category, Kircher discussed stones that contained pictures for other reasons. These included dendrites but also stones with images of saints or the inscription of the cross.[127] He stated that these stones were painted by Nature ('a Naturam depictum') and thereby introduced the concept of Nature as a painter (*natura pictrix*).[128] It followed that Nature had the ability to express herself by means of her art.[129] Kircher therefore regarded the highest level of pictures as worthy of especial attention, interpreting them as stones in which divine providence ('divina providentia') manifested itself.[130] He believed that these stones contained the Divine and described them as medium through which God communicated with human beings.

## Conclusion

Antonio Tempesta can be regarded as one of the leading painters on stone. The paintings on mineral supports which he created in the first half of the 17th century number among the outstanding works in this genre. A distinctive feature of his paintings on stone is the perfect integration of the stone's patterning into the composition, which makes it almost impossible for the beholder to distinguish the boundary between Nature and Art. In comparison to the often miniature-like paintings on patterned and semiprecious stones, his paintings are fairly large. Thus, already the sheer amount of material used in combination with the actual painting made his pictures to exceptional collectors' items. They graced the top-class collections of his time, including Rudolf II's *Kunst-* and *Wunderkammer*, Cosimo II's Uffizi Tribuna as well as the collections of numerous Roman cardinals. On the basis of chronologically ordered sources, the collection context of Tempesta's oeuvre on stone was outlined. Furthermore, three paintings attributed to Tempesta to which scholars have paid only marginal attention are discussed in detail. They show the way of working with the distinctive features of stone supports.

127 "Modus Tertius. Fit ex aliquo singulari accidente, quo figura in lapidibus nunc in hanc in illam formam emergant." Kircher, *Mundus subterraneus,* 2:42.

128 Kircher, *Mundus subterraneus*, 2:30.

129 Robert Felfe's research shows that scientists who developed theories of Nature used different artistic techniques such as painting, sculpture, drawing and printing to describe and explain the origin of stones containing pictures. Thus, palaeolotological interpretations were intimately linked with the knowledge of artistic techniques. Felfe, *Naturform*, p. 152.

130 "Modus Quartus. / De imaginibus lapidum, quibus Divinum quid inesse creditur." Kircher, *Mundus subterraneus*, 2:46.

Finally, the question was raised, why his paintings were so attuned to the taste of his time, thereby generating a great demand for them. With his paintings on stone, Antonio Tempesta catered to collectors whose predilection for patterned and painted stone panels was informed by the complex interrelationship of Art and Nature that was then a central issue not only in the visual arts but also in Science and the study of Nature.

The different theories put forward in the 17th century to explain emphasized the importance of the figuration, can be directly combined with development of paintings on stone.

Questions pertaining to the formation of stones containing pictures and their significance were of crucial importance. The importance of mineral picture supports in which Nature and Art interact with each other is founded in the manifold attempts to explain the existence of pictorial stones. The idea that Nature had her own creative powers and could express herself in skilfully fashioned images which the artist had only to complete, was of fundamental importance. Thus, paintings on patterned stones could be included in a *paragone* discussion—not in the one on the merits of painting and sculpture initiated by Benedetto Varchi, but in one in which Art rivaled Nature.[131] In the 16th to 18th centuries views varied as to whether Art and Nature were equal partners, or whether one or the other was to be given precedence. Nature's great skill was, however, increasingly admired and regarded as superior to Art.[132] The specialization of the sciences in the 18th century led to a separation of domains of Art and Nature, and paintings on stone such as those created by Tempesta and his contemporaries largely ceased to be produced.

## Bibliography

Adamowsky, Natascha, Hartmut Böhme, and Robert Felfe, eds. *Ludi Naturae. Spiele der Natur in Kunst und Wissenschaft*. Munich, 2011.

Alberti, Leon Battista. *Das Standbild. Die Malkunst. Grundlagen der Malerei. De statua. De pictura. Elementa picturae*. Edited by Oskar Bätschmann and Christoph Schäublin. Darmstadt, 2000.

131 Cf. Robert Felfe, "Figuration im Gestein," in *Paragone als Mitstreit*, ed. Joris van Gastel, et al. (Berlin, 2014), pp. 153-55.

132 In his text *The Wisdom of God Manifested in the Works of the Creation* (1691), John Ray, for instance, states: "The most accurate Engravings, or Embossments, seem such rude, bungling, and deform'd Work, as if they had been done with a Mattock, or Trowel; so vast a Difference is there betwixt the Skill of Nature, and the Rudeness and Imperfection of Art." John Ray, *The Wisdom of God Manifested in the Works of the Creation. In two Parts* (London, 1691 and 1727; reprint ed. Charleston, 2010), p. 58.

Alberti, Leon Battista. *On painting and On sculpture: the latin texts of* De pictura *and* De statua. Edited and translated by Cecil Grayson. London, 1972.

Albertus Magnus. *Book of Minerals*. Translated by Dorothy Wyckoff: Oxford, 1967.

Aldrovandi, Ulisse. *Musaeum metallicum in libros IIII distributum bartholomaeum ambrosinus in patrio bono*. Bologna, 1648.

Amendola, Adriano. *La collezione del principe Lelio Orsini nel palazzo di piazza Navona a Roma*. Rome, 2013.

Amico, Bernardino. *Trattato delle piante et imagini dei sacri edificii di Terra Santa disegnate in Gierusalemme secondo le regole della prospettiva, & vera misura della lor grandezza*. Rome, 1609.

Anselmi, Alessandra, ed. *Il diario del viaggio in Spagna del cardinale Francesco Barberini scritto da Cassiano dal Pozzo*. Aranjuez, 2004.

*L'arme e gli amori. La Poesia di Ariosto, Tasso e Guarini nell'arte Fiorentina dei Seicento*. Edited by Elena Fumagalli, Massimiliano, Rossi, and Riccardo Spinelli. Exhibition catalogue, Florence. Florence, 2001.

Aronberg Lavin, Marilyn. *Seventeenth-Century Barberini Documents and Inventories of Art*. New York, 1975.

Baglione, Giovanni. *Le Vite de pittori scultori et architetti. Dal Pontificato di Gregorio XIII. del 1572 in fino a tempi di Papa Urbano Ottauo nel 1642*. Rome, 1642.

Baltrušaitis, Jurgis. *Imaginäre Realitäten. Fiktion und Illusion als produktive Kraft. Tierphysiognomik, Bilder im Stein, Waldarchitektur, Illusionsgärten*. Cologne, 1984.

Barocchi, Paola, and Giovanna Gaeta Bartelà. *Collezionismo Mediceo e storia artistica*. 6 vols. Florence, 2002.

Baudi di Vesme, Alessandro. *La Regia Pinacoteca di Torino*. Rome, 1897.

Bauer, Rotraud, and Herbert Haupt. *Das Kunstinventar Kaiser Rudolfs II. 1607-1611*. Jahrbuch der Kunsthistorischen Sammlungen in Wien, vol. 72. Vienna, 1976.

Bigazzi, Isabelle, ed. *Un memoriale di Piero di Vincenzo Strozzi*. Florence, 1986.

Bocchi, Francesco, and Giovanni Cinelli. *Le Bellezze della Città di Firenze dove a pieno di pittura, di scultura, di sacri templi, di palazzi, i più notabili artifizi, e più preziosi si contengono, scritte già da M. Francesco Bocchi ed ora da M. Giovanni Cinelli ampliate ed accresciute*. Florence, 1677.

Brugnoli, Maria Vittoria. "I primi affreschi nel Palazzo Bassano di Sutri." *Bollettino d'Arte* 42 (1957): 241-54.

Campori, Giuseppe. *Raccolta di cataloghi ed inventari inediti di quadri*. Modena, 1870.

Cappelletti, Francesca, and Laura Testa. *Il Trattenimento dei virtuosi. Le collezioni seicentesche di quadri nei Palazzi Mattei di Roma*. Rome, 1994.

Cavazzini, Patrizia. *Painting as Business in Early Seventeenth-Century Rome*. University Park, 2008.

Cecchi, Alessandro. "Le pêche des perles aux Indes. Une peinture d'Antonio Tempesta." *La Revue du Louvre* 36, no. 1 (1986), 45-57.

Chiarini, Marco. *Teodoro Filippo di Liagno detto Filippo Napoletano. 1589-1626. Vita e opere*. Florence, 2007.

Cifani, Arabella and Monetti, Franco. "'L'Illustrissimo cugino'. Cassiano e Amadeo dal Pozzo. Le relazioni artistiche del Marchese di Voghera e la storia della sua quadreria." In *I segreti di un collezionista*, edited by Francesco Solinas, pp. 29-52. Rome, 2001.

Collomb, Anne-Laure. *Spendeurs d'Italie. La peinture sur pierre à la Renaissance*. Tours and Rennes, 2012.

Della Pergola, Paola. *Galleria Borghese. I dipinti*. 2 vols. Rome, 1955-59.

Della Pergola, Paola. "L'inventario Borghese del 1693 (III)." *Arte Antica e Moderna* 30 (1965), 202-17.

Dorati, Maria Cristina. "Gli scultori della Cappella Paolina di P. Maria Maggiore." *Commentari* 18 (1967), 231-60.

Dunlop, Anne. "On the origin of European painting materials, real and imagined." In *The matter of art. Materials, practices, cultural logics, c. 1250-1750*, edited by Christy Anderson, Anne Dunlop, and Pamela H. Smith, pp. 68-96. Manchester, 2015.

Eusterschulte, Anne. "Naturwunder aus dem Inneren der Erde. Athanasius Kirchers Mundus Subterraneus und die Etablierung einer geokosmischen Wissenschaft." In *Ludi Naturae. Spiele der Natur in Kunst und Wissenschaft*, edited by Natascha Adamowsky, Hartmut Böhme, and Robert Felfe, pp. 177-216. Munich, 2011.

Federici, C., and G. Balbi. *Viaggi di C. Federici e G. Balbi alle Indie Orientali* (Venice 1587). Edited by Olga Pinto. Rome, 1962.

Felfe, Robert. "Figuration im Gestein." In *Paragone als Mitstreit*, edited by Joris van Gastel, Yannis Hadjinicolaou, and Markus Rath, pp. 153-75. Berlin, 2014.

Felfe, Robert. *Naturform und bildnerische Prozesse. Elemente einer Wissensgeschichte in der Kunst des 16. und 17. Jahrhunderts*. Berlin, 2015.

Fischel, Angela. *Natur im Bild. Zeichnung und Naturerkenntnis bei Conrad Gessner und Ulisse Aldrovandi*. Berlin, 2009.

Fučíková, Eliška. "Die Malerei am Hofe Rudolphs II." In *Prag um 1600. Kunst und Kultur am Hofe Kaiser Rudolphs II*, pp. 177-92. Exhibition catalogue, Essen and Vienna. Freren, 1988.

Gesner, Conrad. *De omni rerum fossilium genere, gemmis, lapidibus, matellis, et huius modi, libri aliqot, plerique nunc primum editi*. Zurich, 1565.

Giusti, Annamaria. *Pietra Dura. Bilder aus Stein*. Munich, 2005.

Granata, Belinda. "Appunti e ricerche d'archivio per il cardinal Alessandro Montalto." In *Decorazione e collezionismo a Roma nel Seicento*, edited by Francesca Cappelletti, pp. 37-63. Rome, 2003.

Granata, Belinda. *Le passioni virtuose. Collezionismo e commitenze artistiche a Roma del cardinale Alessandro Perretto Montalto (1571-1623)*. Rome, 2012.

Hadjinicolaou, Yannis. "Malerei auf Stein, steinerne Malerei." In *Paragone als Mitstreit*, edited by Joris van Gastel, Yannis Hadjinicolaou, and Markus Rath, pp. 211-35. Berlin, 2014.

Heikamp, Detlef. "Zur Geschichte der Uffizien-Tribuna und die Kunstschränke in Florenz und Deutschland." *Zeitschrift für Kunstgeschichte* 26 (1963), 193-268.

Herklotz, Ingo. *Cassiano dal Pozzo und die Archäologie des 17. Jahrhunderts*. Munich, 1999.

Höper, Corinna. *Italienische Zeichnungen 1500-1800. Bestandskatalog*, part 2. Stuttgart, 1992.

Janson, Horst W. "The Image Made by Chance in Renaissance Thought." In *De Artibus opuscula XL. Essays in Honor of Erwin Panofsky*, edited by Millard Meiss, pp. 254-66. New York, 1961.

Janson, Horst W. "Chance images." In *The Dictionary of the History of Ideas. Studies of Selected Pivotal Ideas*, vol. 1, edited by Philip P. Wiener, pp. 340-53. New York, 1973.

Jonietz, Fabian. "Le conchiglie della Tribuna." In *La tribuna del principe*, edited by Antonio Natali, Alessandro Nova, and Massimiliano Rossi, pp. 191-203. Florence, 2014.

Kircher, Athanasius. *Mundus subterraneus in XII libros digestus*. Edited by Gian Battista Vai. 2 vols. Reprint of 1678 edition. Bologna, 2004.

Labrot, Gérard. *Collections of Paintings in Naples 1600-1780*. Munich, 1982.

*La tribuna del principe*. Edited by Antonio Natali, Alessandro Nova, and Massimiliano Rossi. Florence, 2014.

*Lapislazzuli. Magia del blu*. Edited by Maria Sframeli. Exhibition catalogue, Florence. Florence, 2015.

Lastri, Marco. *L'Etruria pittrice ovvero Storia della pittura toscana dedotta dai suoi monumenti che si esibiscono in stampa dal secolo X fino al presente*. 2 vols.. Florence, 1791-95.

Leonardi, Camillo. *Speculum Lapidum* [...]. Venice, 1502.

Leonardo da Vinci. *Treatise on Painting*. Rev. ed. Translated by John Francis Rigaud. London, 1877.

Leuschner, Eckhard. "Antonio Tempesta's Drawings for a 'Gerusalemme liberata'." *Masterdrawings* 37, no. 2 (1999), 138-155.

Leuschner, Eckhard. "Antonio Tempesta's 'New' Subjects in the Context of Roman Printmaking ca. 1600." In *The Art Market in Italy 15th– 17th Centuries*, edited by Marcello Fantoni, Louisa C. Matthew, and Sara F. Matthews-Grieco, pp. 65– 73. Modena, 2003.

Leuschner, Eckhard. *Antonio Tempesta. Ein Bahnbrecher des römischen Barock und seine europäische Wirkung*. Studien zur internationalen Architektur und Kunstgeschichte, vol. 26. Petersberg, 2005.

Leuschner, Eckhard, ed. *The Illustrated Bartsch*. Vol. 35, no. 1: Commentary. New York, 2007.

Leuschner, Eckhard, and Yaha Kouroshi. "Topographie und Parteilichkeit. Bernadino Amicos *Terra Santa* (Rom 1609, Florenz 1620) als Dokumentation der christlichen Stätten unter osmanischer Herrschaft." *Revista de Historiografía* 21 (2014), 67-86.

Lloyd, Karen. "Adopted Papal Kin as Art Patrons in Early Modern Rome (1592-1676)." PhD diss., Rutgers University, 2010. (<https://rucore.librariep.rutgerp.edu/rutgers-lib/26547/pdf/1/>) (2016)

Lohff, Johanna Beate. *Malerei auf Stein. Antonio Tempestas Bilder auf Stein im Kontext der Kunst- und Naturtheorie seiner Zeit*. Munich, 2015.

Lohff, Johanna Beate. "Bilder auf Stein als Diplomatengeschenke und ihre Verbreitung in Europa durch den diplomatischen Verkehr." In *Materielle Grundlagen der Diplomatie. Schenken, Sammeln und Verhandeln in Spätmittelalter und Früher Neuzeit*, edited by Mark Häberlein and Christof Jeggle, pp. 209-25. Irseer Schriften N.F., vol. 9. Constance, 2012.

Manilli, Giacomo. *Villa Borghese fuori di Porta Pinciana descritta da Iacomo Manilli Romano. Guardaroba di detta Villa*. Rome, 1650.

Massinelli, Anna. "Aggiunte al tesoro della Tribuna degli Uffizi." In *La tribuna del principe*, edited by Antonio Natali, Alessandro Nova, and Massimiliano Rossi, pp. 205-15. Florence, 2014.

O'Malley, Michelle. *The Business of Art: Contracts and the Commissioning Process in Renaissance Italy*. New Haven and London, 2005.

*Paintings on Stone of the 16th & 17th Centuries*. Edited by Alessandra Di Castro and Alexis Kugel. Exhibition catalogue, New York. Château de Saint-Rémy-en-l'Eau, 2017.

*Pietra dipinta. Tesori nascosti del '500 e del '600 da una collezione privata milanese*. Edited by Marco Bona Castellotti. Exhibition catalogue, Milan. Milan, 2000.

Poupin, Philip. "La ricerca del lapislazzuli." In *Lapislazzuli. Magia del blu*, edited by Maria Sframeli, pp. 13-19. Exhibition catalogue, Florence. Florence, 2015.

Ray, John. *The Wisdom of God Manifested in the Works of the Creation. In two Parts*. London, 1691 and 1727; reprint edition, Charleston, 2010.

Rubsamen, Gisela. *The Orsini Inventories*. Malibu, 1980.

Salerno, Luigi. "The Picture Gallery of Vincenzo Giustiniani (II)." *Burlington Magazine* 102 (1960), 93-104.

Schütze, Sebastian, and Enrico Colle. "I Sacchetti. Forme e funzioni, strategie ed intenzioni del mecenatismo. Un'introduzione." In *Palazzo Sacchetti*, ed. by Sebastian Schütze, pp. 21-43. Rome, 2003.

Seifertová, Hana. *Malby na kameni. umelecký experiment v 16. a na zacátku 17. století. Painting on Stone. An Artistic Experiment in the 16th and Early 17th Centuries*. Prague, 2007.

Sickel, Lothar. *Caravaggios Rom. Annäherungen an ein dissonantes Milieu*. Emsdetten and Berlin, 2003.

Sparti, Donatella Livia. "The dal Pozzo Collection Again. The Inventories of 1689 and 1695 and the Family Archive." *Burlington Magazine*, 132, no. 1049 (1990), 551-70.

Spinelli, Riccardo. "Documenti artistici dall'archivio Niccolini di Camugliano—I. Marmi antichi e 'moderne pitture' di Giovanni di Agnolo (Firenze, 1544-1611)." *Paragone Arte* 61, no. 663 (2005), 80-103.

Spinelli, Riccardo. "Giovanni Antonio Dosio e il progetto della cappella Niccolini in Santa Croce a Firenze. Quarantatre lettere inedite. *Rivista d'arte* 44 (1992), 199-296.

*Stradanus 1523-1605. Court Artist of the Medici.* Edited by Alessandra Baroni and Manfred Sellink. Exhibition catalogue London and Amsterdam. Turnhout, 2012.

Vannugli, Antonio. "Antonio Tempesta. Un retablo portátil en la catedral de Sevogia y otras pinturas sobre piedra." In *In Sapientia libertas. Escritos en homenaje al profesor Alfonso E. Pérez Sánchez*, pp. 230-46. Madrid, 2007.

CHAPTER 6

# Glances into Stone: Hans von Aachen's Paintings on Stone

*Susanne Wegmann*

On 25 June 1611, Philipp Hainhofer (1578-1647)—the enterprising Augsburg merchant, art agent, and diplomat closely tied to the ruling houses of his time—in a letter to Philip II, Duke of Pomerania, makes the following remarks with regard to a black stone tablet painted on by Hans von Aachen (1552-1615):

> (...) but those who do not understand art will not respect it, just like (...) the little tablet on black stone, a painting by Hans von Aachen, of Christ lying in the grave, served by two angels. The way the two candles are placed, however, has been added in a papist manner and there is nothing written of the two lights at the entombment of Christ in the Holy Scripture.[1]

Hainhofer, whose critique in this passage of the 'papist' iconographic formula clearly demarcates him as a bible-abiding Protestant, describes here a tablet which scholarship has unequivocally related to a black marble plate painted on by Hans von Aachen that is part of the art collection of the Benedictine Kremsmünster Abbey [**Plate 23**].[2] The plate—measuring 16.5 by 22 cm, signed

1 Philipp Hainhofer, Letter to Philip II, Duke of Pomerania, cf. Oscar Doering, *Des Augsburger Patriciers Philipp Hainhofer Beziehungen zum Herzog Philipp II. von Pommern-Stettin, Correspondenzen aus den Jahren 1610-1619*, Quellenschriften für Kunstgeschichte und Kunsttechnik des Mittelalters und der Neuzeit, Neue Folge, vol. 6 (Vienna, 1894), p. 148: "wer aber die Kunst nit verstehet wurds nit achten, so wol (...) daß Täflin auf schwartzen stain von hans von Ach gemahlt, daß ist wie Christus im grab liget, vnd 2 Engel Ihme aufwarten, die 2 liechter aber so darbei stehen ist auf Papistische art hinzu gemachet worden, und würd von den liechtern bei der begrebtnuß Christi in hailiger schrifft sonst nichts gelesen."

2 Joachim Jacoby, *Hans von Aachen (1552-1615)* (Munich and Berlin, 2000), pp. 116-18, cat. 24. Hana Seifertová, *Malby na kameni—umělecký experiment v 16. a na začátku 17. století/Paintings on Stone—An Artistic Experiment in the 16th and Early 17th Centuries* (Prague, 2007), pp. 76-77. *Hans von Aachen (1552-1615): Hofkünstler in Europa*, ed. Thomas Fusenig, exhibition catalogue, Suermondt-Ludwig-Museum, Aachen (Berlin and Munich, 2010), pp. 152-53, cat. 34 (Joachim Jacoby).

 | DOI 10.1163/9789004361492_010

by Hans von Aachen on the bottom right, the date given as 1594 ("HVA. 94")—was bought for the monastery's art collection on the Augsburg art market in 1877. It is not possible to trace its provenance further back.[3] From the correspondence between Hainhofer and Philip II it remains unclear whether he means to offer the plate to the Duke as a possible purchase or whether he is drawing up an inventory of the works on their way to the latter.[4]

Hans von Aachen, born in Cologne, might be considered the most important representative north of the Alps of painting on stone materials. His connections to Italy, where the Medici were among his patrons, to the Fugger family in Augsburg, and to William V, Duke of Bavaria, certainly commended him for the appointment as court painter for Rudolf II in Prague. While he had made a name for himself primarily as portraitist, he proved to be among the most versatile court painters at the Court in Prague. In the following, particular focus will be placed on his experiments with the materiality of various pictorial grounds.[5]

The marble plate in Kremsmünster presents an unusual iconography, considered by Joachim Jacoby as Hans von Aachen's innovation. Not only would Hans von Aachen repeat it in different media and on different support materials, but it was also taken up by other artists.[6] The observer is granted a view

3 *Die Kunstdenkmäler des Benediktinerstiftes Kremsmünster. Teil 2: Die stiftlichen Sammlungen und die Bibliothek*, ed. Hans Gertele-Grenadenberg. Österreichische Kunsttopographie, vol. 43 (Vienna. 1977), p. 98, n. 28.

4 The correspondence is presented only in excerpts in Doering's edition, cf. Doering, *Des Augsburger Patriciers*.

5 Concerning Hans von Aachen's artistic context and his patrons prior to his appointment at the Prague Court: Jacoby, *Hans von Aachen (1552-1615)*, pp. 8-44. Eliška Fučikova, "Das Leben," in *Hans von Aachen (1552-1615): Hofkünstler in Europa*, edited by Thomas Fusenig, exhibition catalogue, Suermondt-Ludwig-Museum, Aachen (Berlin and Munich, 2010), pp. 3-8. Fučikova, "Die Malerei," in *Hans von Aachen (1552-1615): Hofkünstler in Europa*, pp. 13-26.

6 Jacoby, *Hans von Aachen (1552-1615)*, pp. 116-20, cat. 24-25. The composition was disseminated through an engraving by Raphael Sadeler [Joachim Jacoby, *Hans von Aachen* (Rotterdam, 1995), cat. 38]. Also see the different iconographic repetitions and variations on a large-format (115 × 255 cm) slate plate by Paolo Piazza, Rome, Pinacoteca Capitolina; on a small (21 × 28 cm) oval slate plate (Milano, Sgl. Giulini); on amethyst by a Venetian (or Prague) master in the Milano collection Giulini, cf.: *Pietra Dipinta. Tesori nascosti del '500 e del '600 da una collezione privata milanese* ed. Marco Bona Castellotti, exhibition catalogue, Palazzo Reale, Milan (Milan, 2000), p. 173, cat. 139. Anne-Laure Collomb, "La Peinture sur pierre en Italie 1530-1630" (PhD diss., University of Lyon, 2006), pp. 30, 39, fig. 10, 129. In 1601, Paolo Piazza visited Bohemia to work at the Court of Rudolf II, cf. Seifertová, *Malby na kameni*, p. 67; hence the slate plate from after 1611 ascribed to him might be influenced directly by Hans von Aachen's composition.

into the still closed sepulcher, in which the body of Christ, mourned by two angels, lies in the scene's foreground on a layering of stones and altar steps; these are covered with a thin, seemingly transparent cloth. The upper body is resting against the altar block, on the *mensa* on which the head is laid, further supported by a smaller block of stones. Two candle holders with burning candles are placed in the background. The coloring of the image appears reduced; the cave's contours, the altar steps, the altar block, and the corpse are represented as grisaille, white on the black stone with white veining. Only small sections are executed in color; hence the flames of the burning candles are glowing in bright yellow, the wick within it glimmering in red. The faces and the flesh tones of the angels are also intensified by red, part of the candles' reflection. The stigmata have only the slightest hue of red, and the garments are contrasted from the stone plate's black background only by means of an almost black, subtly dark-green shading.

Hans von Aachen's hitherto little-noticed *Christ in the Sepulcher* is a striking example of content, form, and material being inextricably related. The material and its particular qualities in image, iconography, and painterly form-giving are explicitly addressed. Questions concerning how the materiality of the pictorial ground comes to bear on Hans von Aachen's works, how the image carriers become effective in terms of an aesthetics both of production and reception, and how the subject matter is clarified, made visible, and graspable for the recipient in the *kunstkammer* are further examined in the reflections that follow.

## Experiments with the Pictorial Surface in Hans von Aachen's Work: Influences from Italy

As Joachim Jacoby has observed, a special treatment and aestheticization of the pictorial ground are already discernible in Hans von Aachen's early work.[7] The double-portrait *Two Laughing Men* at the Archiepiscopal Chateau at Kroměříž and the self-portrait at the Wallraf-Richartz-Museum in Cologne are both painted on a support made of oak wood. Neither painting exhibits a cohesive application of paint. Not only does the brush style remain visible, but the brown base of the wooden ground shines through and is taken into account in the representation.[8] Jacoby considers the structures and color transitions a means of giving particular animation to the faces. This form of painting he

7 Jacoby, *Hans von Aachen (1552-1615)*, p. 11.

8 Cf. Helen Langdon's essay in this volume.

evaluates as unique in the context of Hans von Aachen's environment in Cologne in the years between 1565 and 1575. Aside from the effect the integration of the pictorial surface into the composition has on the image reception, as emphasized by Jacoby, it is similarly important to emphasize in terms of an aesthetics of production the resulting disclosure of the work process. The construction of the image from its ground to the layers of paint, the visibility of the artist's signature in the visibility of the brush strokes—these testify to the artist's *maniera*, render recognizable *his* art and procedures.[9] The wooden materiality of the image carrier is in this case not so much addressed as part of the subject matter, but primarily effective aesthetically.

Yet, as Kia Vahland has demonstrated in relation to Angolo Bronzino's portrait of Laura Battiferi, even the common wooden image carrier can become significant in its materiality.[10] She refers to a sonnet by Bronzino relating the pictorial base and subject matter by means of word play. He compares Laura as portrayed by him on wood with the nymph Daphne, fleeing from Apollo's advances and transforming into a laurel tree. By means of the image composed by the artist, Laura Battiferi performs the same metamorphosis. In the process of creation, she literally transforms into wood—in Italian, *laura* signifies laurel. Bronzino's poem primarily alludes to the sonnets by Francesco Petrarca, considered as models and widely read and addressed among humanist scholars and artists. By constructing his Laura's transformation into wood, Bronzino positions her as unreachable. Poetically describing this act, the painter not only displays consciousness of the unremarkable, common wooden tablet—rather, in a metaphorical idealization—but also renders the painting's support a kind of laurel, thereby ennobling himself, the artist.

It is hard to say whether Hans von Aachen could have been acquainted with this metaphorical use of the painting's support playing with the name of the person portrayed, but unquestionably his later experiments with the pictorial surface and the support materials were informed by Italian influences. The range of materials used by him for his paintings' supports is particularly remarkable. Not only does he frequently employ the copper plates that became popular in the late 16th century for miniature-like images to be

9 Concerning the visibility of the grounding in the image and its thematic use in Rembrandt's work, cf. Nicola Suthor, "Transparenz der Mittel. Zur Sichtbarkeit der Imprimitur in einigen Werken Rembrandts," in *Der Grund. Das Feld des Sichtbaren*, ed. Gottfried Boehm and Matteo Burioni (Munich, 2012).

10 Kia Vahland, *Lorbeeren für Laura. Sebastiano del Piombos lyrische Bildnisse schöner Frauen* (Berlin, 2011), p. 143, describes the relation as a "Wortspiel, das im Frauenbildnis Malgrund und Sujet ineinander aufgehen lässt" ("word play collapsing in the female portrait surface and subject").

viewed from up close in a collection context,[11] but also, according to Karel van Mander, tin and lead surfaces.[12] In terms of stone-based supports, aside from the black marble he also makes use of lighter forms. Among his most elaborate and appealing works are his paintings on so-called oriental alabaster and on jasper agate. The colored stratification, the crystalline veins, and the alteration between opaque and translucent parts are virtuosically integrated into the compositions. Aside from the extant known works, written sources attest to further paintings on stones by Hans von Aachen. The inventory drawn up between 1607 and 1611 of Rudolf II's art chamber lists five of his paintings on stone that do not seem to have survived: these include a painting of a shipwreck occurring in a thunderstorm, executed on grey marble, as well as a marble plate on which hell, magic, and witches' work was to be seen. A black marble tablet formed the ground of a candle-lit night scene with two boys. A small alabaster plate showed Pluto and Proserpina with many "fladergeistern." The inventory further records an alabaster plate painted on two sides, described as "Our Lady in the sun"—hence probably the Woman of the Apocalypse and the War in Heaven.[13] It seems that Hans von Aachen encoun-

11 Cf. Andreas Henning, "The new technique of painting on copper," in *Captured emotions. Baroque painting in Bologna, 1575-1725*, ed. Andreas Henning and Scott Schaefer, exhibition catalogue, J. Paul Getty Museum, Los Angeles (Los Angeles, 2008). Edgar Peters Bowron, "A Brief History of European Oil Paintings on Copper, 1560-1775," in *Copper as Canvas. Two Centuries of Masterpiece Paintings on Copper 1575-1775*, exhibition catalogue, Phoenix Art Museum (New York, 1999).

12 Jacoby, *Hans von Aachen (1552-1615)*, pp. 89-91, cat. 9. Jürgen Müller, *Concordia Pragensis. Karel van Manders Kunsttheorie im Schilder-Boek. Ein Beitrag zur Rhetorisierung von Kunst und Leben am Beispiel der rudolfinischen Hofkünstler* (Munich, 1993), p. 201.

13 The exact wording according to the edition of the inventory: Rotraud Bauer and Herbert Haupt, eds., *Das Kunstkammerinventar Kaiser Rudolfs II., 1607-1611*, Jahrbuch der Kunsthistorischen Sammlungen in Wien, vol. 72 (Vienna, 1976), pp. 139-40: "Truhen N°. 20 Gemeld auf alabaster or: und andere stain gemaltt / 2800. N°. 949. Auff wassergrawen marmorstain, darauf H.V.A. ein tempest und ungewitter mit plitz und schiffbruch gemalt. / 2801. N°. 950. Ein ander gleichmessiger marmelstein, darauf H.V.A. ein höll, zauberey und Hexenwerckh gemalt. / (...) 2803. 957. Von H.V.A. ein doppelt gemalter alabaster or:, uff einer seitten ist unser fraw in der sonnen, uff der anderen, wie der trach auß dem himel under sich herabgestürtzt wirt. / 2804. 954. Von H.V.A. uff schwartzen marmor gemalt, wie bey der nacht 2 knaben von einem brandt ein liecht uffblasen, der stain ist 8-egget. / (...) 2806. 944. Ein klein quaderlin von H.V.A. gemalt auf or: alabaster, historа von Pluto und Proserpina mit vilen fladergeistern und höll." Cf. Seifertová, *Malby na kameni*, pp. 136-37; Seifertová, "Pittura su pietra der rudolfinischen Zeit und ein Vanitasgemälde von Christoph Gertner," in *Rudolf II, Prague and the World. Papers from the International Conference Prague 2-4 September 1997*, ed. Lobomír Konečný (Prague, 1998), p. 95.

tered the techniques and the different applications of oil painting on metal and stone grounds on his travels to Italy. During his stays in Venice, Rome, and Florence from 1574 onwards, he was able to see in particular examples of oil painting on slate or *pietra paragone*. It was not only the Florentine Vasari who ascribed its invention to the Venetian Sebastiano del Piombo; in 1530 already, the Venice-born papal valet de chambre Vincenzo Soranzo had proudly praised this artistic invention. Even Vasari himself made use of the technique which, according to his own art-theoretical grounds, promised durability and seemed to secure lasting glory for the artist.[14]

It is Hans von Aachen's representation of the body of Christ on black marble ground in particular that seems reminiscent of works by Sebastiano del Piombo. The latter took up the black surface in his paintings on slate, even if the stone mostly did not remain visible without treatment. Themes of the passion of Christ—the procession, the lamentation, the entombment—formed frequently executed iconographies; Sebastiano knew how to augment the dramatic quality by allowing the pallid gloom of the surface to shine through. Similarly, he rendered palpable the stone's tectonics as a sculptural element to his paintings.[15]

## Artistic and Scholarly Encounters with Material at the Court of Rudolf II

Having recognized the influence of Italian developments on Hans von Aachen's work, the significance of his environment at the Court of Rudolf II in Prague in terms of his paintings on stone and of his referencing of the material base of

---

14 Concerning the development of painting on stone in Italy and its art theoretical reflection, cf.: Christiane J. Hessler, "Maler und Bildhauer im sophistischen Tauziehen. Der Paragone in der italienischen Kunstliteratur des 16. Jahrhunderts," in *Wettstreit der Künste. Malerei und Skulptur von Dürer bis Daumier*, ed. Ekkehard Mai and Kurt Wettengl, exhibition catalogue, Haus der Kunst, Munich (Wolfratshausen, 2002), p. 93. Seifertová, *Malby na kameni*, pp. 17-46. Claudio Strinati, "Sebastiano del Piombo und die Renaissance in Rom," in *Raffaels Grazie–Michelangelos Furor. Sebastiano del Piombo (1584-1547)*, ed. Wolfgang Lindemann and Claudio Strinati, exhibition catalogue, Museen Gemäldegalerie, Berlin, and Palazzo Venezia, Rome (Berlin, 2008), p. 16. Concerning Sebastiano del Piombo, see the relevant contributions in this volume.

15 See for example Sebastiano del Piombo, Pieta, oil on slate, 120 × 107 cm, 1533-40, Seville, Casa de Pilatos. Cf. Seifertová, *Malby na kameni*, pp. 22-27, fig. 3. Elena Calvillo, "Authoritative Copies and Divine Originals: Lucretian Metaphor, Painting on Stone, and the Problem of Originality in Michelangelo's Rome," *Renaissance Quarterly* 66, no. 2 (2013), pp. 480-91.

the image is not to be underestimated. In 1592, he was appointed official court painter ('Kammermaler von Haus aus'); after several shorter stays at court, he settled in Prague in 1596 and quickly became one of Rudolf II's most significant and productive court painters, not least of all for his works' quality as diplomatic gifts, successfully representing the court abroad.[16] The encounter with other court artists experimenting with metals and minerals as painting surfaces, such as Bartholomäus Spranger and Joseph Heintz, is certainly no coincidence. The latter especially had specialized on metal surfaces. Of the forty-six works which Jürgen Zimmer ascribes to Joseph Heintz in his still definitive monograph from 1971, fifteen are painted on copper plates.[17] The inventory of Rudolf II's art chamber ascribes to Heintz an alabaster plate painted on two sides with the Annunciation to Mary and the birth of Christ.[18] Painting on copper seems to have played a lesser role for Bartholomäus Spranger. His representation of Apollo with the muses executed on marble appears to be an isolated exception as part of his body of work; yet the fact that this tablet is fractured suggests a high rate of loss for these fragile pieces.[19]

The court of Rudolf II offered to painters a productive environment in which they could experiment with a variety of arts, working materials, and

16 Thomas DaCosta Kaufmann, *The School of Prague. Painting at the Court of Rudolf II* (Chicago and London, 1988), p. 33.

17 Jürgen Zimmer, *Joseph Heintz der Ältere als Maler* (Weissenhorn, 1971).

18 Bauer and Haupt, *Das Kunstkammerinventar*, p. 140: "2802. 952. Von O. Heintzen uff or: alabaster gemalt, uff einer seitten ein englischer gruß, uff der andern seiten ein weihnachten." A work, a two-sided alabaster image, currently held in Vienna (Kunsthistorisches Museum, Inv. Nr. 2687) corresponds to this description; yet, research has doubted its ascription to Heintz, assigning it instead to Paolo Piazza, cf. Seifertova, *Malby na kameni*, p. 137, leaves open the ascription. DaCosta Kaufmann, *The School of Prague*, p. 201, fundamentally rejects the ascription to Joseph Heintz.

19 Michael Henning, *Die Tafelbilder Bartholomäus Sprangers (1546-1611). Höfische Malerei zwischen "Manierismus" und "Barock"* (Essen, 1987), pp. 177-92, records for Spranger sixty-seven certified originals, among them eight on copper, one on marble (Apollo, Minerva and the Muses, Vienna, Kunsthistorisches Museum) and one on brass. Sally Metzler, *Bartholomeus Spranger. Splendor and Eroticism in Imperial Prague. The Complete Works* (New Haven, London, and New York, 2014), catalogues for Spranger eighty-eight paintings, among them thirteen on copper and one on marble (pp. 132-33, cat. 60). Concerning Spranger's painting *Apollo and the Muses* on marble, cf. recently: Yannis Hadjinicolaou, "Malerei auf Stein, Steinerne Malerei. Die Farbgestaltung bei Spranger und de Gelder," in *Paragone als Mitstreit*, ed. Joris van Gastel, et al. (Berlin, 2014), pp. 211-15. Concerning Christoph Gertner, who worked in Prague around 1600, cf. Seifertová, "Pittura su pietra." Gertner, court painter for Henry Julius, Duke of Brunswick-Lüneburg, was one of only few painters north of the Alps specialized on slate surfaces.

techniques. There was work for sculptors, bronze founders, goldsmiths, copperplate engravers, medallists, wax sculptors, lapidaries, and glass cutters. Many projects were carried out as collaborations between different arts. Hence painters and sculptors worked hand in hand as a matter of course, and works by painters and sculptors were copied and disseminated by the copperplate engravers. The stone cuts of the Miseroni were treated by goldsmiths and silversmiths.[20] For a diplomatic present from Rudolf to Christian II, Elector of Saxony, the Castrucci specialized on the Commessi di Pietre Dure, united with Hans von Aachen. While the Saxon coat of arms was executed by the Italian artist as a virtuosic and detailed *pietra dura*, Hans von Aachen painted on the reverse, a jasper agate plate. On this he showed an allegorical representation of the government of Christian II, as well as the victory over the Turks achieved together with Rudolf.[21] The stone intarsia, emulating painting in its coloring and illusionist effect, is closely linked to a painting technique which takes up in its use of stone the sculptor's material and the advantages of its durability. Hence the front and back of the work unite two aspects of the *paragone* discourse, challenging the observer to compare the differences between images.

The stone cut and the stone intarsias in particular encountered similar challenges as painting on stone: the stone's natural colors and structures directed and inspired the artist's work. Hence Giorgio Vasari praises the abilities of the stonecutter Matteo del Nassaro in reacting to the unexpected qualities concealed in a stone by nature and in incorporating and referencing these in the compositions. Red particles in the jasper turn into the wounds of Christ; and a chalcedony, characterized by its differently colored stone layers, is treated so skillfully by Matteo that he is able to make use for the representation of a Deianira with the lionskin of a red stripe for the back surface of the flayed fur, of a different spot for the hair and of a white layer for face and breast.[22]

In Rudolf's collection, the painters were able to study not only the works of the lapidaries and vessel cutters and their engagement with the challenges given by the materials, but also model paintings on stone which Rudolf II had acquired from Italy. In 1595, Rudolf attempted to buy the famous work on slate

20 This is also referred to by Joris van Gastel et al., "Paragone als Mitstreit," in *Paragone als Mitstreit*, ed. Joris van Gastel, et al. (Berlin, 2014), p. 24.

21 Ulrike Weinhold, "Die Habsburger und die frühe Dresdner Kunstkammer," in *Die kurfürstliche-sächsische Kunstkammer in Dresden. Geschichte einer Sammlung*, ed. Dirk Syndram and Martina Minning (Dresden, 2012), p. 72, fig. 6-7; Jacoby, *Hans von Aachen (1552-1615)*, pp. 180-81, cat. 59.

22 Giorgio Vasari, *Die Leben der ausgezeichneten Gemmenschneider, Glas- und Miniaturmaler Valerio Belli, Guillaume de Marcillat und Giulio Clovio*, trans. Victoria Lorini and ed. Anja Zeller (Berlin, 2006), pp. 26, 28.

painted by Daniele da Volterra on two sides, giving two views of David's victory over Goliath. Through the choice of material and the design on the front and back, Daniele da Volterra, too, positions himself argumentatively in the *paragone* debate. For not only does he make use of the sculptor's material to bestow durability to his work, at least insofar as that durability was ascribed to the material by art theory, he also proves painting's capacity for the multiple view. While Rudolf was not able to acquire the image himself—explained among other reasons by transport problems given the actual fragility of the stone plates—he commissioned his court painter Joseph Heintz to manufacture a copy.[23] Rudolf was more successful with Antonio Tempesta's smaller limestone dendrite plates. These are exemplary in terms of an artist's co-creation with nature.[24] A natural landscape of trees, shrubbery and grass is formed by the dendrites; he places two hunting scenes into this landscape, a bear and a boar hunt. According to the inventory of Rudolf's court chamber, the two plates are kept in a trunk together with the stone paintings of his court painters. On the inventory, the Commessi di Pietre Dure by Cosimo and Giovanni Castrucci follow immediately upon the paintings on stone.[25] The proximity of the two forms of art in the inventory's taxonomy attests to their connectedness, both presenting to the observer a play of *ars* and *natura*.[26]

---

23 Daniele da Volterra, *David and Goliath*, oil on slate, 130 × 172 cm, 1555/1556, Paris Louvre. Seifertová, *Malby na kameni*, p. 46, fig. 15, 16. For an interpretation as a contribution to the *paragone* controversy: Marina Haiduk, "Daniele da Volterras 'David und Goliath'. Ein Wettstreit auf mehreren Ebenen," *Wiener Jahrbuch* 57 (1999). Ben Thomas, "'The lantern of painting': Michelangelo, Daniele da Volterra and the paragone," *Apollo* 154, no. 474 (2001), 46-53.

24 Cf. Johanna Beate Lohff's essay in this volume. Also see Johanna Beate Lohff, *Malerei auf Stein. Antonio Tempestas Bilder auf Stein im Kontextder Kunst- und Naturtheorie seiner Zeit* (Munich, 2015), pp. 104-18 (for an art-theoretical discussion of accidental images, images made by nature and the interplay of art and nature), p. 69, cat. 4.1.-2 (concerning Tempesta's paintings on dendrite as part of Rudolfs II's collection).

25 Bauer and Haupt, *Das Kunstkammerinventar*, p. 140. "2805. 955. Zwey grosse stuckh stain, so von Roma kommen, darein wie bäumb oder landschafften natürlich gewachsen von Anthon. Tempest jaglandtschafften darein gemalt"; Seifertová, *Malby na kameni*, pp. 116-19, fig. 59, 60; *Prag um 1600. Kunst und Kultur am Hofe Rudolfs II*, exhibition catalogue, Villa Hügel, Essen (Freren, 1988), p. 297, cat. 171-72 (Eliška Fučíková). It is not entirely clear from the inventory whether the stone intarsias were stored in chest 20 with the paintings on stone.

26 Oscar Doering, *Des Augsburger Patriciers Philipp Hainhofer Reisen nach Innsbruck und Dresden* (Vienna, 1901), p. 117. "Jetzt erzehlte historiae sein gar erkantlich vnd subtil von öhlfarben auf die stain gemahlt, so das ars vnd natura mit ain ander spilen." Hainhofer describes stone as created by nature, so-called Florentine landscape marble with its

At the Court of Rudolf II, the arts were placed in constant exchange with scholars. Astronomers, astrologers, mathematicians, and physicists—including illustrious characters such as Johannes Kepler and Tycho Brahe—were active at the Prague Court. In terms of painting on stone, the collection of minerals and its scientific treatment on the part of the physicist, geologist, and mineralogist Anselmus Boethius de Boodt is of special importance.[27] While he was Rudolf II's private physician since 1583, he was primarily responsible for the construction and maintenance of his comprehensive collection of minerals. In 1609, he published his scientific magnus opus, the *Gemmarum et lapidum historia*. Hans von Aachen's works mirror scientific discoveries and discourses concerning the genesis and qualities of minerals, and de Boodt's direct influence seems to suggest itself. Knowledge from natural history and scholarly insight had a productive effect on art and was turned into its subject matter. At the same time, the arts' influence is visible in the mineralogist's work.[28] Following the model of Pliny's *Naturalis Historia*, de Boodt references the use of minerals in the arts, listing famous works since antiquity and emphasising in particular the examples assembled at the Rudolfine court. The most famous work, remarked and reflected on several times by de Boodt, is certainly the *pietre dure* table that Rudolf II had manufactured with Bohemian stone in Florence.[29] In the dedication to his book, de Boodt gives to the emperor something of an instruction of how to view this marvel of stone-cutting, calling it the eighth wonder of the world:

---

bizarre landscapes and cities, as well as the paintings executed on agate, carnelian, chalcedony, and jasper at the art cabinet which Archduke Leopold of Austria ordered as a present for the Great Duke of Tuscany. Cf. Vgl. Jurgis Baltrušaitis, *Imaginäre Realitäten. Fiktion und Illusion als produktive Kraft* (Cologne, 1984), p. 59.

27 Cf.: Joaneath Spicer, "The Role of 'Invention' in Art and Science at the Court of Rudolf II," *Studia Rudolphina* 5 (2005); Christiane Mannheim, "Edelsteine und Mineralien. Symbole theologischer Spekulationen und mystischer Naturverehrung werden zu Objekten wissenschaftlicher Forschungen," in *Erkenntnis, Erfindung, Konstruktion. Studien zur Bildgeschichte von Naturwissenschaften und Technik vom 16. bis zum 19. Jahrhundert*, ed. Hans Holländer (Berlin, 2000), pp. 204-06.

28 Concerning the competition of *natura* and *ars*, cf. recently Robert Felfe, "Figurationen im Gestein. Ko-Produktionen zwischen Kunst und Natur," in *Paragone als Mitstreit*, ed. Joris van Gastel et al. (Berlin, 2014). Felfe explicitly describes the relation between nature and art not in terms of a one-sided influence, but emphasizes instead the mutual challenge.

29 The table plate praised by de Boodt as the eighth world wonder: Karla Langedijk, "The Table in Pietre Dura for the Emperor. A New Understanding of Rudolph II as a Collector," *Mitteilungen des Kunsthistorischen Institutes in Florenz* 42 (1999), 358-82.

> ... in them [the gems and stones] you may contemplate the perfection of God and his ineffable power, which seems to have brought together the beauty of the whole world and to have enclosed the power of all other matter in such minuscule bodies, whereby you may ever have before your eyes something of the light and appearance of the Divinity.[30]

A similar aesthetics of perception is mirrored in the report of Carlo Francesco Manfredi, ambassador to the Duke of Savoy, upon his visit to Rudolf's art chamber. He is shown a large agate bowl, on the bottom of which one could read in Greek letters the word 'Christ' as written into the stone by *natura*, as if it was God himself speaking to the emperor through this work of nature.[31] The stones as treated by the artists give the emperor the possibility for contemplation and self-reflection, uniting the beauty of the world as a whole and rendering visible the divine. The ruler is mirrored in the stones' divine, sublime glow, and the stones cast on him the reflected splendor of the almightiness of God.

## The Creative Power of the Four Elements

The two alabaster panels originating around 1600—painted on both sides, depicting the triumphal procession of Amor and Bacchus, the fall of Phaethon, the releasing of the winds by Aeolus, and the liberation of Andromeda—are certainly to be numbered among Hans von Aachen's principal works.[32] While

30 Quoted from Langedijk, "The Table in Pietre Dura," p. 376. Anselmus Boethius de Boodt, *Gemmarum et Lapidum Historia* (Hanau, 1609), p. 7: "(...) sed vt in iis Dei excellentiam, eiusq(ue) vim ineffabilem, qui in tam exiguis corpusculis totius mundi pulchritudinem coegisse, ac omnium aliarum rerum vires inclusisse videtur, contempletur, ac ante oculos diuinitatis quandam lucem & speciem perpetuo habeat." Annamaria Giusti, *Pietra Dura. Bilder aus Stein* (Munich, 2005), pp. 115-17; Rudolf Distelberger, *Die Kunst des Steinschnitts. Prunkgefäße, Kameen und Commessi aus der Kunstkammer* (Vienna, 2002), pp. 248-49.

31 Achatschale, 4th century, Constantinople, Vienna, Kunsthistorisches Museum, Weltliche Schatzkammer; Paula Findlen, "Cabinets, Collecting and Natural Philosophy," in *Rudolf II and Prague. The Imperial Court and Residential City as the Cultural and Spiritual Heart of Central Europe*, ed. Eliška Fučiková (Prague, London, and Milan, 1997), p. 212. Also see: Seifertová, *Malby na kameni*, p. 127; Susanne König-Lein, "Ein Spiel der Natur? Bildersteine und Steinbilder," in *SpielKunstGlück. Die Wette als Leitlinie der Entscheidung. Beispiele aus Vergangenheit und Gegenwart in Kunst, Wissenschaft, Wirtschaft*, ed. Johann Konrad Eberlein (Vienna, Berlin, and Münster, 2011), p. 137.

32 38 × 45 cm, Vienna, Kunsthistorisches Museum, Sammlungen Schloss Ambras. Kayo Hirakawa, "Narrative and Material: Paintings on Rare Metal, Stone and Fabrik in the Late Sixteenth and Early Seventeenth Centuries," in *Aspects of narrative in art history.*

the works are first traceable as part of the estate of Emperor Matthias in Vienna in 1619, their provenance from Rudolf II's art collection appears likely. Presumably they became part of Matthias' property after his brother's—Rudolf's—death. The 1619 inventory describes them like this: "On a base of ebony an oblong, quadrangular, large plate of agate, framed by ebony, painted on both sides, on one of these Phaethon, on the other side Bacchus, Venus and other Gods and Goddesses" and "a large quadrangular plate of agate, on one side Perseus and Andromeda, a shipwreck on the part of Aeneas."[33] The description of the base and frame made of ebony suggests that the plates were exhibited free-standing in a room.

One of the plates shows the boy Amor standing triumphant on a carriage [**Fig. 6.1**], followed by Bacchus riding on a donkey. They are accompanied by Venus holding a burning heart and by Ceres and the cornucopia. The procession is led by cherubs holding up hearts on arrows like standards, as well as bow and arrow and a winged heart—while amoretti scatter flowers across them from heaven. Placed in heaven, too, Mercury, Apollo, and Juno watch the festive procession. A round temple becomes visible in the background to the right; in front of it, we see a kneeling woman, milk running from her breasts, identified by Jacoby as the "embodiment of natura."[34] The triumphal procession of Amor goes back to motifs in Ovid's *Amores* (1.2.21-52); the reverse side with its representation of the fall of Phaethon [**Plate 24**], son of the sun god Apollo, refers to Ovid's *Metamorphoses* (2.5.1-700). Apollo is depicted in a round temple in heaven, surrounded by a bright nimbus and an aureole. Phaethon is kneeling in front of the temple. In order to dissipate Phaethon's doubt with regard to his divine provenance, Apollo promises to fulfil all his

---

*Proceedings of the international workshop for young researchers, held at the Graduate School of Letters, Kyoto University, Kyoto 2-3 December 2013*, ed. Kayo Hirakawa (Kyoto, 2014), pp. 39-43; *Hans von Aachen (1552-1615): Hofkünstler in Europa*, pp. 196-99, cat. 65-66 (Christof Metzger); Seifertová, *Malby na kameni*, pp. 132-36; *Die Entdeckung der Natur. Naturalien in den Kunstkammern des 16. und 17. Jahrhunderts*, ed. Wilfried Seipel, exhibition catalogue, Schloss Ambras, Innsbruck, and Kunsthistorisches Museum, Vienna (Vienna, 2006), pp. 171-74, cat. 3.14 (Hana Seifertová); Jacoby, *Hans von Aachen (1552-1615)*, pp. 141-45, cat. 40-41; DaCosta Kaufmann, *The School of Prague*, pp. 140-41, cat. 1.17; p. 149, cat. 1.47.

33 "Auf ainem ebenen fuess ain ablang viereckhete grosse tafel von agata, in eben eingefast, auf beiden seiten gemahlt, als auf ainer Phaeton, auf der andern seithen Bacchus, Venus und andere götter und göttinnen," "Ain grosse gefierte taffel von agata, auf einer seiten Perseus und andromeda, ein Schiffbruch Aeneae." Quoted from Jacoby, *Hans von Aachen (1552-1615)*, pp. 141, 143.

34 Jacoby, *Hans von Aachen (1552-1615)*, p. 142.

wishes. Hence, he also has to agree to Phaethon's audacious desire to conduct the sun chariot. Jupiter, flying on an eagle, hurls a lightning bolt on Phaethon in order to prevent his setting ablaze earth and heavens. Phaethon falls from the chariot; on the bank of the river Eridanos, his sisters, the Heliadae, are transformed into trees (weeping willows) by grief, his half-brother Cygnus into a swan.[35]

The second plate shows on one side the releasing of the winds by Aeolus [**Fig. 6.2**], which was to thwart Aeneas' flight across the sea. In the foreground, Juno, in pursuit of the Trojan hero, asks the god of wind for help, in exchange for which he is to receive a nymph, the latter being sat on a rock to the right. Aeolus stands in front of a mountain, from which he frees the winds, blowing cherubs. In the sea ruffled by storm, Aeneas' fleet threatens to sink. To protect Aeneas, Venus appears above it in her chariot pulled by doves. The reverse depicts the freeing of Andromeda according to Ovid's *Metamorphoses* [**Fig. 6.3**]. After his adventure with the Gorgons, Perseus, flying on Pegasus, catches sight of Andromeda on the seashore. Chained to a rock, she is to be sacrificed to a sea monster in order to appease Neptune. Hans von Aachen depicts the moment in which Perseus, his sword drawn, descends rapidly upon the monster emerging from the sea.

Hans von Aachen impressively displays his virtuosity in incorporating into the compositions stone structure, coloring, layering, and veining. For the liberation of Andromeda he places the onrushing hero on the winged horse into a glowing light-yellow part with red particles and dark-grey veiling. The dramatic red insinuates the bloody struggle to follow. The clouded structures of the stone are used by the painter for the sea made turbulent by the releasing of the winds. As part of Amor's triumphal procession, the stone structure predetermines the mountain range formation in the background. The sky's red coloring lights the hearts on the cherubs' 'standards,' as well as the burning torch.

Jacoby has considered the images as a cycle of the four elements: the earth is represented by the triumphal procession of Amor and Bacchus, the air by the winds of Aeolus, water by the freeing of Andromeda, and fire by the fall of Phaethon.[36] In referring to the elements, the images also address the development of the minerals and hence their own origins, traditionally ascribed by the

35 The composition refers to an invention by Michelangelo, cf. Jacoby, *Hans von Aachen (1552-1615)*, p. 142. A drawing by Hans von Aachen (Basel, Kunstmuseum) shows the process of how the composition and the given structures of the stone are combined, *Hans von Aachen (1552-1615): Hofkünstler in Europa*, pp. 200-01, cat. 67 (Christof Metzger).

36 Jacoby, *Hans von Aachen (1552-1615)*, pp. 143-44.

natural sciences to an interaction between the four elements. From the mixing of the four elements, gems and stones form into primary matter. Different mixture relations effect different qualities, such as translucence, clarity, weight, or hardness. Anselmus de Boodt, too, emphasizes in his above-mentioned treatise that this nature-philosophical perspective, if flawed in some details, is principally accurate.[37]

Hans von Aachen's images, however, not only allegorize the four elements as an allusion to the origin and the ur-substances of stone, his chosen support; on different levels they illustrate the process of metamorphosis, occurring as a result of the mixing of elements and the petrification of ur-matter. Hence Phaethon's unfortunate ride with the sun chariot hardens the earth through the impact of fire. It becomes fissured, the cracks on the ground allowing the penetration of light into Tartarus (Ovid, *Metamorphoses,* 2.5.260-61). The myth describes the petrification of the earth by means of fire and the quality of the material's translucency. It is not only the representation of myth that recalls this process, it is also the image surface itself, materially constituting petrified earth. Furthermore, the bloodied red around the fighting hero Perseus is the color of the stone, representing blood within the image. The other way round, from the head of Medusa—which Perseus carries with him during the struggle with the sea monster—blood is flowing into the sea, turning into corals. As reported by Ovid, the coral is still characterized by the quality of forming a shrub in the sea and hardening into stone on land. Finally, it is well-known that Medusa's gaze itself had the power of turning everything into stone; this is how the Atlas mountains, the petrified form of the titan Atlas, originated.[38]

Rock and the forces inherent to it is the subject of the winds of Aeolus. The force entrapped in and liberated from the stone, air, is capable of turning the sea into a wild vortex, formed by the alabaster stone and its swirling structures itself. Consequently, the painter is able to explain the origin of these structures, showing the power of air and water that have affected the forms of the painting surface. Andromeda is tied to the opposite side of the same mountain in which the winds are imprisoned. Similar to Hans von Aachen's image is the contrast

37 Cf. Johannes E. Hiller, "Die Mineralogie Anselmus Boetius de Boodts," *Quellen und Studien zur Geschichte der Naturwissenschaften und der Medizin* 8 (1942), pp. 42-43. With regard to conceptions common around 1400 concerning the origin and development of the painter's materials (pigments, metals and minerals, gems and stones) and related writing in the field of natural history (e.g. Marbode de Rennes, Albertus Magnus), see Anne Dunlop, "On the origins of European painting materials, real and imagined," in *The matter of art. Materials, practices, cultural logics, c. 1250-1750*, ed. Christy Anderson, Anne Dunlop, and Pamela H. Smith (Manchester, 2016), pp. 83-87.

38 See Christopher Nygren's contribution in this volume, pp. 153-54.

between stone, life, and movement set into play in Ovid's *Metamorphoses*: if the hero had not noticed the windblown hair, the tears running from her eyes, he would have considered Andromeda, in her beauty, a marble image. Hans von Aachen contrasts the perfection of the female nude resembling a marble statue with the dark-red rock. The marvel with which Perseus looks at this image apparently made of stone, yet filled with life, certainly corresponds to the admiration hoped for by Hans von Aachen in the reception of his painterly transformations of stone into life-like characters.[39]

Hans von Aachen succeeds in illustrating in his work not only the genesis of the painting's surface ground from the elements and the process of transformation from primary matter into stone,[40] but also the characteristic qualities of alabaster. He draws attention to them, even gives them influence over his invention and composition. In particular, the alabaster 'lives' off of its translucency; it is the quality which de Boodt mentions as particularly praiseworthy in the agate.[41] For the reception of the artwork, this characteristic is especially important as changing light also changes the appearance of the image: some sections might flash up, others be extinguished. Image elements are emphasized with clarity or sink back into the image ground. Hence light contributes to the animation of the image, giving life to the figures and scenes formed in and of stone.

The semi-transparency of stone is usually described as a challenge to the painter. Haindorfer already emphasizes the special quality of an alabaster tablet painted on both sides, the respective compositions coordinated in a way so

39 Ovid, *Metamorphosen. Lateinisch / Deutsch*, trans. and ed. Michael von Albrecht (Stuttgart, 2010), 4, 5.673-77: "quam simul ad duras religatam bracchia cautes / vidit Abantiades (nisi quod levis aura capillos / moverat et tepido manabant lumina fletu, / marmoreum ratus esset opus), trahit inscius ignes / et stupet et visae correptus imagine formae / paene suas quatere est oblitus in aëre pennas."

40 Hans Holländer, "'Denkwürdigkeiten der Welt oder sogenannte Relationes Curiosae': Über Kunst- und Wunderkammern," in *Die Erfindung der Natur. Max Ernst, Paul Klee, Wols und das surreale Universum*, ed. Karin Orchard and Jörg Zimmermann, exhibition catalogue, Sprengel-Museum, Hannover (Freiburg im Breisgau,1994), p. 41, speaks of an alchemical "principle of the potential transformation of all things" ("Prinzip der potentiellen Verwandlungsfähigkeit aller Dinge") which is shown via these metamorphoses on the respective levels of material, narrative and representation. In relation to the works by Bernard Pallisy, Lisbet Tarp, "Images on trial. Nature's Role in Late Renaissance Decorative Arts," in *Paragone als Mitstreit*, ed. Joris van Gastel, et al. (Berlin: 2014,) pp. 148-49, refers to the fact "that the work of art was associated and aligned with natural petrifaction."

41 The 1619 inventory of the estate notably does not describe the stone as oriental alabaster but as agate. Concerning the translucency of agate, see de Boodt, *Gemmarum et Lapidum Historia*, p. 125: "Interdum Achates semitranslucet, quod in ea laudatur."

they do not disrupt each other.[42] Hans von Aachen, however, draws attention precisely to the translucency of the material, demonstrating his virtuosity in meeting the difficult task of painting on two sides. It is striking to see that the front and back of the two plates, respectively, refer to one another: some objects, figures and animals can be observed on both sides. In both Amor's triumphal procession and Phaethon's fall, a chariot plays, so to speak, the 'lead role'; they are composed to match one another. Similarly, swans and a round temple appear on both sides; and, in both scenes, river gods with their amphorae are positioned in the foreground. On the second plate, a sea monster emerges from the floods on both sides. A naked woman sits or stands, respectively, on the rocky shore. Ships are to be seen in the ruffled sea. The quality of translucency is intensified not only by means of the artful device of moving the figures and props forwards and backwards as a result of the changing light incidence, but these are even constructed so as to penetrate the stone, forming a new scene. This is particularly the case for the sea monster, diving through the stony waves of the sea, spreading fear and dread on both sides.

## Made from Earth and Animated by God

Two matching alabaster plates by Hans von Aachen depicting the Creation of Eve [**Fig. 6.4**] and the Expulsion from Paradise are dated around 1605.[43] While the unique representation of a solar eclipse in the Expulsion has drawn some scholarly attention, as proof of the exchange between art and sciences at the Court of Rudolf II, the Creation scene has been little remarked on.

Again the composition is closely attuned to the stone structures. The celestial bodies, sun, and moon, as well as a flying angel, are encircled by a light

42 It is the alabaster painting by an unnamed artist which Philipp Hainhofer also describes in his letter to Philip II, Duke of Pomerania, from 26 June 1611. Doering, *Des Augsburger Patriciers Philipp Hainhofer Beziehungen*, p. 148: "Mehr ist vorhanden ein schöner durchsichtiger Romanscher stain, auf den auf ainer seithen daß Manna wie es das Israelitisch volckh in der wüesten es aufleset, auf der anderen seithen daß Pharonisch heer wie es mit dem König im rothen meer ersaufft, gemahlet, und wie ich verstehe von Clemente octauo ist herauß geschickhet worden, schön daß der Stain durchsichtig vnd doch kain gemahl daß ander hindert (...)".

43 The tables are painted on one side only, but cut from the same block of stone. 25 × 21.5 cm, Vienna, Kunsthistorisches Museum, Sammlungen Schloss Ambras. Seifertová, *Malby na kameni*, pp. 128-30; *Die Entdeckung der Natur*, pp. 169-71, cat. 3.13 (Hana Seifertová); Jacoby, *Hans von Aachen (1552-1615)*, pp. 78-80, cat. 1-2; DaCosta Kaufmann, *The School of Prague*, pp. 148-49, cat. 1.45-1.46.

aureole formed by the stone; due to the translucency of the alabaster, the aureole lights up and shines under changing light circumstances. The landscape, too, is co-determined by the stone layers, forming mountains and valleys. While God, in the image center, turns with a raised right hand towards Eve as emerging from Adam's side, seemingly pointing towards her with a stick (?) in his left, on the right image margin a male figure is stuck in the ground up to his hip. Jacoby has suggested this might be the "embodiment of the budding tree of knowledge."[44] It seems more probable, however, that the figure stuck in the earth and hence seemingly formed from the earth is an allusion to the creation of Adam from soil. At the same time, the earthen man depicted in this scene also relates to God's creation around him. His hand extended, he points to the act of creation. Man, as formed from earth in this image, is simultaneously a figure invented by the painter on the stone tablet, hence created from stone; this figure is still part of the earth—which is simultaneously the material which constitutes the image surface itself. Both stone and man are made of earth, the artist himself ultimately creating his figures from earth. His art is based on the material's genesis.[45] The figure thus references the origin of all creatures and things, prompting a comparison of different acts of creation by means of the pointing of the finger. The animation of earth-based creatures, however, is only given to God. de Boodt correspondingly defines stone as a "mixed, inanimate, hard (...) body, formed by nature from mere earth (...)".[46] The capacity of forming stone (*virtus lapifica*) was added to the primary material, it is "prescribed, arranged and created by God as it were as a natural facility (*seminaria*) of all things."[47] Divine omnipotence gives life to the essence of stone, as generated by *natura* and *ars*. Hence the figure stuck in soil and stone surface remains stone and soil. Correspondingly, by means of its greyish colorlessness, the painter differentiates it from the animated and ensouled, colorful work of God. If, according to de Boodt, divine omnipotence is enclosed in stone, it was to be discovered by the observer within and on the stone; Hans von Aachen adds to this nature-philosophical observation the artist's view on his own creative activity.

44 Jacoby, *Hans von Aachen (1552-1615)*, p. 79. Seifertová, *Malby na kameni*, pp. 128-30, does not address the figure.

45 Cf. Felfe, "Figurationen im Gestein," pp. 155-62.

46 Cf. Felfe, "Figurationen im Gestein," pp. 155-62.

47 See Hiller, "Die Mineralogie," pp. 46-47. de Boodt, *Gemmarum et Lapidum Historia*, pp. 11-12: "Causa formalis lapidum vel gemmarum nihil aliud est, quam virtus lapidifica, quae vel materiae praeparata insidet, vel efficie(n)ti causa accedere debet, ut lapis fiat. Haec a Deo Opt. Max. vt aliarum rerum seminiaria, ordinata, constituta & creata est."

### Overcoming Dead Materiality

An early modern observer might also have recognized or intuited divine omnipotence when regarding the painting on black marble described in the introduction to this essay. This work depicts a scene preceding Christ's resurrection. Already in this early example of painting on stone as part of Hans von Aachen's work, the artist responds to the shapes given by nature, making use in particular of the structure of a cave suggested by the stone. Consequently, he addresses the materiality of stone in the image, painting stone on the stone support, layered, dressed, and raw. The stone-based image carrier is, correspondingly, stone in the image: the stone of the burial cave and the stone on which Jesus is laid out. The painter's work is already inextricably linked to nature through the material. The treatment of the material by nature and by the painter become an inseparable unit: the subtly used thin white lines, the veiled parts letting the black of the marble shine through, the hardly noticeable colored hues imitating the natural appearance of marble, literally granting the observer a glance into the stone. This, too, occurs in a double sense, for, as the gaze of the observer penetrates the black marble, he is able to see the image concealed therein,[48] as well as penetrate the stone surrounding the sepulcher.[49] Through the work of the artist, the observer is able to see the body of Christ even before the latter will penetrate the closed grave in order to overcome death through the resurrection. Here, too, only divine omnipotence is capable of bringing to life dead, inanimate life.

In this case, too, the particular qualities of the material are effective in terms of an aesthetics of reception. While in the case of the alabaster panels, this especially concerned their play with translucency, the marble work is to be apprehended primarily by its haptic characteristics. The *kunstkammer*

48 The image concealed in stone is an art topos already raised repeatedly in Pliny's *Natural History*. Cf.: König-Lein, "Ein Spiel der Natur," pp. 135-36; Horst W. Janson, "The 'Image made by chance' in Renaissance thought," in *De artibus Opuscula XL. Essays in Honor of Erwin Panofsky*, ed. Millard Meiss, vol. 1 (New York, 1961). The motif is also taken up for the *Paragone* of painters and sculptors—the sculptor with laborious craftsmanship uncovering the image concealed in the stone. Cf. Hessler, "Maler und Bildhauer," pp. 87-88.

49 The gaze precisely does not—as described by Jacoby, *Hans von Aachen (1552-1615)*, p. 116—fall "into the grave from the outside, through the opened grave" ("von außen, durch das geöffnete Grab in die Höhle"); rather, the viewer sees that which is hidden and concealed inside the stone. Cf. *Hans von Aachen (1552-1615): Hofkünstler in Europa*, p. 153 (Joachim Jacoby); while here he emphasizes that the image represents something "unreachable, even unlawful" ("Unerreichbares, ja Unerlaubtes"), he sees the grave plate pushed into the image below.

rendered possible such a reception as the works could be held, touched, and felt. Similarly, the categorization in de Boodt's text aims at a perception of the material with all the senses. He emphasizes the fact that differences between the rocks are perceived by the five senses, correspondingly ordering them according to visual, haptic, and olfactory characteristics. They are described as cold, warm, dry, colored, heavy, light, transparent, or dark. The smell can resemble that of the iris or be reminiscent of musk, thyme, must, wine, or milk.[50] The mineralogist's classification aids are similarly relevant for the reception of the artwork, which is not only to be seen, but also to be perceived in its weight and temperature, and with which particular smells are associated. Looking at the image of the dead body of Christ in the stone tomb is related to the haptic impression of the hard, dead material.

The stone's hardness, effective in terms of an aesthetics of production and reception, is also crucial for an understanding of the representation of the body of Christ in the sepulcher. In terms of the dimension of production, the manner of how hard stone, especially diamond or porphyry, are to be mastered is already described by Pliny, as well as by Theophilus, Vasari, and de Boodt. The effective means is buck's blood, on the one hand hardening the tool, on the other hand softening the stone. The recipe is already made use of by Augustine for the illustration of the Christian salvation history. According to Alcuin, it is only the blood of Christ that softens the sinner, just as the hard diamond is softened neither by fire nor hammer, but by the buck's blood.[51] The iconography criticized in the art agent Hainhofer's letter as mentioned in the beginning—the representation of the altar with burning candles in the sepulcher—refers to the sacrificial lamb Christ, the sacrifice of which was to be repeated at the catholic altars during the mass. This sacrifice is prerequisite for the overcoming of death. In this manner, too, the penetration and softening of the hard material is addressed, for the blood of Christ softens the stone of the sepulcher just as the buck's blood softens the hard stone. The material guides the seeing and feeling observer towards insight into Christ's sacrifice and the wonder of the resurrection through the enclosed grave made of stone.

The shining polished surface becomes similarly important in terms of questions of reception. The surface's glow connects the image of Christ as lit by the glow of the painted candles with the reflections cast by their surroundings. The black polished marble mirrors its environment. Hainhofer describes in terms of his visit to the hall of antiquities in Munich the black marble wall covering

50 Hiller, "Die Mineralogie," pp. 37-38.

51 Cf. Friedrich Ohly, *Diamant und Bocksblut. Zur Traditions- und Auslegungsgeschichte eines Naturvorgangs von der Antike bis in die Moderne* (Berlin: 1976), esp. pp. 72-73.

reflecting the entire room surrounding it.[52] In the stone to be observed close-up, the viewer himself is inevitably mirrored as shadowy [**Fig. 6.5**] as the painted figures themselves; hence he becomes part of the image, his 'penetration' of the stone performed not only through his gaze but occurring in a real and quasi-corporeal manner through the reflection. The observer is now in the image and hence in the sepulcher of Christ. As both Jacoby and Seifertová have argued, this painting is certainly a devotional image. The work suggests impressively that the reception context of the *kunstkammer*—the discourse concerning the rivalry between the arts situated therein and their 'sophistic tug of war'[53] with nature—does not exclude the communication of religious content, but makes use of it in order to render tangible with all senses the devotional perception and contemplation of the passion.

## Bibliography

Baltrušaitis, Jurgis. *Imaginäre Realitäten. Fiktion und Illusion als produktive Kraft*. Cologne, 1984.

Bauer, Rotraud, and Herbert Haupt, eds. *Das Kunstkammerinventar Kaiser Rudolfs II., 1607-1611*. Jahrbuch der Kunsthistorischen Sammlungen in Wien, vol. 72. Vienna, 1976.

Bowron, Edgar Peters. "A Brief History of European Oil Paintings on Copper, 1560-1775." In *Copper as Canvas. Two Centuries of Masterpiece Paintings on Copper 1575-1775*, pp. 9-30. Exhibition catalogue, Phoenix Art Museum. New York, 1999.

Calvillo, Elena. "Authoritative Copies and Divine Originals: Lucretian Metaphor, Painting on Stone, and the Problem of Originality in Michelangelos Rome." *Renaissance Quarterly* 66, no. 2 (2013), 453-508.

Collomb, Anne-Laure. "La Peinture sur pierre en Italie 1530-1630." PhD diss., University of Lyon, 2006.

DaCosta Kaufmann, Thomas. *The School of Prague. Painting at the Court of Rudolf II*. Chicago and London, 1988.

de Boodt, Anselmus Boethius. *Gemmarum et Lapidum Historia*, Hanau, 1609. <urn:nbn:de:bvb:12-bsb10156506-6>.

Distelberger, Rudolf. *Die Kunst des Steinschnitts. Prunkgefäße, Kameen und Commessi aus der Kunstkammer*. Vienna, 2002.

52 Anne Langenkamp, "Philipp Hainhofers Münchner Reisebeschreibungen. Eine kritische Ausgabe" (PhD diss., Technische Universitat Berlin, 1990), p. 151: "Überall umb die gesimbs herumb ist schwarzer, glenzender marmelstain, das man das ganz antiquarium darin sehen kan."

53 Hessler, "Maler und Bildhauer."

Doering, Oscar. *Des Augsburger Patriciers Philipp Hainhofer Beziehungen zum Herzog Philipp II. von Pommern-Stettin, Correspondenzen aus den Jahren 1610-1619*. Quellenschriften für Kunstgeschichte und Kunsttechnik des Mittelalters und der Neuzeit, Neue Folge, vol. 6. Vienna, 1894.

Doering, Oscar. *Des Augsburger Patriciers Philipp Hainhofer Reisen nach Innsbruck und Dresden*. Vienna, 1901.

Dunlop, Anne. "On the origins of European painting materials, real and imagined." In *The matter of art. Materials, practices, cultural logics, c. 1250-1750*, edited by Christy Anderson, Anne Dunlop, and Pamela H. Smith, pp. 68-96. Manchester, 2016,

*Die Entdeckung der Natur. Naturalien in den Kunstkammern des 16. und 17. Jahrhunderts*. Edited by Wilfried Seipel. Exhibition catalogue, Schloss Ambras, Innsbruck, and Kunsthistorisches Museum, Vienna. Vienna, 2006.

Felfe, Robert. "Figurationen im Gestein. Ko-Produktionen zwischen Kunst und Natur," *In Paragone als Mitstreit*, edited by Joris van Gassel, Yannis Hadjinicolaou, and Markus Rath, pp. 153-75, Berlin, 2014.

Findlen, Paula. "Cabinets, Collecting and Natural Philosophy." In *Rudolf II and Prague. The Imperial Court and Residential City as the Cultural and Spiritual Heart of Central Europe*, edited by Eliška Fučiková, pp. 209-19. Prague, London, and Milan, 1997.

Fučiková, Eliška, "Das Leben." In *Hans von Aachen (1552-1615): Hofkünstler in Europa*, edited by Thomas Fusenig, pp 3-11. Exhibition catalogue, Suermondt-Ludwig-Museum, Aachen. Berlin and Munich, 2010.

Fucikova, Eliška. "Die Malerei." In *Hans von Aachen (1552-1615): Hofkünstler in Europa*, edited by Thomas Fusenig, pp. 13-31. Exhibition catalogue, Suermondt-Ludwig-Museum, Aachen. Berlin and Munich, 2010.

Gastel, Joris van, Yannis Hadjinicolaou, and Markus Rath. "Paragone als Mitstreit." In *Paragone als Mitstreit*, edited by Joris van Gastel, Yannis Hadjincalaou, and Markus Rath, pp. 15-32. Berlin 2014.

Giusti, Annamaria. *Pietra Dura. Bilder aus Stein*. Munich, 2005.

Hadjinicolaou, Yannis. "Malerei auf Stein, Steinerne Malerei. Die Farbgestaltung bei Spranger und de Gelder." In *Paragone als Mitstreit*, edited by Joris van Gastel, Yannis Hadjincalaou, and Markus Rath, pp. 211-35. Berlin 2014.

Haiduk, Marina. "Daniele da Volterras 'David und Goliath'. Ein Wettstreit auf mehreren Ebenen." *Wiener Jahrbuch* 57 (2008), 97-131.

*Hans von Aachen (1552-1615): Hofkünstler in Europa*. Edited by Thomas Fusenig. Exhibition catalogue, Suermondt-Ludwig-Museum, Aachen. Berlin and Munich, 2010.

Henning, Andreas. "The new technique of painting on copper." In *Captured emotions. Baroque painting in Bologna, 1575-1725*, edited by Andreas Henning and Scott Schaefer, pp. 25-35. Exhibition catalogue, J. Paul Getty Museum, Los Angeles. Los Angeles, 2008.

Henning, Michael. *Die Tafelbilder Bartholomäus Sprangers (1546-1611). Höfische Malerei zwischen "Manierismus" und "Barock."* Essen, 1987.

Hessler, Christiane J. "Maler und Bildhauer im sophistischen Tauziehen. Der Paragone in der italienischen Kunstliteratur des 16. Jahrhunderts." In *Wettstreit der Künste.*

*Malerei und Skulptur von Dürer bis Daumier*, edited by Ekkehard Mai and Kurt Wettengl, pp. 83-97. Exhibition catalogue, Haus der Kunst, Munich. Wolfratshausen, 2002.

Hiller, Johannes E. "Die Mineralogie Anselmus Boetius de Boodts." *Quellen und Studien zur Geschichte der Naturwissenschaften und der Medizin* 8 (1942), 1-215.

Hirakawa, Kayo. "Narrative and Material: Paintings on Rare Metal, Stone and Fabrik in the Late Sixteenth and Early Seventeenth Centuries." In *Aspects of narrative in art history. Proceedings of the international workshop for young researchers, held at the Graduate School of Letters, Kyoto University, Kyoto 2-3 December 2013*, edited by Kayo Hirakawa, pp. 33-45. Kyoto, 2014.

Holländer, Hans. "'Denkwürdigkeiten der Welt oder sogenannte Relationes Curiosae': Über Kunst- und Wunderkammern." In *Die Erfindung der Natur. Max Ernst, Paul Klee, Wols und das surreale Universum*, edited by Karin Orchard and Jörg Zimmermann, pp. 34-45. Exhibition catalogue, Sprengel-Museum, Hannover. Freiburg im Breisgau, 1994.

Jacoby, Joachim. *Hans von Aachen*. Rotterdam, 1995. (The new Hollstein German engravings, etchings and woodcuts).

Jacoby, Joachim. *Hans von Aachen (1552-1615)*. Munich and Berlin, 2000.

Janson, Horst W. "The 'Image made by chance' in Renaissance thought." In *De artibus Opuscula XL. Essays in Honor of Erwin Panofsky*, edited by Millard Meiss, 1:254-66. New York, 1961.

König-Lein, Susanne. "Ein Spiel der Natur? Bildersteine und Steinbilder." In *SpielKunst-Glück. Die Wette als Leitlinie der Entscheidung. Beispiele aus Vergangenheit und Gegenwart in Kunst, Wissenschaft, Wirtschaft*, ed. Johann Konrad Eberlein, pp. 133-44. Vienna, Berlin, and Münster, 2011.

*Die Kunstdenkmäler des Benediktinerstiftes Kremsmünster. Teil 2: Die stiftlichen Sammlungen und die Bibliothek*. Edited by Hans Gertele-Grenadenberg. Österreichische Kunsttopographie, vol. 43. Vienna, 1977.

Langedijk, Karla. "The Table in Pietre Dura for the Emperor. A New Understanding of Rudolph II as a Collector." *Mitteilungen des Kunsthistorischen Institutes in Florenz* 42, 1998 (1999), 358-82.

Langenkamp, Anne. "Philipp Hainhofers Münchner Reisebeschreibungen. Eine kritische Ausgabe." PhD diss., Technische Universitat Berlin, 1990.

Lohff, Johanna Beate. *Malerei auf Stein. Antonio Tempestas Bilder auf Stein im Kontext der Kunst- und Naturtheorie seiner Zeit*. Munich, 2015.

Mannheim, Christiane. "Edelsteine und Mineralien. Symbole theologischer Spekulationen und mystischer Naturverehrung werden zu Objekten wissenschaftlicher Forschungen." In *Erkenntnis, Erfindung, Konstruktion. Studien zur Bildgeschichte von Naturwissenschaften und Technik vom 16. bis zum 19. Jahrhundert*, edited Hans Holländer, pp. 197-222. Berlin, 2000.

Metzler, Sally. *Bartholomeus Spranger. Splendor and Eroticism in Imperial Prague. The Complete Works*. New Haven and London, 2014.

Müller, Jürgen. *Concordia Pragensis. Karel van Manders Kunsttheorie im Schilder-Boek. Ein Beitrag zur Rhetorisierung von Kunst und Leben am Beispiel der rudolfinischen Hofkünstler*. Munich, 1993.

Ohly, Friedrich. *Diamant und Bocksblut. Zur Traditions- und Auslegungsgeschichte eines Naturvorgangs von der Antike bis in die Moderne*. Berlin, 1976.

Ovid (P. Ovidius Naso). *Metamorphosen. Lateinisch / Deutsch*. Translated and edited by Michael von Albrecht. Stuttgart, 2010.

*Pietra Dipinta. Tesori nacosti del '500 e del '600 da una collezione privata milanese*. Edited by Marco Bona Castellotti. Exhibition catalogue, Palazzo Reale, Milan. Milan, 2001.

*Prag um 1600. Kunst und Kultur am Hofe Rudolfs II*. Exhibition catalogue, Villa Hügel, Essen. Freren, 1988.

Seifertová, Hana. "Pittura su pietra der rudolfinischen Zeit und ein Vanitasgemälde von Christoph Gertner." In *Rudolf II, Prague and the World. Papers from the International Conference Prague 2-4 September 1997*, edited by Lobomír Konečný, pp. 95-98. Prague, 1998.

Seifertová, Hana. *Malby na kameni—umělecký experiment v 16. a na začátku 17. století/ Paintings on Stone—An Artistic Experiment in the 16th and Early 17th Centuries*. Prague, 2007.

Spicer, Joaneath. "The Role of 'Invention' in Art and Science at the Court of Rudolf II." *Studia Rudolphina* 5 (2005), 7-16.

Strinati, Claudio. "Sebastiano del Piombo und die Renaissance in Rom." In *Raffaels Grazie–Michelangelos Furor. Sebastiano del Piombo (1584—1547)*, edited by Wolfgang Lindemann and Claudio Strinati, pp. 15-21. Exhibition catalogue, Museen Gemäldegalerie, Berlin, and Palazzo Venezia, Rome. Berlin, 2008.

Suthor, Nicola. "Transparenz der Mittel. Zur Sichtbarkeit der Imprimitur in einigen Werken Rembrandts" In *Der Grund. Das Feld des Sichtbaren*, edited by Gottfried Boehm and Matteo Burioni, pp. 223-49. Munich, 2012.

Tarp, Lisbet. "Images on trial. Nature's Role in Late Renaissance Decorative Arts." In *Paragone als Mitstreit*, edited by Joris van Gastel, Yannis Hadjincalaou, and Markus Rath, pp. 141-52. Berlin, 2014.

Thomas, Ben. "'The lantern of painting': Michelangelo, Daniele da Volterra and the paragone," *Apollo* 154, no. 474 (2001), 46-53.

Vahland, Kia. *Lorbeeren für Laura. Sebastiano del Piombos lyrische Bildnisse schöner Frauen*. Berlin, 2011.

Vasari, Giorgio. *Die Leben der ausgezeichneten Gemmenschneider, Glas- und Miniaturmaler Valerio Belli, Guillaume de Marcillat und Giulio Clovio*. Translated by Victoria Lorini and edited by Anja Zeller. Berlin, 2006.

Weinhold, Ulrike. "Die Habsburger und die frühe Dresdner Kunstkammer." In *Die kurfürstliche-sächsische Kunstkammer in Dresden. Geschichte einer Sammlung*, edited by Dirk Syndram and Martina Minning, pp. 62-77. Dresden, 2012.

Zimmer, Jürgen. *Joseph Heintz der Ältere als Maler*. Weissenhorn, 1971.

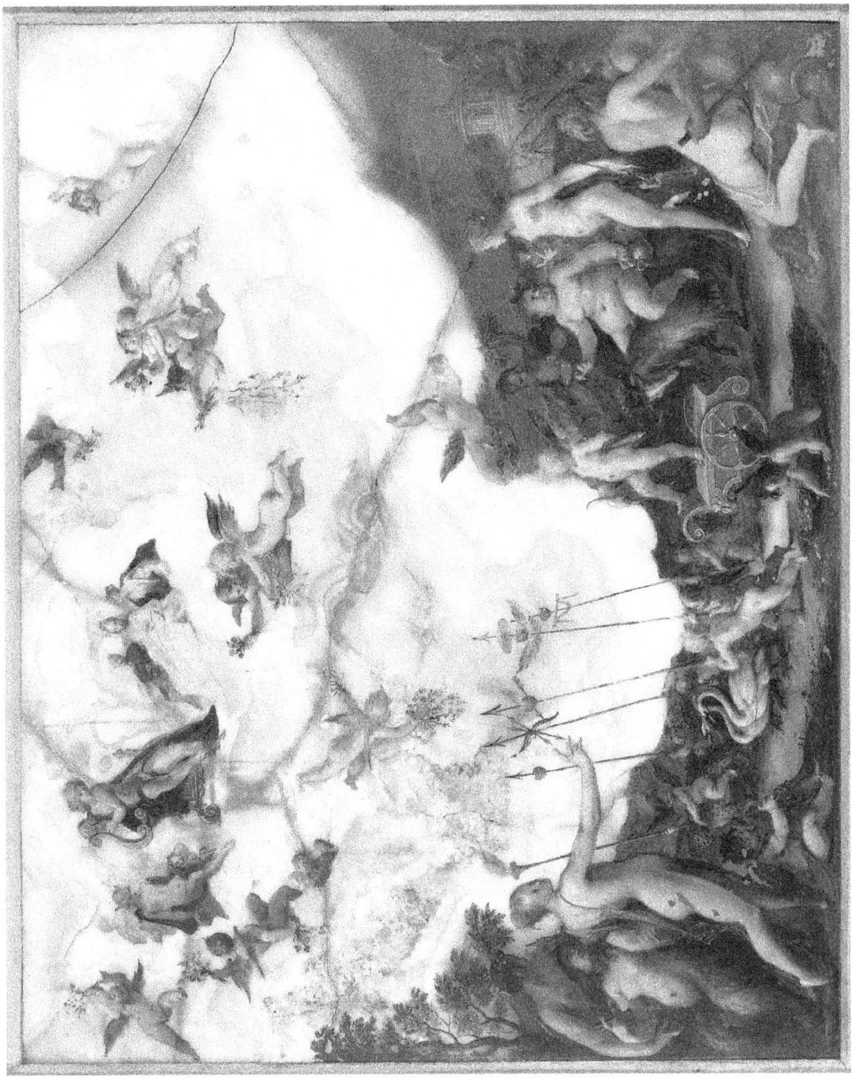

FIGURE 6.1 *Hans von Aachen,* The Triumph of Amor and Bacchus, *oil on alabaster, ca. 1600, Innsbruck, Ambras Castle* (PHOTO: KUNSTHISTORISCHES MUSEUM, VIENNA).

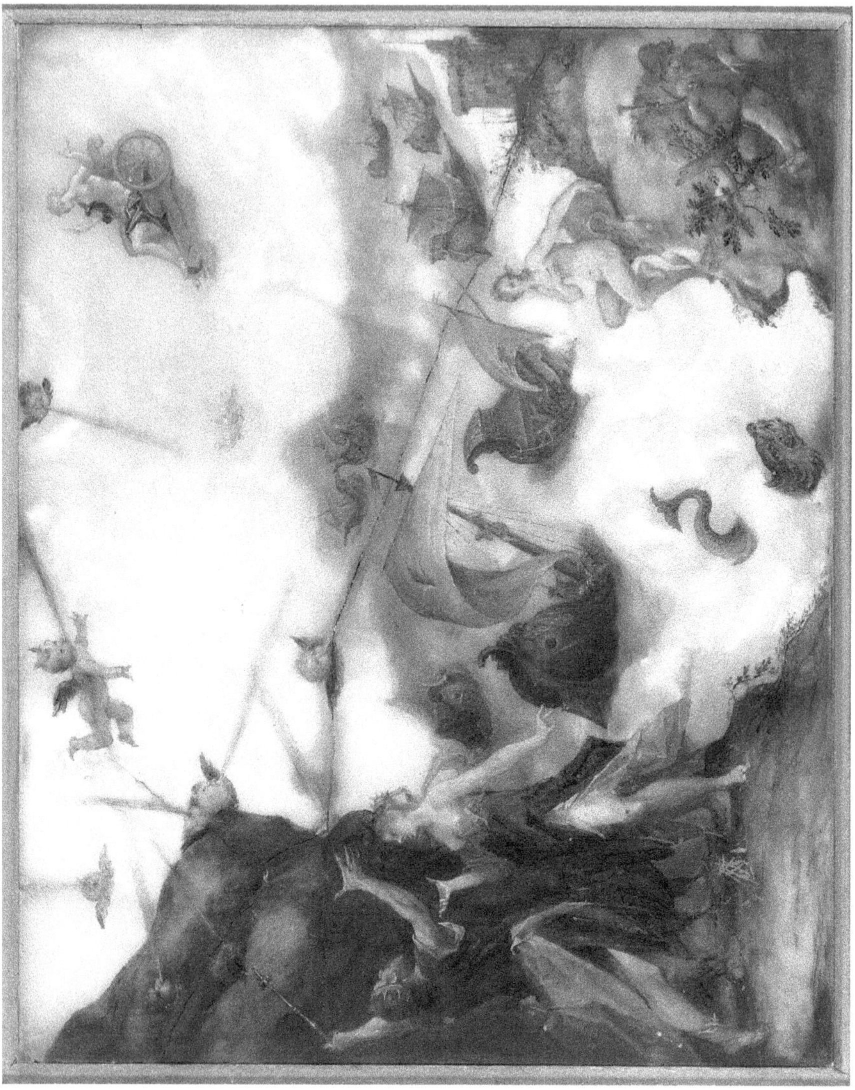

FIGURE 6.2 *Hans von Aachen,* The Unleashing of the Winds, *oil on alabaster, ca. 1600, Innsbruck, Ambras Castle* (PHOTO: KUNSTHISTORISCHES MUSEUM, VIENNA).

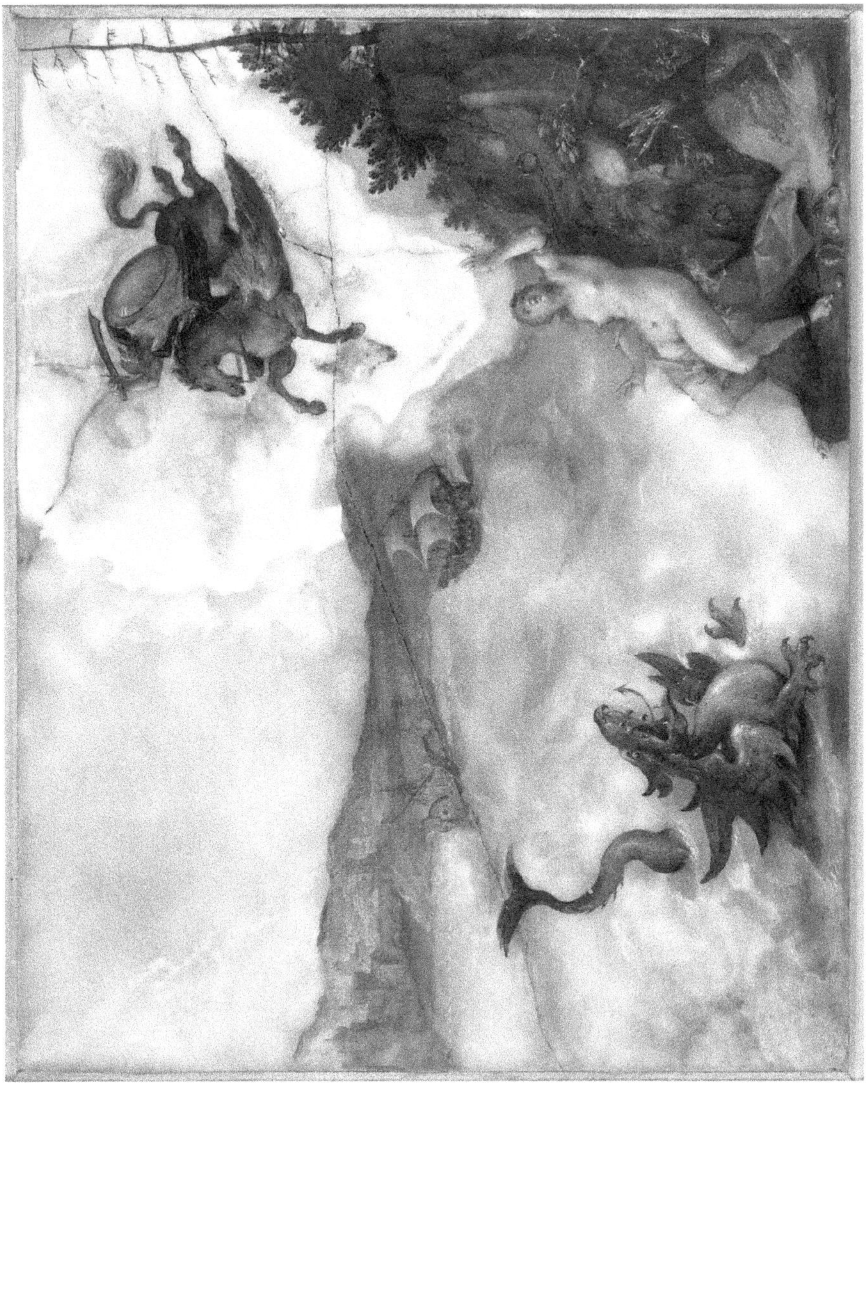

FIGURE 6.3 *Hans von Aachen,* The Freeing of Andromeda, *oil on alabaster, ca. 1600, Innsbruck, Ambras Castle* (PHOTO: KUNSTHISTORISCHES MUSEUM, VIENNA).

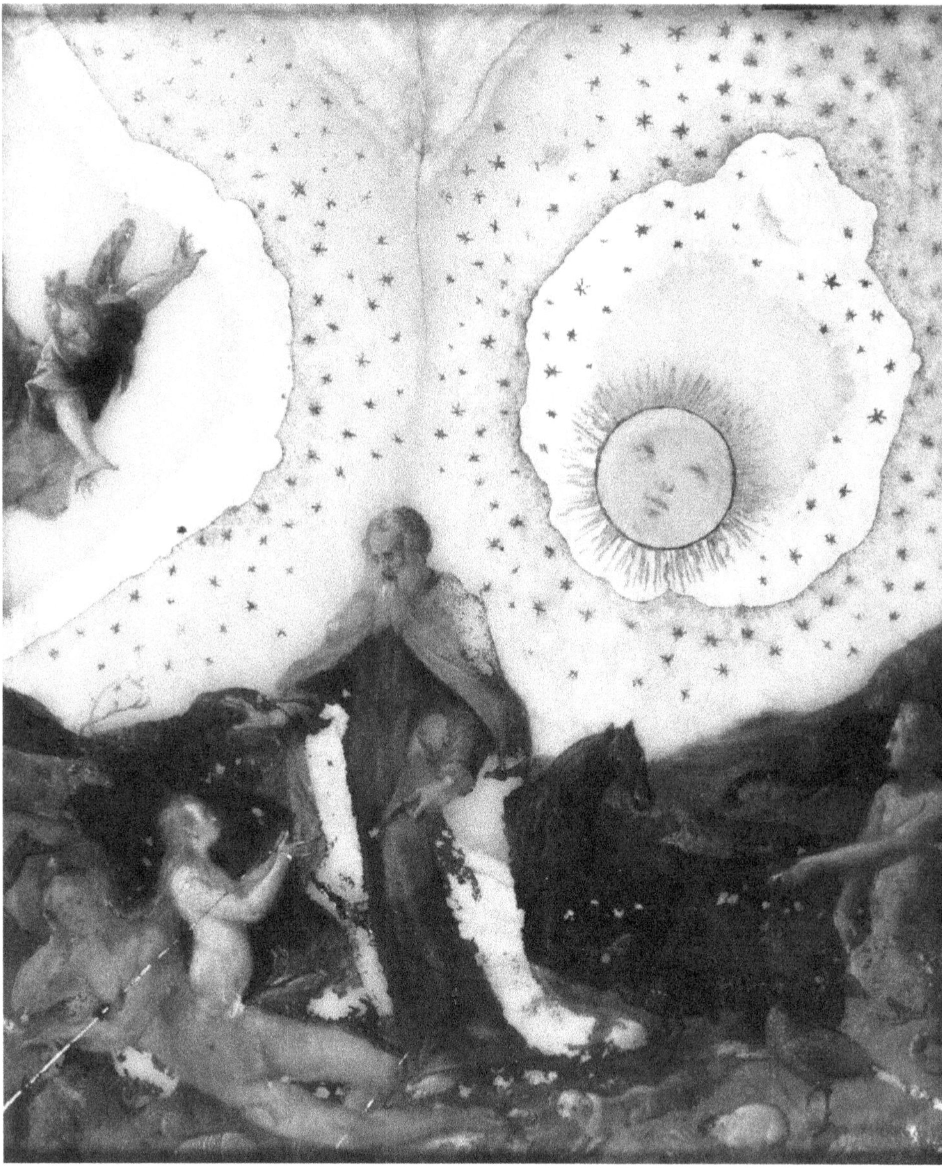

FIGURE 6.4 *Hans von Aachen,* The Creation of Eve, *oil on alabaster, ca. 1605, Innsbruck, Ambras Castle* (PHOTO: KUNSTHISTORISCHES MUSEUM, VIENNA).

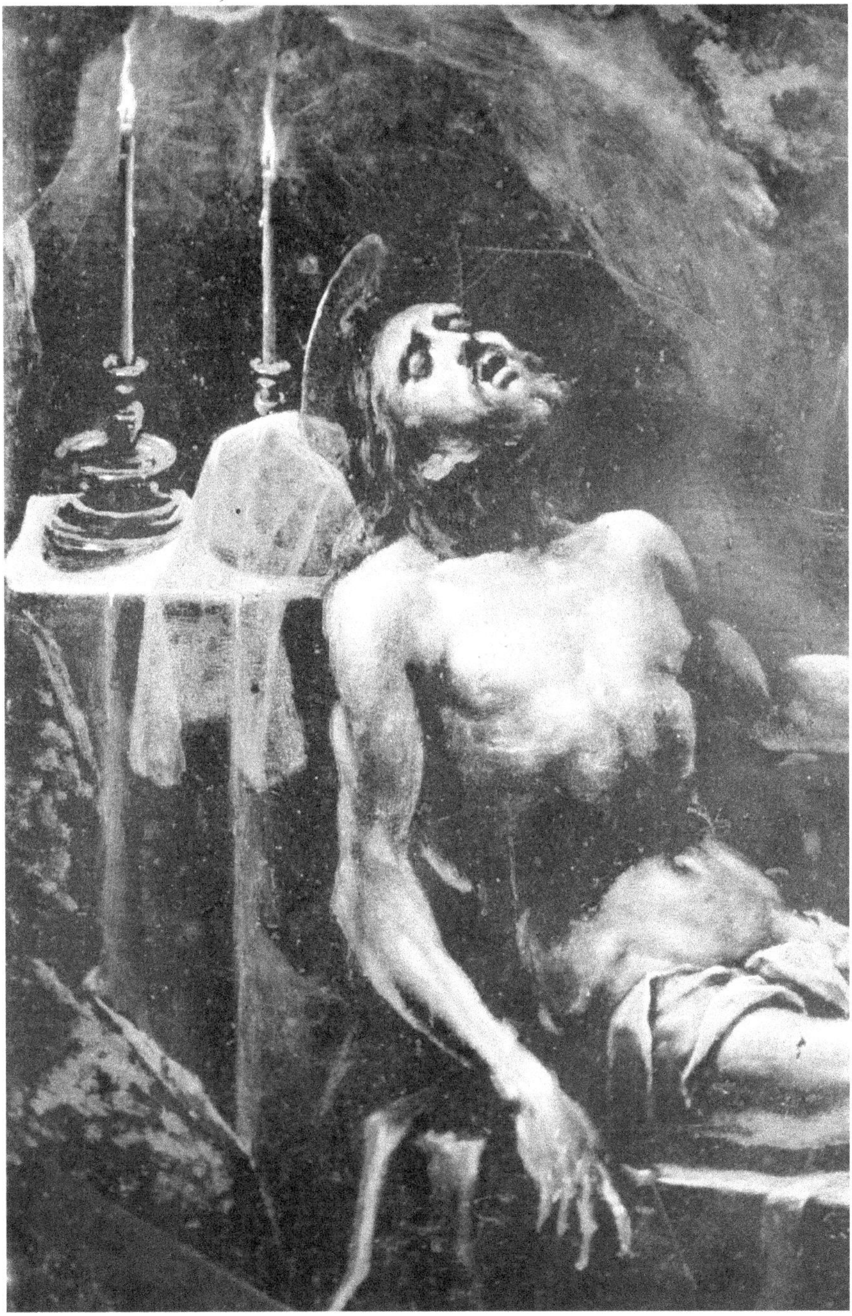

FIGURE 6.5 *Hans von Aachen,* Pietà, *oil on black marble, 1594, Kremsmünster Abbey, detail* (PHOTO: AUTHOR).

CHAPTER 7

# Painting on Stone and Metal: Material Meaning and Innovation in Early Modern Northern European Art

*Nadia Baadj*

In the years following the consecration of the Jesuit Church of Antwerp (Saint Carolus Borromeuskerk) in 1621, a number of Flemish artists including Sebastian Vrancx, Pieter Neefs I and II, and Wilhelm Schubert van Ehrenberg depicted its opulent Baroque interior, which featured a spectacular material interplay between sculpture, stucco, multiple varieties of marble, and oil paintings.[1] In contrast to the majority of these compositions, which were painted on panel or canvas, Ehrenberg's painting on marble, dated to 1668, represents an innovative approach to picturing the multi-sensorial experience evoked by the exceptional materiality of the "Marmortempel" (marble temple), as the church was often called [**Plate 25**].[2] By portraying the interior of the church on Carrara marble, one of the chief varieties from which it was constructed, Ehrenberg underscores the novelty of this material in Netherlandish architecture and also creates a direct and tangible link between the visual effects of the stone's surface—its subtle translucency, polish, and veining—and its substantial weight and tactile qualities, such as its smoothness, coldness, and hardness.

In addition to the obvious allusion to the predominantly gray marble interior, the decision to paint this particular architectural structure on marble as opposed to more conventional supports also resonates with the materiality of one of the church chapels. The Houtappel Chapel was commissioned in 1635 by the three Houtappel sisters, 'spiritual daughters' of the Jesuit order, and their cousin, Anna 's Grevens.[3] In addition to an altar painting of the

1 On the decoration of the church interior, see Léon Lock, "Rubens and the Sculpture and Marble Decoration," in *Innovation and Experience in the Early Baroque in the Southern Netherlands: The Case of the Jesuit Church in Antwerp*, ed. Piet Lombaerde (Turnhout, 2008), pp. 155-74.

2 Hanna Seifertova, *Malby na kameni: Uměcký experiment v 16. Na začátku 17 století/ Painting on Stone: An Artistic Experiment in the 16th and Early 17th Centuries* (Prague, 2007), p. 162.

3 On the decoration and commission of the Houtappel family chapel, see Bert Timmermans, "The Chapel of the Houtappel Family and the Privatisation of the Church in Seventeenth-Century Antwerp," in Lombaerde, *Innovation and Experience*, pp, 175-86.

 | DOI 10.1163/9789004361492_011

Assumption of the Virgin copied after a work by Peter Paul Rubens, stucco ceiling reliefs designed by Rubens, white marble sculptures of the patron saints of the three sisters and their cousin, and a plethora of white marble ornament, the Houtappel chapel also contains a panoply of nearly thirty different kinds of local and imported colored marbles.[4] Following the practice of Italian and Central European forerunners, like Hans von Aachen, who manipulated the natural veining of marble and alabaster to suggest compositional motifs such as rocky outcroppings, the Antwerp artist Hendrick van Balen painted a series of scenes from the Life of the Virgin on white marble panels flanking the chapel's altar [**Plate 26**].[5] Juxtaposed conspicuously with the oil paintings on marble are several unpainted colored marble panels that are framed suggestively, eliciting a comparison between painted marble and the natural figurations [**Figs. 7.1 and 7.2**]. The latter's vivid colors compete with Van Balen's rich palette, and the forms evoked by the stone's natural striations echo the whimsical white marble garlands, festoons, vegetal ornament, and masks that punctuate the entire chapel.

Ehrenberg's painting of the Jesuit Church on marble generated material meaning for the early modern viewer in a number of ways. First, Ehrenberg's use of the same type of marble from which the Church was built underscores the commercial as well as socio-cultural significance of that particular stone, and of the numerous foreign species of colored marbles used to decorate the church interior, to the Antwerp community, which placed an extremely high value on the local collecting and crafting of imported exotic raw materials in the late 16th and 17th centuries.[6] Second, by choosing to paint the architectural and sculptural elements of the church interior in such a veristic manner on

4 See Lock, "Rubens and the Sculpture and Marble Decoration," pp. 156-58. The different varieties of marble are also illustrated in Lombaerde, *Innovation and Experience*, pp. 223-24.

5 For further discussion of Van Balen's painted series, see Anna C. Knaap, "Marvels and Marbles in the Antwerp Jesuit Church: Hendrick van Balen's Stone Paintings of the Life of the Virgin (1621)," in *Jesuit Image Theory in Europe and the Overseas Missions*, ed. Walter Melion, Wietse de Boer, and Karl Enenkel (Leiden, 2016), pp. 352-93.

6 The central role of exotic goods and raw materials in early modern Antwerp has been explored recently by Michael Limberger, "'No town in the world provides more advantages': Economies of Agglomeration and the Golden Age of Antwerp," in *Urban Achievement in Early Modern Europe: Golden Ages in Antwerp, Amsterdam and London*, ed. Patrick O'Brien et al. (Cambridge, 2001), pp. 39-62; Sven Dupré, "Trading Luxury Glass, Picturing Collections and Consuming Objects of Knowledge in Early Seventeenth-Century Antwerp," *Intellectual History Review* 20 (2010), 53-78; Christine Göttler, "The Place of the 'Exotic' in Early-Seventeenth-Century Antwerp," in *Looking East. Rubens's Encounter with Asia*, ed. Stephanie Schrader (Los Angeles, 2013), pp. 88-107.

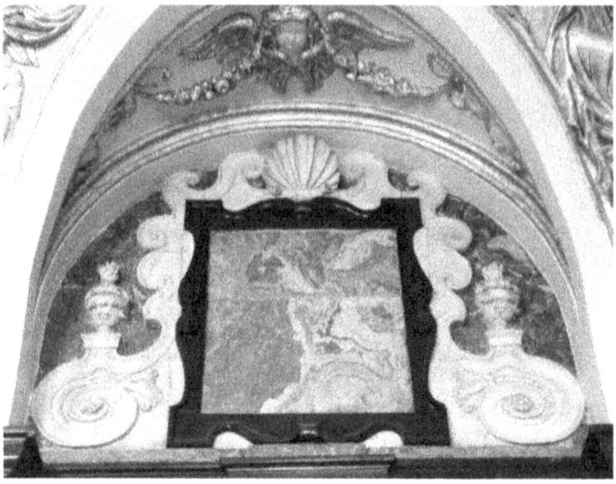

FIGURE 7.1 *Houtappel Chapel, Antwerp Sint-Carolus Borromeuskerk* (PHOTO: AUTHOR).

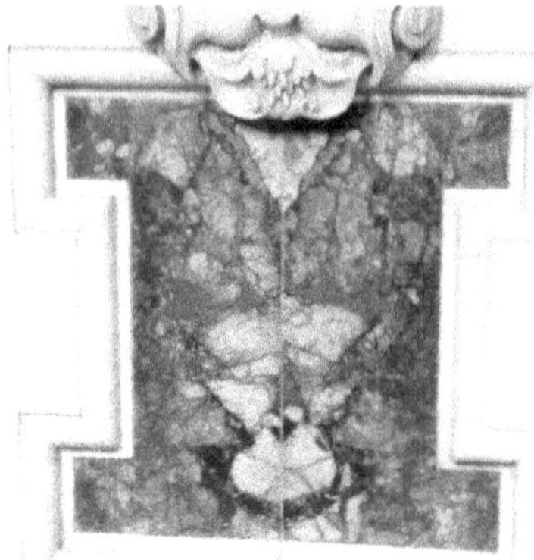

FIGURE 7.2 *Houtappel Chapel. Antwerp Sint-Carolus Borromeuskerk* (PHOTO: AUTHOR).

marble, Ehrenberg takes ownership of the sculptor's material and demonstrates oil paint's powerful capability to create mimetically the illusion of the edifice, without the need for the sculptor's tools or physical strain. Third, Ehrenberg's painting of the church interior on stone, which depicts the entrance to the Houtappel chapel at the far left, parallels the competition between *ars* and *natura* engendered by the artworks inside the chapel, namely Hendrick van Balen's paintings on marble and the framed 'natural' compositions of colored marble beside them.

Although the technique of painting with oil on stone may have originated in Italy, the practice later spread to Northern Europe. Following Sebastiano del Piombo's innovation around 1530, a number of Netherlandish and German artists who visited or worked in Italy, such as the Fleming Bartholomaeus Spranger and his German contemporary Hans von Aachen, adopted the stone medium. In the 1590s, the court environments of Vienna and Prague in particular provided these and other Northern artists ample opportunities for experimenting with and expanding the boundaries of stone's creative potential. From 1600 on, painting on stone became increasingly popular in the Northern and Southern Netherlands, Germany, and France.

A crucial, and often overlooked, aspect of painting on stone supports is the technique's inextricable link to painting on metallic surfaces. This material connection is noted by Giorgio Vasari, who suggests that Sebastiano's experimentation with stone supports paved the way for the practice of painting on silver, copper, tin, and other metals.[7] Painting on copper flourished in the Low

7 Giorgio Vasari, *Lives of the Most Eminent Painters, Sculptors, and Architects*, trans. by Gaston Du C. de Vere, 10 vols. (London, 1912-15), 6:183.

Countries between 1575 and 1650 to such an extent that the Dutch painter and art theorist Karel van Mander identified it as *the* working method of Netherlandish painters.[8] Examining the intertwined histories of these two practices indicates that their simultaneous emergence and efflorescence in this region resulted from the artisanal response to a decidedly Northern 'period eye.'[9] Although artistic exchange between Northern and Southern Europe complicates the identification of regional differences in technique and reception, certain elements generally associated with Northern art, including the predilection for pictorial illusion and the imitation of precious materials, the rise of small-format 'cabinet' pictures, admiration of God's handiwork, and the empirical investigation of the natural world and sustained scrutiny of objects and specimens, are especially suited to painting on stone and metallic surfaces. These characteristics can also be linked to a distinctive culture of curiosity that emerged in the late 16th and early 17th centuries. The multivalent connotations of curiosity in the early modern period, ranging from rarity, miniaturization, and secrecy, to wonder, strangeness, and meticulousness, are visible in the form and content of many cabinet paintings on metal and stone supports.

The technique of painting on these economically and culturally valuable materials also forged perceptible links between oil painting and various other media, effectively mirroring the fluidity and frequent cross-fertilization of artisanal and craft practices in Northern Europe. The use of copper plates for painting, for example, made a tangible allusion to printmaking, enameling, and smithing; the use of marble and alabaster elicited comparisons with

8 In Van Mander's Life of Hans Rottenhammer, he notes that the German Rotenhammer "began to paint on copperplates, in the manner of the Dutch" (*begaf hem op platen te schilderen, gelijck de wijse der Nederlanders is*). Karel van Mander, *Het Schilder-boeck* (Haarlem, 1604; reprint edition, Utrecht, 1969), fol. 296v. On the practice of painting on copper in early modern Europe, see especially Johannes van der Graaf, "Development of Oil-Paint and the Use of Metal-Plates as a Support," paper presented at the Congress on Conservation of Paintings and the Graphic Arts, Lisbon, 2972 (London: International Institute for Conservation of Historic and Artistic Works, 1972), pp. 139-52; Michael Komanecky, "Antwerp Artists and the Practice of Painting on Copper," in *Painting Techniques, History, Materials, and Studio Practice*, ed. Ashok Roy and Perry Smith (Dublin ), pp. 136-39; Christine Göttler, "Affectionate Gifts: Rubens's Small Curiosities on Metallic Supports," in *Munuscula Amicorum: Contributions on Rubens and his Context in Honour of Hans Vlieghe*, ed. Katlijne Van der Stighelen (Turnhout, 2006), pp. 47-66; Michael Komanecky et al., eds., *Copper as Canvas: Two Centuries of Masterpiece Paintings on Copper, 1575-1775*, exhibition catalogue, Phoenix Art Museum; Nelson-Atkins Museum of Art, Kansas City; Mauritshuis, The Hague (New York, 1999).

9 On the concept of a 'period eye,' meaning the social and cultural practices that shape visual form within a given culture, see Michael Baxandall, *Painting and Experience in Fifteenth-Century Italy: A Primer in the Social History of Pictorial Style* (Oxford, 1988).

sculpture and responded to the *paragone* of the arts; and the practice of painting on precious and semi-precious stones central to contemporary mineralogical debates engaged with the rivalry perceived between *ars* and *natura*. As Koenraad Jonckheere has suggested, the materials of Netherlandish paintings produced in the late 16th and early 17th centuries can also be profitably examined relative to contemporary debates about the physicality and materiality of art.[10] Following the iconoclastic period of *c.* 1550-85, both Catholic and Protestant writings regularly referred to the various materials--gold, marble, wood, paint—employed in the creation of artworks, and artists and artisans in turn devoted significant attention to this aspect of their production. For example, the aim of depicting the human body 'accurately' in a lifelike manner, as opposed to a stony sculptural one, was suited to painting on smooth, non-porous supports that afforded greater mimetic capabilities.

In an effort to explore the theory and practice of painting on these less conventional supports beyond Italy, this chapter focuses on the contributions of Northern European artists, particularly those from the Low Countries. It examines the rise of painting on stone and metal at a moment characterized by major economic, social, political, and religious change, as well as changing beliefs, value systems, and artisanal environments. Whereas much of the discussion concerning paintings on stone from Italy centers on their presentation as durable and eternal objects, this chapter reveals that a constellation of factors in Northern Europe, such as the rise of Antwerp and later Amsterdam as global hubs for foreign materials, goods, skills, and knowledge; a distinctive culture of curiosity, collecting, and connoisseurship; and a heightened pursuit of natural and empirical knowledge, prompted novel uses and meanings of materials from around 1600.

## Precious Stones for Collectors, Craftsmen, and Connoisseurs

In the early modern period gem- and hardstones played a central role in numerous different arenas, including global trade, the natural sciences, medicine, and the decorative arts. Both the stones and the array of specialized skills recruited to work their materials moved fluidly between different disciplines and communities. Not only diamond-cutters, goldsmiths, and jewelers, but also merchants, natural philosophers, collector-connoisseurs, and artists were

10 Koenraad Jonckheere, "Images of Stone: The Physicality of Art and the Image Debates of the Sixteenth Century," in *Meaning in Materials: Netherlandish Art 1400-1800*, ed. H. Perry Chapman, Ann-Sophie Lehmann, and Frits Scholten, Nederlands Kunsthistorisch Jaarboek 62 (Leiden, 2013), pp. 117-46.

involved in their circulation, manipulation, and evaluation. The increasingly global reach of cities like Amsterdam and Antwerp, thanks to vast international trading routes and the activities of the Dutch East India Company (VOC) and Dutch West India Company (WIC), also stimulated the importation of and widespread interest in new kinds of stones, metals, and other exotic materials. Beginning in the 16th century, Antwerp became a storehouse for raw materials and goods from around the world.[11] The city on the Scheldt was home to a large population of goldsmiths, jewelers, and stone grinders and polishers whose products were admired and avidly collected by local residents, as attested to in the numerous Antwerp probate inventories and related legal documents transcribed and published by Erik Duverger.[12] The city itself was even likened to the diamond in the Netherlandish ring of the world in Carolus Scribani's *Antverpia* (1610), a metaphor that was reiterated in Cornelis de Bie's *Het gulden cabinet* of 1662.[13]

The popularity of precious and semi-precious stones in early modern Antwerp, and in the Low Countries more generally, also resonates with the multivalent conception of curiosity in this period. As small-scale objects that were perceived as exotic, rare, and strange, and also, through transformative craft processes, exhibited meticulous handling, and diligent description that appealed to the intertwined sensibilities of collectors, connoisseurs, natural-

11 The Florentine merchant and humanist writer Ludovico Guicciardini (1521-89) in his widely-circulated *Descrittione di tutti i Paesi Bassi* (Descriptions of the Low Countries, 1567), commented extensively on Antwerp's vast riches, foreign inhabitants, and cosmopolitanism. For further reading on Guicciardini and Antwerp's role as a global hub for raw materials and skilled craftsmen, see Göttler, "The Place of the 'Exotic'" and Arjan van Dixhoorn, "The values of Antwerp and the Prosperity of Belgica: Political Economy in Guicciardini's *Descrittione di tutti I Paesi Bassi* (1567)," in *Trading Values in Early Modern Antwerp*, ed. Christine Göttler, Bart Ramakers, and Joanna Woodall, Nederlands Kunsthistorisch Jaarboek 64 (Leiden, 2014) pp. 76-107. On Antwerp's trade in precious stones in particular, see John Everaert, "Antwerp, Turning Table of the International Circuit in Precious Stones and Jewels (16th-17th Centuries)," *Zeitschrift für Kunsttechnologie und Konservierung* 1 (2004), 121-42.

12 Erik Duverger, *Antwerpse Kunstinventarissen uit de Zeventiende Eeuw*, 14 vols (Brussels, 1984-2009).

13 On Scribani's metaphor, see Raingard Esser, "The Diamond of the Netherlands: Histories of Antwerp in the Seventeenth Century,", in *Trading Values in Early Modern Antwerp*, pp. 349-69. Cornelis de Bie, *Het gulden cabinet van de edel vry schilderconst* (Antwerp, 1662; reprint edition, Soest, 1971), p. 179.

ists, and alchemists among others, they engaged with a broad range of criteria associated with curiosity.[14]

The rarity and exoticism of materials, from individual gemstones to the varieties of imported colored marbles decorating the interior of the Houtappel Chapel, enhanced their value and made them even more appealing to local collectors and connoisseurs. The 1637 Antwerp inventory of the household goods of the widow of a wealthy Portuguese converso merchant describes an abundance of gemstones and other precious materials, noting their foreign origins, including diamonds and rubies from India, coral, cochineal from New Spain, Indian shellac, and ebony cabinets.[15] Stones described as 'oriental' were desirable due to their foreign origins, even if those origins were not always identified with geographic precision.

In addition, the miniature size, intriguing tactile and optical qualities, and portability of gemstones also attracted the interest of early modern collectors and craftsmen. The collections of Antwerp artisans involved in the trade and transformation of gemstones reveal how these stones were integrated into both the plastic and pictorial arts. Gemstones, both unworked and those which were combined with other materials and transformed into valuable luxury items, are abundant in 17th-century inventories of Antwerp collections, including the 1681 probate inventory of the silversmith, *pietre dure* specialist, jeweler, and prolific art collector, Jan Gillis.[16] His multifaceted role as creator, collector, evaluator, and salesman of precious objects makes his collection an exemplary case study for investigating the complex networks forged by the gemstone industry in Antwerp. In addition to his expertise as a silversmith, Gillis's exquisite *pietre dure* work garnered attention from all over Europe. He capitalized on

---

14 In the past two decades, there has been an abundance of scholarship on the diverse meanings and uses of curiosity in early modern art, science, and literature. See, for example: Krzysztof Pomian, *Collectors and Curiosities: Paris and Venice 1500-1800*, trans. Elizabeth Wiles-Portier (Cambridge, 1990); Lorraine Daston, "Curiosity in Early Modern Science," *Word and Image* 11 (1995), 391-404; Christopher Wood, "'Curious Pictures' and the Art of Description," *Word and Image* 11 (1995), 332-52; Neil Kenny, *Curiosity in Early Modern Europe: Word Histories* (Wiesbaden, 1998); Barbara Benedict, *Curiosity: A Cultural History of Early Modern Inquiry* (Chicago, 2001); Peter Harrison, "Curiosity, Forbidden Knowledge, and the Reformation of Natural Philosophy," *Isis* 92:2 (2001): 265-90; Neil Kenny, *The Uses of Curiosity in Early Modern France and Germany* (Oxford, 2004); Alexander Marr and R.J.W. Evans, eds., *Curiosity and Wonder from the Renaissance to the Enlightenment* (Aldershot, 2006).

15 Duverger, *Antwerpse Kunstinventarissen*, 4:130-31.

16 For a transcription of the main contents of the inventory, see Duverger, *Antwerpse Kunstinventarissen*, 11:166-73.

the rapidly growing interest in and demand for gemstones, expanding his technical virtuosity to include both metal and stonework. Using precious and semi-precious stones, especially agate, Gillis created inlaid surfaces on a variety of luxury objects including multimedia art cabinets (*kunstkasten*), sculptures, metalwork, ceramics, paintings, and picture frames. As evidenced by the inventory of his residence, the luxury objects collected and produced by Gillis frequently merged together some or all of these different media. He worked extensively for the Forchondt art dealership, contributing agatework to cabinets and other luxury furniture, which was sold in Antwerp and other major European markets. His use of agate in particular reflects the fact that this stone was especially sought after by 16th- and 17th-century naturalists and artisans, who valorized its intrinsic ability to emulate natural forms and stimulate the creative imagination.

As was the case with many artists and artisans in 17th-century Antwerp, Gillis's private residence, workshop, and collection were highly fluid spaces that intersected. The inventory of his residence, *De Spiegel* (The Mirror), on Antwerp's Grote Markt contains the following objects: a series of portrait busts of Roman emperors and Cleopatra, all carved from agate and further decorated with intaglio agate stones, rubies, diamonds, silver, gold, and copper; a shell made from agate with a foot garnished with gold; and several small agate paintings (referred to as both "schilderijkens van agate" and "agate schilderykens") depicting birds and other small animals, encased in gilded copper frames.[17] It is unclear whether these so-called 'agate paintings' were *pietre dure* panels or oil paintings on stone panels, as both were used to decorate cabinet facades.[18] These kinds of precious objects are representative of the compositeness—in terms of materials, media, and craft specializations—of objects represented in elite Antwerp collections.

Small-format, finely painted, highly detailed pictures also appealed to the local culture of collecting, which in the Netherlands was not dominated by ruling nobility, as in Italy or Spain, but rather by the city's large population of *liefhebbers*, a category of art collectors who presented themselves as possessing superior connoisseurial intuition and skills.[19] Although these types of collec-

17 Duverger, *Antwerpse Kunstinventarissen*, 11:166-73.

18 The paintings are referred to in the inventory simply as "schilderijkens van agate" or "agate schilderykens." According to Ria Fabri, in a conversation in September 2014, both *pietre dure* panels and oil paintings on agate panels were used to decorate Antwerp *kunstkasten*.

19 On the emergence and characteristics of *liefhebbers* in the 17th century, see Zirka Filipczak, *Picturing Art in Antwerp* (Princeton, 1988), esp. pp. 51-53; Marten Jan Bok, "Art-Lovers and their Paintings: Van Mander's Schilder-boeck as a Source for the History of the Art

tions originated in aristocratic contexts, such as the imperial *Kunstkammer* of Rudolf II or the *studiolo* of Francesco I de' Medici, the commercial elite in the Southern Netherlands sought to emulate these well-known and documented collections by acquiring similar works. Paintings on stone and copper held great appeal for these ambitious connoisseurs, who saw art collecting as a means to demonstrate their knowledge and elevate their social status.

Close inspection and intimate viewing, both associated with curiosity, were especially encouraged by paintings on stone and metal, which were typically limited in size by their materials and which encouraged viewers to hold them, turn them around, and scrutinize their miniscule brushstrokes and details. Copper plates, which were more commonly used for printmaking in the 16th and 17th centuries, were usually delivered to painters in the standard sizes associated with prints. Unlike canvas or wood panel, metal could also be easily formed into small or irregularly-shaped painting supports, and both metal and stone were ideal for painting detailed, intimate subjects such as nature studies, portraits, devotional images, and erotic scenes intended for private enjoyment. Such paintings were widely collected and displayed in curiosity cabinets alongside other objects that appealed to the culture of collecting in 16th- and 17th-century Europe, which favored objects that were small, wondrous, precious, and rare, and integrated natural and manmade artifice. The intimate nature of miniature paintings on stone and metal is also indicated by their frequent integration into domestic furnishings, such as art cabinets. As already noted, oil paintings on both stone and copper were commonly inlaid in the facades of the ebony *kunstkasten* that were produced in great numbers in 17th-century Antwerp.

For the savvy connoisseur, paintings on copper presented the challenge of distinguishing individual hands, since the medium afforded artists the opportunity to experiment with a wider variety of techniques than were possible when working on canvas or panel. In the 17th century, the concept of curiosity was relevant not only to the content of a painting, but also its technique and the artist's manual control. It was especially linked to meticulous brushwork,

---

Market in the Northern Netherlands," in *Dawn of the Golden Age: Northern Netherlandish Art 1580-1620*, ed. Ger Luijten and Ariane van Suchtelen (Amsterdam, 1993), p. 143; Bert Timmermans, "The Seventeenth-Century Antwerp Elite and Status Honour: The Presentation of Self and the Manipulation of Social Perception," in *On the Edge of Truth and Honesty: Principles and Strategies of Fraud and Deceit in the Early Modern Period*, eds. Jan L. De Jong, Toon van Houdt, and Zoran Kwak (Leiden, 2002), p. 151; Bert Timmermans, *Patronen van Patronage in het Zeventiende-Eeuwse Antwerpen: Een Elite als Actor Binnen een Kunstwereld* (Amsterdam, 2008), pp. 229-31.

minute descriptive detail, and a finely-wrought appearance, a manner characteristic of the Leiden *fijnschilders*, such as Gerard Dou and Frans van Mieris.[20] Seventeenth-century descriptions of paintings on copper frequently include the adjective 'curious,' in reference to these properties.[21] Much more frequently than paintings on canvas or panel, paintings on copper are described in contemporary sources in terms of the unique pictorial effects afforded by the metallic medium. In his lives of Netherlandish and German artists, Karel van Mander frequently praises those who painted on copper, singling out their ability to paint with speed and agility while simultaneously maintaining diligent attention to detail. Van Mander employs the term *neerstichcheyt* to encompass the speed, detail, and precision required to produce small-format cabinet paintings.[22] This combination of intense detail, miniaturization, and extreme naturalism on the one hand, and quickness and fluidity of handling on the other hand typified a 'curious' manner of painting that enhanced and rendered ordinary objects strange and wondrous. Moreover, its appearance of nonchalance appealed to contemporary connoisseurs and served to counterbalance overwrought fussiness or stiffness, which were among the more negative connotations of curiosity and the fine manner of painting.[23]

## Material Mimesis and Manipulation

In addition to their foreign origins, precious qualities, and status as collectibles, gemstones were prized by artisans, collectors, and naturalists for their

20 Wood, "Curious Pictures."

21 An examination of the numerous auction catalogues from the late 17th and first half of the 18th century assembled by Gerard Hoet reveals a consistent tendency to qualify particularly fine works, such as those painted on copper. For example, "Een zeer Curieus Perspectieff op Coper," (a very curious perspective on copper), and "Een curieus landscap op een kopere plaet," (a curious landscape on a copper plate), in Gerard Hoet, *Catalogus of naamlyst van schildereyen, met derzelver pryzen: zedert een lange reeks zoo In Holland als op andere plaatzen in het openbaar verkogt*, 2 vols. (The Hague, 1752; reprint edition, Soest 1976), 1:464, 33.

22 Van Mander, *Het Schilder-boeck*. On the meaning of the term 'neerstichheyt', see E. Verwijs and J. Verdam, *Middelnederlandsch Woordenboek III* (The Hague, 1894), cols. 2353-54. The range of meanings associated with the early modern use of the term 'diligence' is discussed in Lucy C. Cutler, "Virtue and Diligence: Jan Brueghel I and Federico Borromeo," in *Virtus: Virtuositeit en kunstliefhebbers in de Nederlanden 1500-1700;* Nederlands Kunsthistorisch Jaarboek 54 (Leiden, 2003): pp. 203-27, esp. pp. 207-08.

23 Wood, "Curious Pictures," esp. pp. 346-50.

pictorial and optical qualities, which were not only studied and admired, but also imitated. Comparisons, imitations, and forgeries of materials were common in early modern Northern Europe: the fictitious creation of diamonds, rubies, amethysts, emeralds, and other gemstones occurred frequently, as did the staining or painting of common woods to imitate expensive imported woods like ebony.[24] Both classical and medieval texts as well as contemporary recipes and craft manuals provided early modern artists and artisans with detailed instructions for reproducing the highly sought-after qualities of gemstones and other precious materials. In his famous treatise on gems, metals, and minerals, *De Mineralibus* (*c.* 1250), Albertus Magnus makes a direct correlation between the preciousness of stones and their varying degrees of transparency and translucency, and he also writes about the practice of creating fictitious gems.[25] In a related vein, instructions for altering and enhancing the natural color of gemstones by applying oil is discussed in the encyclopedic texts of Pliny the Elder and Isidore of Seville, while the technique of painting with oil on polished metallic surfaces, such as tin and copper, in order to imitate gemstones and gold is recounted by Theophilus.[26]

Additionally, the development and refinement of oil painting techniques in the 15th century by Jan van Eyck and other Netherlandish painters further stimulated artistic interest in studying and reproducing the optical qualities of different materials such as metal and stone. The desire to imitate precious materials is evident in the Dutch painter Wilhelmus Beurs's treatise on oil painting published in 1692, *De groote waereld in 't kleen geschildert* (The Great World Painted in Small), which provides detailed instructions for depicting materials such as metals, precious stones, and glass.[27] In a section entitled "Of Pearls and costly stones," Beurs gives directions for painting pearls, diamonds,

24 On the imitation or faking of precious materials in the early modern period, see Marjolijn Bol, "Coloring Topaz, Crystal, and Moonstone: Factitious Gems and the Imitation of Art and Nature, 300-1500," in *Fakes!?: Hoaxes, Counterfeits and Deception in Early Modern Science*, Second Watson Seminar in the History of Material and Visual Science, Museo Galileo, Florence, June 7, 2013, ed. Marco Beretta et al. (Sagamore Beach, 2014), pp. 108-29; Ann-Sophie Lehmann, "How Materials Make Meaning," in *Meaning in Materials*, pp. 20-22; Rachel King, "To Counterfeit Such Precious Stones as you Desire: Amber and Amber Imitations in Early Modern Europe," in Birgit Ulrike Münch et al., *Fälschung—Plagiat—Kopie: Künstlerische Praktiken der Vormodernen* (Petersberg, 2014), 87-97.

25 Marjolijn Bol, "Oil and the Translucent: Varnishing and Glazing in Practice, Recipes, and Historiography, 1100-1600" (PhD diss., University of Utrecht, 2012), pp. 93-94.

26 See Bol, "Oil and the Translucent," p. 101, and Theophilus, *De diversis artibus*, trans. and ed. C.R. Dodwell (London, 1961), p. 25.

27 Wilhelmus Beurs, *De groote waereld in 't kleen geschildert, of schilderagtig tafereel van 's Weerelds schilderyen* [...] (Amsterdam, 1692).

rubies, and other stones, emphasizing their color and reflective properties.[28] Additionally, he advises painters to "practice diligence" in order to make their painted stones finer and more valuable.[29]

Related artisanal practices also laid the groundwork for the development of painting on stone and metallic supports in 16th- and 17th-century Northern Europe. Many Netherlandish and German painters were trained as goldsmiths, enamellers, glass painters, and engravers, and these crafts provided them with extensive knowledge of minerals and metals.[30] The craft of enameling, which flourished up until the 17th century in present-day France, Belgium, the Netherlands and Germany, likely provided an important precursor for painting on copper given its use of copper plates covered with glass and tin foil to create richly embellished surfaces. In light of these technical developments and the widespread interest among Northern European painters in manipulating the natural properties of materials to achieve stunning examples of artistic virtuosity, the stage was set by the 16th century for the transition from imitating precious stones and metals to painting on them directly.

An increasing knowledge of and interest in the properties of gemstones and metals and the methods for imitating them provided fertile ground for painters to experiment with stone and metal as supports for oil painting by the mid-16th century. The diverse craft specializations of many of these painters also undoubtedly contributed to their methods for manipulating the surfaces of stone and metallic materials. Mannerist painters such as Joachim Wtewael, who trained initially as a glass engraver in Utrecht, took advantage of copper's smooth, non-porous surface to depict the translucency and sensuality of flesh, as well as a wide range of materials and *naturalia* on a small scale.[31] Classical allegories and myths, with their complex narratives and incorporation of nude bodies and multiple tiny details, were ideally suited to small metal or stone supports amenable to fine brushwork and jewel-like colors.

Emperor Rudolf II's court in Prague was a key site for Netherlandish and German Mannerist painters who worked on metal and stone. The court artists Bartholomaeus Spranger and Hans von Aachen integrated the natural

28 Beurs, *De groote waereld*, pp. 121-24.

29 Beurs, *De groote waereld*, p. 124.

30 For example, Albrecht Dürer apprenticed with his father, who was a goldsmith, while the Utrecht painter Joachim Wtewael began his career as a glassmaker and glass engraver in his father's workshop, according to Karel van Mander.

31 On Wtewael's oeuvre see James Clifton, Liesbeth Helmus, and Arthur K. Wheelock, eds., *Pleasure and Piety: The Art of Joachim Wtewael*, exhibition catatalogue, Centraal Museum, Utrecht; National Gallery of Art, Washington; Museum of Fine Arts, Houston (Princeton, 2015).

striations and swirling colors found in onyx, alabaster, and marble into their compositions. Von Aachen, who was born in Cologne but also worked in Antwerp, Florence, Rome, Augsburg, and Munich before coming to Prague, took advantage of the availability of a wide range of materials in the stonecutters' workshops at Prague Castle, ranging from marble and alabaster to slate and copper.[32]

As demonstrated by a pair of two-sided mythological paintings on alabaster, dated to around 1600, Von Aachen mastered the technique of integrating the natural marbling and tones of the stone into his paintings, adding to their value and aligning them with the precious objects displayed in the Imperial *Kunstkammer*. The first alabaster tablet depicts on the recto Ovid's tale of Perseus freeing Andromeda [**Fig. 6.3** ] and on the verso Virgil's story of Aeolus releasing the winds to prevent Aeneas' escape across the sea [**Fig. 6.2**]. The natural striations of the stone form the clouds on which the gods rest, while the alabaster's translucency allows Von Aachen's painted image to incorporate elements of both sides of the stone when held up to the light. The second tablet contains two Ovidian myths: the Fall of Phaeton on the recto [**Plate 24**] and the Triumph of Amor and Bacchus on the verso [**Fig. 6.1**]. In an inventory of the collection of Rudolf's successor, drawn up in 1619, the second tablet is described as having been framed in ebony so that both recto and verso could be viewed simultaneously, enhancing the effects of the stone's translucency and internal texture as well as the viewer's cognizance of the extent to which the painter had to harness the natural veining of the stone and skillfully arrange the painted motifs so that they correspond not only to these striations but also to the inverse figurations and painted composition.[33] For viewers of these works, the implications of curiosity were two-fold: firstly, the natural figurations of the alabaster itself inspired wonder and awe at nature's artifice and God's handiwork, and secondly, the painter's finely-wrought brushwork and exquisite detail on a small scale would have been greatly admired by connoisseurs who were compelled to try and distinguish painted from figured stone.

32 On Hans von Aachen's career and production, see T. Fusening, ed., *Hans von Aachen (1552-1615): Court Artist in Europe* exhibition catalgoue, Suermondt-Ludwig-Museum, Aachen (Berlin, 2010).

33 Fusening, ed., *Hans von Aachen*, pp. 196-97. Rudolf II's inventory contained many more two-sided paintings on stone, and the manner of display is seldomly noted. In addition to the mirror-like framing mentioned in the 1619 inventory, Seifertova suggests that other paintings on stone remained unframed and were intended to be handled and turned over by the viewer. See Seifertova, *Painting on Stone*, p. 132.

## Meaning and Metaphor in Forges, Mines, and Fires

In addition to their role within early modern cultures of collecting and material experimentation, paintings on stone and metallic supports also possessed a number of metaphorical and symbolic associations. Especially common was the use of metals and dark-colored stones to depict nocturnal scenes and images of fires, furnaces, forges, and the underworld. In the introductory book of *Het Schilder-boek*, *Den grondt der edel vry schilder-const* (The foundations of the noble and free art of painting), Karel van Mander describes paintings of nocturnal scenes in which the color of the stone served as the darkest background tone.[34] Dark stone and metallic surfaces provided a dramatic backdrop for the glowing, radiant effects of natural and artificial light, as seen in several night scenes painted on slate in the 1620s by the Dutch artist Leonard Bramer, as well as in works on copper and silver by the German painters Adam Elsheimer and Hans Rottenhammer. These images permitted Northern painters to explore the behavior of different kinds of light, both natural and artificial, on a broad range of materials, such as metal, glass, skin, and stone.

Not only do such scenes draw attention to the optical properties of fire and other light sources, but they also underscore connections between the transmutation of metal and stone through processes of mining, metalworking, forging, and alchemy on one hand, and the transformative craft of painting on stone and metal on the other hand.[35] Images of alchemists' workshops, mines, and forges make implicit the transformative potential of stone and metal and they draw attention to contemporary interests in alchemy and debates about the transformation, animation, and petrification of stone and other natural materials.[36] Painting on stone and metal forged a tangible connection between art and nature, while also asserting the painter's skills and knowledge of the natural world. The ability to not only represent nature, but also to transform it resonates with Pamela Smith's concept of the artisanal cognition and the

34 Karel van Mander, *Den grondt der edel vry schilder-const*, ed. Hessel Miedema (Utrecht, 1973), part 1, pp. 196-97.

35 For further discussion of fire and mining landscapes in Netherlandish art, see Leopoldine Prosperetti, *Landscape and Philosophy in the Art of Jan Brueghel the Elder (1568-1625)* (Burlington, 2009), esp. Chapter 3. On the association of fire and alchemy with transformative craftsmanship, particularly in regards to Antwerp, see Christine Göttler, "The Alchemist, the Painter, and the 'Indian Bird': Joining Arts and Cultures in Seventeenth-Century Antwerp: Adriaen van Utrecht's *Allegory of Fire* in the Royal Museums of Fine Arts in Brussels," in *Synergies in Visual Culture / Bildkulturen im Dialog: Festschrift für Gerhard Wolf*, ed. Manuela De Giorgi et al. (Munich, 2013), pp. 499-512.

36 Sven Dupré et al., eds., *Art and Alchemy: The Mystery of Transformation* (Munich, 2014).

acquisition of natural knowledge through bodily experience.[37] Through making art from and about such materials, these painters ostensibly acquired a deeper understanding of their origins, behaviors, and properties.

The proliferation of mineralogical treatises in the 16th and 17th centuries also stimulated interest in painting on stone. Georgius Agricola's *De re metallica* (1556) and Conrad Gesner's *De omni rerum fossilium genere* (1565) were pivotal in promoting the ordering of stones according to visual criteria. Portrayals of forging and mining activities also illustrate the increasingly positive associations with these occupations as well novel developments and techniques. Jan Brueghel the Elder's highly-detailed depictions of forges, several of which are painted on copper, provide near encyclopedic accounts of contemporary metallurgical tools and techniques [**Fig. 7.3**].[38] Not coincidentally, it was these very developments that made copper more affordable and accessible to Netherlandish printmakers and painters.

New scientific knowledge and technologies, as well as a heightened interest in the empirical investigation and representation of nature, played a crucial role in the rise of painting on stone and metal in 16th- and 17th-century Europe. Across different media, artists sought to visually describe flora, fauna, and other natural phenomena, leading to the emergence of new pictorial subjects and techniques for representing them. In the Netherlands, still life emerged as an independent pictorial genre that enabled artists to explore natural subjects previously confined to the margins or background of compositions. Jan Brueghel played a key role in the development of the flower still life as a discrete and extremely popular type, especially among collectors who relished the opportunity to test their botanical knowledge and identify local and exotic species, many of which were available in their own gardens or specimen collections. In a similar vein, Brueghel's grandson, Jan van Kessel the Elder, established insects as an independent subject of oil painting [**Fig. 7.4**].[39] Many

37 Pamela H. Smith, *The Body of the Artisan: Art and Experience in the Scientific Revolution* (Chicago, 2004).

38 For further discussion of the iconography and materiality of Jan Brueghel's paintings of fiery forges and their potential connection to contemporary alchemical knowledge, see Prosperetti, *Landscape and Philosophy*, esp. Chapter 3; Göttler, "The Alchemist, the Painter, and the 'Indian Bird,'" esp. pp. 507-509; Göttler, "Allegories of Fire," in Dupré et al., *Art and Alchemy*, pp. 134-45, and Göttler, "Vulcan's Forge: The Sphere of Art in Early Modern Antwerp," in Sven Dupré and Christine Göttler, eds. *Cultures of Knowledge in Early Modern Europe* (forthcoming).

39 Van Kessel's insect paintings on copper supports are discussed in greater depth in Nadia Baadj, *Jan van Kessel I (1626-1679): Crafting a Natural History of Art in Early Modern Antwerp* (Turnhout, 2016).

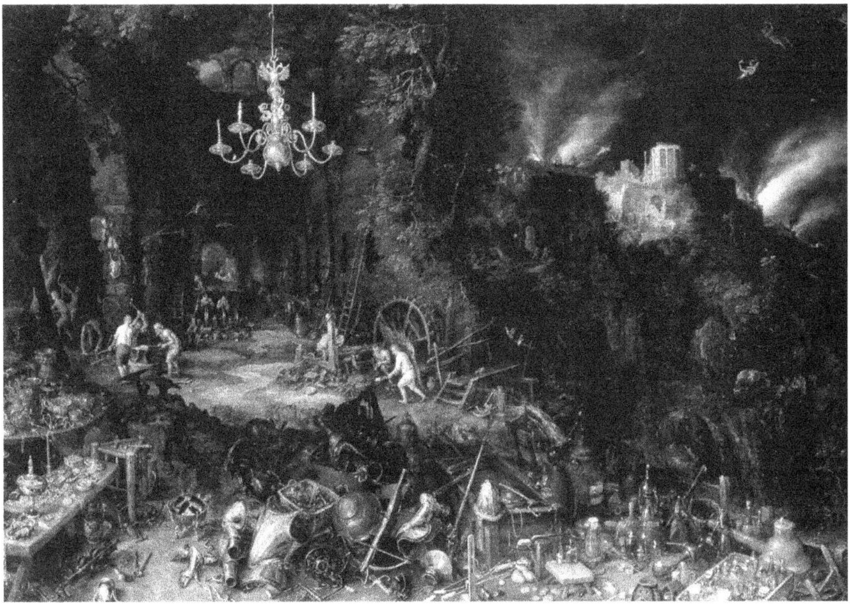

FIGURE 7.3 *Jan Brueghel the Elder,* Allegory of Fire, *oil on copper, ca. 1608, Milan, Pinacoteca Ambrosiana* (PHOTO: ©VENERANDA BIBLIOTECA AMBROSIANA – MILANO/DE AGOSTINI PICTURE LIBRARY).

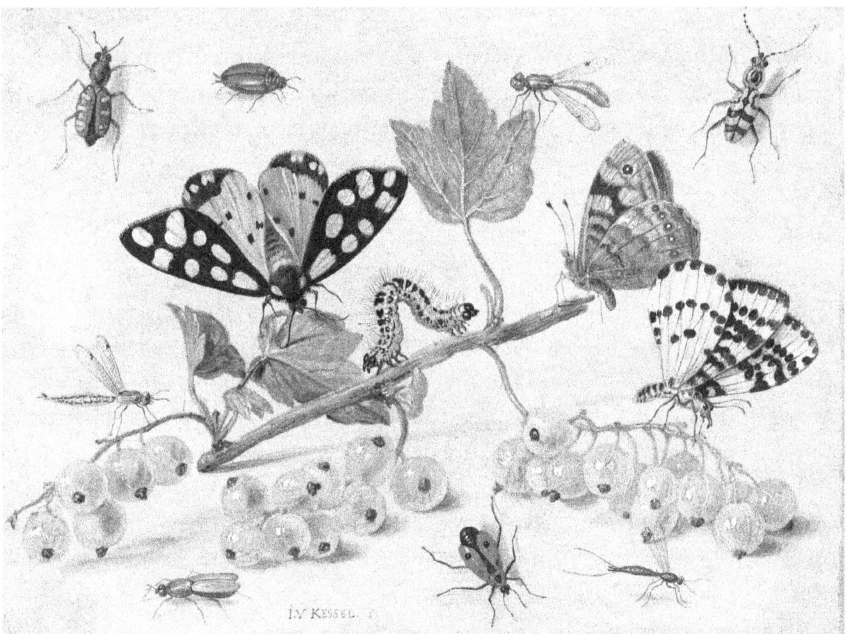

FIGURE 7.4 *Jan van Kessel I,* Insects and Fruit, *oil on copper, ca. 1660-1665, Amsterdam, Rijksmuseum* (PHOTO: RIJKSMUSEUM).

still-life painters, including Brueghel and Van Kessel, favored copper supports that allowed them to render meticulously even the tiniest details of their natural subjects.

Van Kessel's miniature insect paintings were innovative due to their translation of subject matter that was previously restricted to watercolor or graphic media into precious oil paintings on copper. The tiny scale of insects enhanced the precious quality of his already jewel-like paintings, as did their intricate anatomies, which were only recently beginning to be explored due to advancements in visual technology such as the microscope. In his insect works, Van Kessel exploited the material properties of copper in ways that complemented and became inseparable from his tiny, intricate subjects. Since antiquity, the status of insects was inextricably linked with models of finely-wrought craftsmanship and the production of precious objects. As animals that were themselves exemplars of fine craftsmanship and also possessed the ability to produce valuable materials (silk, cochineal, etc.), it was only natural that the representation of insects in art would emphasize and incorporate precious materials. Van Kessel's elevation of insects as a subject worthy of painting on copper underscores the interconnectedness of metalworking and zoological investigation in the 17th century. Alchemical recipes appealed to both artisanal and natural historical knowledge by including detailed instructions for how to transform insects and other animals into valuable substances.[40] Van Kessel's insect paintings thus provide a parallel to alchemists' transformation of base materials into objects of great value. In both cases, transformative artistry serves as the agent that turns diminutive animals into precious and prestigious works of art.

## Conclusion

In conclusion, I would like to return to the Houtappel Chapel where we began. The most important example of the revival of family chapels in 17th-century Antwerp, it displays an uninhibited range of material virtuosity, integrating painting, architecture, and sculpture in such a way that all three media engage and compete with the natural properties of stone. Hendrick van Balen's painted scenes from the life of the Virgin recruit and manipulate the natural veining of their marble supports in order to transform these figurations into cavernous

40 See for example the life-casting techniques discussed in Pamela H. Smith and Tonny Beentjes, "Nature and Art, Making and Knowing: Reconstructing Sixteenth-Century Life-Casting Techniques," *Renaissance Quarterly* 63 (2010), 128-79.

rock formations (in the *Adoration of the Shepherds* and *Adoration of the Magi*) and marble columns, steps, and floors (in the *Presentation of the Virgin* and *Presentation of Christ in the Temple*). Van Balen's use of marble panels in the chapel altar to depict the illusion of marble inside another (painted) sacred edifice is especially striking. Framing these paintings are the colored marble panels, which engage in a formal dialogue with both Van Balen's cycle and the white marble sculptural decoration by means of their vibrant tones and suggestive figurations. Taken together, the oil paintings on marble, the figurative colored marbles, and the sculpted marble ornament blur the boundaries between nature and artifice and between image and support, compelling the viewer to question whether creative potential and agency originated in the materials or the maker.

Furthermore, this multi-sensorial space presents beholders with a material microcosm of not only the entire church, but also of the city of Antwerp and its global reach in the late 16th and 17th century. The different varieties of marble, stuccowork, oil paintings, and sculpture engender contemplation of the city's status as a hub for raw materials from around the world and as an innovative laboratory for combining, comparing, and transforming these materials through consummate craftsmanship and artistry. On an even smaller scale, Ehrenberg's virtuosic summation of the chapel's striking materiality in oil paint and stone brings into focus its associations with a distinctive local culture of curiosity, connoisseurship, and collecting that prized the exotic, the rare, and the wondrous, as well as empirical inquiry and technical virtuosity.

## Bibliography


Anderson, Christy, Anne Dunlop, and Pamela H. Smith, eds. The Matter of Art: Materials, Technologies, Meanings, *c. 1250-1650*. Manchester, 2014.

Baadj, Nadia. *Jan van Kessel I (1626-1679): Crafting a Natural History of Art in Early Modern Antwerp*. Turnhout, 2016.

Baxandall, Michael. *Painting and Experience in Fifteenth-Century Italy: A Primer in the Social History of Pictorial Style*. Oxford, 1988.

Benedict, Barbara. *Curiosity: A Cultural History of Early Modern Inquiry*. Chicago, 2001.

Beurs, Wilhelmus. *De groote waereld in ,t kleen geschildert, of schilderagtig tafereel van 's Weerelds schilderyen* […]. Amsterdam, 1692.

Bok, Marten Jan. "Art-Lovers and their Paintings: Van Mander's *Schilder-boeck* as a Source for the History of the Art Market in the Northern Netherlands." In *Dawn of the Golden Age: Northern Netherlandish Art 1580-1620*, edited by Ger Luijten and Ariane van Suchtelen, pp. 136-66. Amsterdam, 1993.

Bol, Marjolijn. "Coloring Topaz, Crystal, and Moonstone: Factitious Gems and the Imitation of Art and Nature, 300-1500." In *Fakes!?: Hoaxes, Counterfeits and Deception in Early Modern Science, 2nd Watson Seminar in the History of Material and Visual Science, Museo Galileo, Florence, June 7, 2013*, edited by Marco Beretta and Maria Conforti, pp. 108-29. Sagamore Beach, 2014.

Bol, Marjolijn. "Oil and the Translucent: Varnishing and Glazing in Practice, Recipes, and Historiography, 1100-1600." PhD diss., University of Utrecht, 2012.

Cannon, Jo, Joanna Kirby, and Susie Nash, eds. *Trade in Artists' Materials: Markets and Commerce in Europe to 1700*. London, 2010.

Chapman, H. Perry, Ann-Sophie Lehmann, and Frits Scholten, eds. *Meaning in Materials: Netherlandish Art 1400-1800*. Nederlands Kunsthistorisch Jaarboek, vol. 62. Leiden, 2013.

Clifton, James, Liesbeth Helmus, and Arthur K. Wheelock, eds. *Pleasure and Piety: The Art of Joachim Wtewael*. Exhibition catalogue, Centraal Museum, Utrecht; National Gallery of Art, Washington; Museum of Fine Arts, Houston. Princeton, 2015.

Cook, Harold J., Amy R.W. Meyers, and Pamela H. Smith, eds. *Ways of Making and Knowing: the Material Culture of Empirical Knowledge*. Ann Arbor, 2014.

Cutler, Lucy C. "Virtue and Diligence: Jan Brueghel I and Federico Borromeo." In *Virtus: Virtuositeit en kunstliefhebbers in de Nederlanden 1500-1700*, edited by Jan de Jong, pp. 203-27. Nederlands Kunsthistorisch Jaarboek, vol. 54. Leiden, 2003.

Daston, Lorraine. "Curiosity in Early Modern Science." *Word and Image* 11 (1995), 391-404.

De Bie, Cornelis. *Het gulden cabinet van de edel vry schilderconst*. Antwerp: Jan Meyssens, 1662. Reprint edition, Soest, 1971.

Dixhoorn Arjan van. "The values of Antwerp and the Prosperity of Belgica: Political Economy in Guicciardini's *Descrittione di tutti I Paesi Bassi* (1567)." In *Trading Values in Early Modern Antwerp*, edited by Christine Göttler, Bart Ramakers, and Joanna Woodall, pp. 76-107. Nederlands Kunsthistorisch Jaarboek, vol. 64 Leiden, 2014.

Dupré, Sven, Dedo von Kerssenbrock-Krosigk, and Beat Wismer, eds. *Art and Alchemy: The Mystery of Transformation*. Munich, 2014.

Dupré, Sven. "Trading Luxury Glass, Picturing Collections and Consuming Objects of Knowledge in Early Seventeenth-Century Antwerp." *Intellectual History Review* 20, no. 1 (2010), 53-78.

Duverger, Erik. *Antwerpse kunstinventarissen uit de zeventiende eeuw*, 14 vols. Brussels, 1984-2009.

Esser, Raingard. "The Diamond of the Netherlands: Histories of Antwerp in the Seventeenth Century." In *Trading Values in Early Modern Antwerp*, edited by Christine Göttler, Bart Ramakers, and Joanna Woodall, 349-369. Nederlands Kunsthistorisch Jaarboek, vol. 64. Leiden, 2014.

Everaert, John. "Antwerp, Turning Table of the International Circuit in Precious Stones and Jewels (16th-17th Centuries)." *Zeitschrift für Kunsttechnologie und Konservierung* 1 (2004): 121-42

Filipczak, Zirka. *Picturing Art in Antwerp*. Princeton, 1988.

Fusenig Thomas, ed. *Hans von Aachen (1552-1615): Court Artist in Europe*. Exhibition catalogue Suermondt-Ludwig-Museum, Aachen. Berlin, 2010.

Göttler, Christine, Bart Ramakers, and Joanna Woodall, eds. *Trading Values in Early Modern Antwerp*. Nederlands Kunsthistorisch Jaarboek, vol. 64. Leiden, 2014.

Göttler, Christine. "Affectionate Gifts: Rubens's Small Curiosities on Metallic Supports." In *Munuscula Amicorum: Contributions on Rubens and his Context in Honour of Hans Vlieghe*, edited by Katlijne Van der Stighelen, pp. 47-66. Turnhout, 2006.

Göttler, Christine. "The Alchemist, the Painter, and the 'Indian Bird': Joining Arts and Cultures in Seventeenth-Century Antwerp: Adriaen van Utrecht's *Allegory of Fire* in the Royal Museums of Fine Arts in Brussels." In *Synergies in Visual Culture/Bildkulturen im Dialog: Festschrift für Gerhard Wolf*, edited by Manuela De Giorgi, Annette Hoffman, and Nicola Suthor, pp. 499-512. Munich, 2013.

Göttler, Christine. "Allegories of Fire," In *Art and Alchemy: The Mystery of Transformation*, edited by Sven Dupré, Dedo von Kerssenbrock-Krosigk, and Beat Wismer, pp. 134-45, Munich, 2014.

Göttler, Christine. "The Place of the 'Exotic' in Early-Seventeenth-Century Antwerp." In *Looking East: Rubens's Encounter with Asia*, edited by Stephanie Schrader, pp. 88-107. Los Angeles, 2013.

Göttler, Christine. "Vulcan's Forge: The Sphere of Art in Early Modern Antwerp," In *Cultures of Knowledge in Early Modern Europe* (forthcoming), edited by Sven Dupré and Christine Göttler.

Graaf, Johannes van der. "Development of Oil-Paint and the Use of Metal-Plates as a Support." Paper presented at the Congress on Conservation of Paintings and the Graphic Arts, Lisbon, 1972. London: International Institute for Conservation of Historic and Artistic Works, 1972, pp. 139-52.

Harrison, Peter. "Curiosity, Forbidden Knowledge, and the Reformation of Natural Philosophy." *Isis* 92, no. 2 (2001), 265-90.

Hoet, Gerard. *Catalogus of naamlyst van schildereyen, met derzelver pryzen: zedert een lange reeks zoo In Holland als op andere plaatzen in het openbaar verkogt*, 2 vols. The Hague, 1752. Reprint edition, edited by Pieter Terwesten, Soest, 1976.

Jonckheere, Koenraad. "Images of Stone: The Physicality of Art and the Image Debates of the Sixteenth Century." In *Meaning in Materials: Netherlandish Art 1400-1800*, edited by H. Perry Chapman, Ann-Sophie Lehmann, and Frits Scholten, pp 117-146. Nederlands Kunsthistorisch Jaarboek, vol. 62. Leiden, 2013.

Kenny, Neil. *Curiosity in Early Modern Europe: Word Histories*. Wiesbaden, 1998.

Kenny, Neil. *The Uses of Curiosity in Early Modern France and Germany*. Oxford, 2004.

King, Rachel. "To Counterfeit Such Precious Stones as you Desire: Amber and Amber Imitations in Early Modern Europe." In *Fälschung—Plagiat—Kopie: Künstlerische Praktiken der Vormodernen*, edited by Birgit Ulrike Münch et al., pp. 87-97 Petersberg, 2014.

Knaap, Anna C. "Marvels and Marbles in the Antwerp Jesuit Church: Hendrick van Balen's Stone Paintings of the *Life of the Virgin* (1621)." In *Jesuit Image Theory in Europe and the Overseas Missions*, edited by Walter Melion, Wietse de Boer, and Karl Enenkel, pp, 352-93. Leiden, 2016.

Komanecky, Michael, ed. *Copper as Canvas: Two Centuries of Masterpiece Paintings on Copper, 1575-1775*. Exhibition catalogue, Phoenix Art Museum; Nelson-Atkins Museum of Art, Kansas City; Mauritshuis, The Hague. New York, 1999.

Komanecky, Michael. "Antwerp Artists and the Practice of Painting on Copper." In *Painting Techniques, History, Materials, and Studio Practice*, edited by Ashok Roy and Perry Smith, 136-39. London, 1998.

Lehmann, Ann-Sophie. "How Materials Make Meaning." In *Meaning in Materials: Netherlandish Art 1400-1800*, edited by H. Perry Chapman, Ann-Sophie Lehmann, and Frits Scholten, pp. 6-27. Nederlands Kunsthistorisch Jaarboek, vol. 62. Leiden, 2013.

Limberger, Michael. "'No town in the world provides more advantages': Economies of Agglomeration and the Golden Age of Antwerp." In *Urban Achievement in Early Modern Europe: Golden Ages in Antwerp, Amsterdam and London*, edited by Patrick O'Brien, Derek Keene, Marjolein't Hart, and Hermann van der Wee, pp. 39-62. Cambridge, 2001.

Lock, Léon."Rubens and the Sculpture and Marble Decoration." In *Innovation and Experience in the Southern Netherlands: The Case of the Jesuit Church in Antwerp*, edited by Piet Lombaerde, pp. 155-74. Turnhout, 2008.

Lombaerde, Piet, ed. *Innovation and Experience in the Southern Netherlands: The Case of the Jesuit Church in Antwerp*. Turnhout, 2008.

Mander, Karel van. *Het schilder-boeck*. Haarlem, 1604. Reprint edition, Utrecht, 1969.

Mander, Karel van. *Den grondt der edel vry schilder-const*, edited by Hessel Miedema. Utrecht, 1973.

Marr, Alexander, and R.J.W. Evans, eds. *Curiosity and Wonder from the Renaissance to the Enlightenment*. Aldershot, 2006.

Pomian, Krzysztof. *Collectors and Curiosities: Paris and Venice 1500-1800*. Translated by Elizabeth Wiles-Portier. Cambridge, 1990.

Prosperetti, Leopoldine. *Landscape and Philosophy in the Art of Jan Brueghel the Elder (1568-1625)*. Burlington, 2009.

Seifertova, Hana. *Malby na kameni: Umĕcký experiment v 16. Na začátku 17 století/ Painting on Stone: An Artistic Experiment in the 16th and Early 17th Centuries*. Prague, 2007.

Smith, Pamela H. *The Body of the Artisan: Art and Experience in the Scientific Revolution*. Chicago, 2004.

Smith, Pamela H., and Tonny Beentjes. "Nature and Art, Making and Knowing: Reconstructing Sixteenth-Century Life-Casting Techniques." *Renaissance Quarterly* 63 (2010), 128-79.

Theophilus. *De diversis artibus*. Translated and edited by C.R. Dodwell. London, 1961.

Timmermans, Bert. "The Chapel of the Houtappel Family and the Privatisation of the Church in Seventeenth-Century Antwerp." In *Innovation and Experience in the Southern Netherlands: The Case of the Jesuit Church in Antwerp*, edited by Piet Lombaerde, 155-74. Turnhout, 2008.

Timmermans, Bert. *Patronen van Patronage in het Zeventiende-Eeuwse Antwerpen: Een Elite als Actor Binnen een Kunstwereld*. Amsterdam, 2008.

Timmermans, Bert. "The Seventeenth-Century Antwerp Elite and Status Honour: The Presentation of Self and the Manipulation of Social Perception." In *On the Edge of Truth and Honesty: Principles and Strategies of Fraud and Deceit in the Early* 165. Leiden, 2002.

Vasari, Giorgio. *Lives of the Most Eminent Painters, Sculptors, and Architects*. Translated by Gaston Du C. de Vere, 10 vols. London, 1912-15.

Verwijs, E., and J. Verdam. *Middelnederlandsch Woordenboek III*. The Hague, 1894.

Wood, Christopher. "'Curious Pictures' and the Art of Description." *Word & Image* 11, no. 4 (1995), 332-52.

# PART 3

## *Other Materials, Metaphors, and Inventions*

CHAPTER 8

# 'Painting the Eternal': Micromosaic Materiality and Transubstantiation in an Icon of Christ at Santa Croce in Gerusalemme, Rome*

*Anna Marazuela Kim*

Like the technique of painting on stone, the art of mosaic was praised by Giorgio Vasari as '*quasi eterno*' due to its unsurpassed durability.[1] Historically as a medium, mosaic carried with it associations of eternity for other reasons as well. Throughout the centuries, mosaics were employed in the decoration of Christian churches, in particular, the surfaces of their lofty, domed apses: spaces that symbolized, and visualized, the eternal realm of the Heavenly Jerusalem. The specific materiality of mosaics—their unique reflective capacity, glowing when illuminated, as if divinely lit from within, gem-like in their rich glittering, colors—made them especially suited to not only the depiction of the sacred, but also to its activation by the devotional viewer.[2]

* My thanks to the editors for their invitation to contribute to this volume and for their insightful comments on an earlier draft. The discussion of the micromosaic icon below is based upon a chapter of my dissertation, "Real Presence, Iconic Images, and Iconoclasm from Byzantium to the Reformation" (PhD diss. University of Virginia, 2014) and was delivered as a paper at the 2013 meeting of the Renaissance Society of America. It now forms part of a book project.

1 "E certo è che il musaico è la più durabile pittura che sia, imperò che l'altra col tempo si spegne e questa nello stare fatta di continua s'accende, et inoltre la pittura manca e si consuma per se medesima, ove il musaico per la sua lunghissima vita si può quasi chiamare eterno." "Introduction," Giorgio Vasari, *Le vite de più eccellenti pittori scultori e architettori nelle redazioni del 1550 e 1568*, ed. Rosanna Bettarini, with comments by Paola Barocchi, 8 vols. (Florence, 1966-84), vol. 1, ch. 29, 148. In the Giunta edition of the *Lives*, "Proem," I, 16, Vasari also writes of the "eternità e nobilità di material n' musaici."

2 For the Roman context in particular, see Erik Thunø's, *The Apse Mosaic in Early Medieval Rome* (Cambridge, 2015), which provides further arguments for the spiritual action of mosaics, discussed below. Otto Demus' *Byzantine Mosaic Decoration: Aspects of Monumental Art in Byzantium* (New Rochelle: Caratzas Brothers, 1976), originally published in 1948, is the *locus classicus* for the exploration of mosaic church decoration in its relation to the sacred. Liz James' *Mosaics in the Medieval World From Late Antiquity to the Fifteenth Century* (Cambridge, 2017), promises a much-needed update.

 | DOI 10.1163/9789004361492_012

As a technique, mosaic is described by Renaissance writers and theorists such as Antonio Filarete (1400-69) and Leon Battista Alberti (1404-72) as an art that had been 'lost' (*perduta*) through the ages, with contemporary practitioners difficult to find.[3] The esteem for the medium is attested by the nearly universal praise for Giotto's lost masterpiece, *Navicella* (c. 1300-10), which was known only by surviving fragments. Although earlier precedents, such as Pietro Cavallini's mosaics from the 1290s, were extant in the Renaissance, Giotto's work in mosaic was considered exceptional: in Vasarian terms, it was created by the first generation of 'modern' artists. Generally, mosaic was considered an art of antiquity, one especially associated with ancient Rome, and also the Byzantines. In addition to church mosaics dating from the early Christian Empire, Byzantine icons would have constituted the most significant exemplars of this art. And these mosaics had a double value, as works of precious or wondrous facture and also as sacred artifacts that claimed an authoritative connection to the life of Christ: holy portraits from life, charged with spiritual potency.

From surviving inventories in Italy, we know that mosaic icons were valued highly by the first major collectors of art. The 1457 inventory of the Venetian Cardinal Pietro Barbo, later Pope Paul II (r. 1464-71), records a remarkable twenty-five examples. Among these are icons described as crafted of the smallest mosaic ("*cum musaico parvenissimmo*"), others finely ornamented with enamel or silver, and still others containing relics.[4] The collection of Lorenzo de' Medici (r. 1442-92), inventoried upon his death in 1492, also lists no fewer than eleven miniature mosaic icons, and specifically describes a Byzantine icon of Christ as 'precious' because of its mosaic technique.[5] A mosaic icon of

---

3 See for example Filarete, *Trattato di Architettura*, vol. 2, eds. Anna Maria Finoli and Lilana Grassi (Milan, 1972), pp. 671-72: "Questa arte, come è detto, è perduta, che da Giotto in qua poco s'è usata." On the scarcity of skilled practitioners, see Carlo Bertelli, "Rinascimento del Mosaico," in *Il Mosaico*, ed. Carlo Bertelli (Milan 1997), p. 232: "no si trovando chi sappia altri."

4 Arne Effenberger, "Images of Personal Devotion," in *Byzantium: Faith and Power (1261-1557)*. ed. Helen Evans (New York and New Haven, 2004), 210. The Italian humanist Niccolo Perotti also describes a collection of "*molte cose di mosaici in tavolette*" (many mosaic panels); see Gabriele Barucca, "I Reliquari da Niccolo Perotti a Sassoferrato," *Studi Umanistici Piaceni* 12 (1992), 16.

5 Anthony Cutler, "From Loot to Scholarship: Changing Modes in the Italian Response to Byzantine Artifacts, ca. 1200-1750," in "Symposium on Byzantium and the Italians, 13th-15th Centuries," special issue, *Dumbarton Oaks Papers*, 49 (1995), 252. On Lorenzo's collection, see Laurie Fusco and Gino Corta, *Lorenzo de'Medici, Collector of Antiquities* (Cambridge, 2006); for a recent analysis of the valuation of the icons, see Rembrandt

Christ, along with mosaic icons of three saints, was displayed in the Palazzo Medici, a showcase for Lorenzo's collection of antiquities and a visual repertoire for Florentine artists under his patronage whose works were included among them.[6] Among these collections, icons in micromosaic—the rarest, most technically difficult of forms, constructed through the skillful insertion of hundreds of minute tesserae—represented a truly lost and inimitable art that must have appeared nearly miraculous to its beholders. It is precisely this 'miraculous' potential of micromosaic—its doubling of artistic and spiritual authority—that this paper aims to explore.[7]

As Miguel Falomir has noted, the paintings on stone inaugurated by Sebastiano del Piombo (1485-1547) during the 1530s in Rome [**Plates 1-4**] would have recalled venerable icons in the medium of mosaic.[8] Given that the ancient Roman art of mosaic originally involved the use of stone tesserae, before the transition in the 3rd to 1st centuries to *smalto*, or colored glass, it is worth engaging the discussion in this volume of painting on stone with a consideration of its overlapping significance with mosaic. For, much like the painterly technique praised by Vasari, mosaic is a form of art that creates a durable, stone-like surface; it preserves a radiance of color that rivals oil paint on stone. Moreover, mosaic is itself constituted of stone-like particles, which, through a technically wondrous procedure, creates works that both refer to and exceed their constitutive elements. So constituted, I will argue, the specific materiality and facture of these works contribute to the activation of their sacred subjects.[9] Finally, although mosaic seems categorically distinct as an artistic

---

Duits, "Byzantine Icons in the Medici Collection," in *Byzantine Art and Renaissance Europe* ed. Angeliki Lymberopoulou and Rembrandt Duits (Farnham and Burlington, 2013), pp. 157-88.

6 In the "camera della sala grande detta di Lorenzo," or the alcove; see Effenberger, "Images of Personal Devotion," p. 212.

7 On this double authority, particularly with respect to Sebastiano del Piombo's stone altarpiece, the *Úbeda Pietà*, see the compelling insights by Elena Calvillo, in "Authoritative Copies and Divine Originals: Lucretian Metaphor, Painting on Stone, and the Problem of Originality in Michelangelo's Rome," *Renaissance Quarterly* 66, no. 2 (2013), 453-508. While relevant to the argument I pursue below, Calvillo's arrives at a different conclusion: she highlights the disjunction, rather than union, between technical artifice and sacred subject matter, one that signals an ontological difference to the devotional viewer between image and its divine prototype.

8 Miguel Falomir, "Sebastiano and 'Spanish Taste'," in *Sebastiano del Piombo 1485-1547*, ed. Giuseppe Scandini, Claudio Strinati, and Bernd Wolfgang Lindemann (Milan, 2008), p. 71, n. 60; cited in Calvillo, "Authoritative Copies," p. 63, n. 65.

9 Carol Walker Bynum has written extensively on the dynamic relation between materiality and sacred subject matter with respect to medieval art; see most recently *Christian*

process, throughout the 16th century it is commonly referred to as 'painting' (*pittura*). How might this earlier medium and technique, which undergoes a notable revival in the Renaissance of Sebastiano's time, give a conceptual and historical grounding for understanding the painterly innovations that followed, and the range of their associative meanings?

In the following essay, I propose a number of parallels between painting on stone and the art of mosaic, articulating their rich material associations and performative possibilities. I begin by briefly situating sacred mosaics within the context of their renewed appreciation in the Renaissance. I then turn to a legendary, Byzantine micromosaic icon of Christ [**Plate 27**] that bears special comparison with Sebastiano's altarpiece on stone which features a similar subject.[10] Here I want to argue that painting the eternal, Eucharistic body of Christ in media with 'eternal' associations, stone altarpiece and mosaic icon stage a similar doubling of artistic and spiritual authority. In the case of the icon, the specific materiality of micromosaic—activated by an unusual reliquary frame and interactive layers of mimetic materials within its setting in the church of Santa Croce in Gerusalemme—enacts a further feat. Not only does it authoritatively represent the Eucharistic body of Christ to the viewer: by means of its "miraculous" facture, the icon enacts the very mystery of transubstantiation it depicts.[11]

## Painting with Stones

If one were to look for textual evidence of a distinction between mosaic and painting, it would likely be found in the treatises and writings that form the basis for emerging art theory in the Renaissance. Yet Alberti and Vasari, both of whom set about to develop finer definitions and categories regarding the specific character of the arts, each refer to mosaic as *painting*, in the *Della Pittura* (1435) and the *Lives* (1550, 1568) respectively.[12] Indeed, in his *vita* of Domenico

---

*Materiality: An Essay on Religion in Late Medieval Europe* (New York, 2011). See also the seminal work of Cynthia Hahn, discussed below in the context of reliquaries.

10 The micromosaic icon is noted as an exemplar for Titian by Christopher Nygren in his essay for this volume.

11 In a complementary and compelling argument, Nygren also invokes 'transubstantiation' in his characterization of Titian's stone paintings in his essay for this volume.

12 As noted above (n. 2), Vasari uses the term "pittura" to describe mosaic, and Alberti, like many writers of the time, refers to Giotto's *Navicella* mosaic as "dipinta"; Leon Battista Alberti, *Della pittura*, in *Opere Volgare*, ed. Cecil Grayson (Bari, 1973), 3:74-75 (bk. 2, section 42).

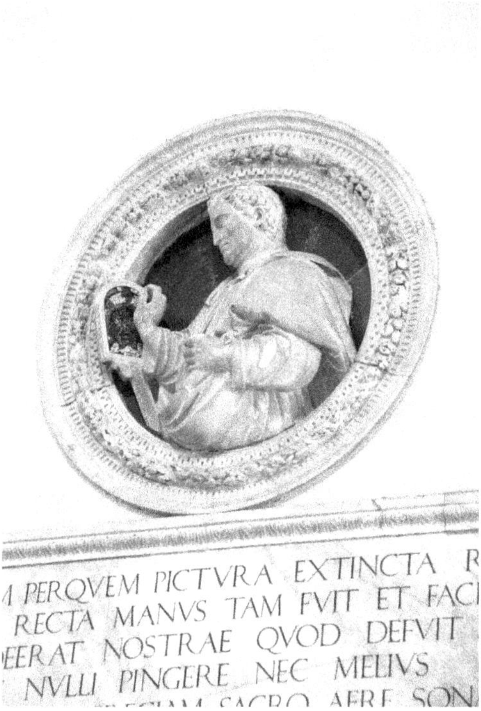


FIGURE 8.1
*Benedetto da Maiano, Monument to Giotto (detail), 1490. Florence, Cathedral* (PHOTO: MICHAEL WATERS).

Ghirlandaio, Vasari relates the claim of mosaic as "la *vera pittura* per la eternità" (italics mine).[13] This categorization warrants further investigation, as the technique of inserting individual tesserae into a bed of adhesive material seems very different from the application of pigment to a surface by the stroke of a brush. The intersection of the two techniques is highlighted most famously in the figure of Giotto who, as mentioned earlier, was unusual as a skilled practitioner of both. As Alexander Nagel and Christopher Wood have recently discussed, the conflation of techniques in Giotto's case presents something of a conundrum within the history of art.[14] This is evident in a commemorative monument to the artist by Benedetto da Maiano [**Fig. 8.1**], commissioned by Lorenzo de' Medici for the cathedral of Florence. On the one hand, the monument celebrates Giotto's canonical claim to fame, earlier described by Lorenzo Ghiberti in *I Commentari* and reiterated by Vasari (and indeed subsequent narratives of Renaissance art into the present). Just below the artist's portrait, an

13 Vasari, *Le vite*, vol. 3, p. 494; cited in Alexander Nagel and Christopher S. Wood, *Anachronic Renaissance* (New York, 2010), p. 399, n. 18.

14 Nagel and Wood, "Authorship and *Acheiropoieton*," in *Anachronic Renaissance*, pp. 123-34.

epitaph by the poet Angelo Poliziano, here translated from Latin, reads: "I AM HE THROUGH WHOM PAINTING, DEAD (EXTINCTA)/ RETURNED TO LIFE (REVIXIT)." The inscription continues:

> AND WHOSE HAND (MANUS) WAS AS SURE AS IT WAS ADEPT.
> WHAT MY ART LACKED WAS LACKING IN NATURE HERSELF.
> TO NO ONE WAS IT GIVEN TO PAINT (PINGERE) BETTER OR MORE.[15]

Giotto's revival of the art of painting, achieved through a return to the direct observation of nature, is central to the historiography of Renaissance art, and by extension to art histories that mark a shift from the Middle Ages to the Renaissance at the watershed of pictorial naturalism. The pivotal moment in this story is Giotto's turn away from the "crude painting" of the Greeks or Byzantines: their practice of merely copying, rather than inventing, authoritative pictures. Presumably the 'non-naturalistic,' hieratic iconographies rejected here are those of the iconic tradition. And yet in Benedetto's portrait, Giotto is depicted not as a painter, as celebrated by Poliziano or these narratives. He is instead imaged as a *mosaicist*, depicted in the act of carefully inserting a tessera into a small panel.[16] Even more surprising—indeed, seemingly at odds with the very premise of a monument to artistic invention—is his work on what appears to be an *icon*. Furthermore, not only is it an icon, but one that clearly makes reference to a venerable Byzantine prototype: the Mandylion, the legendary, acheiropoeitic image of Christ's Holy Face, an image miraculously made *without* human hands.

In addressing the paradox of the artist celebrated for his painter's hand (MANUS), here depicted as working on an image made without one (*acheiropoieton*), Nagel and Wood overlook the interesting overlap in the Renaissance between mosaic and painting. In referring to Alberti's use of 'dipinta' for mosaic, they surmise that either the artist forgot it was mosaic or perhaps "considered mosaic a modality of painting."[17] They instead interpret inscription and portrait as staging a contest between two different modes of image making and authorship. Giotto's work of art—as artistic invention—defines itself and emerges in relation to a longer tradition of cultural transmission of substitutive chains that refer backwards in time to historic prototypes, such as Byzantine icons. A complementary reading of the monument, I would suggest, sees its

15 Nagel and Wood, *Anachronic Renaissance*, pp. 123 and 397, n. 1.

16 The inscription of an earlier tondo portrait of Giotto by Benozzo Gozzoli in San Francesco, Montefalco also describes the artist as *pictor*.

17 Nagel and Wood, *Anachronic Renaissance*, p. 398, n. 11.

celebration of Giotto as the rare artist who excelled in two forms of painting at a moment of their historic intersection: mosaic, the art of icons, which signified the eternal and the durability of the past, and a revitalized (REVIXIT), 'living' painting that founded its own, modern future.

Historically, the spiritual potency and authority of the Mandylion lay in its miraculous generation, relic-like, through the impress of fluids from Christ's face to cloth. This is a process at once mysterious, but also akin to the application of paint to a surface.[18] Therefore it seems significant that the depicted Mandylion to which the artist applies his hand in the monument is rendered as mosaic, and not paint, effectively blurring the boundary between two kinds of images and processes. To add one's hand to an icon of divine facture, as Giotto does in this image, is at once a modest and a bold procedure: it is to insert oneself, tessera-like, into religious and artistic history. The 'double portrait' of Giotto, as painter *and* mosaicist, refers both to the authoritative, miraculous Mandylion and a new one of the painter's invention, produced through the equally miraculous art of mosaic.[19] Here is evidence of the doubling of artistic and spiritual authority to which we will turn.

## Mosaic Revivals in the Renaissance

The monument to Giotto constituted part of a more comprehensive project of Lorenzo at the Duomo involving mosaics: an ambitious, unrealized plan to decorate the interior of Brunelleschi's enormous cupola. The new decoration would have effectively linked the Cathedral to the historic Baptistery directly across, a building thought to be of prestigious, ancient origin and notable for the spectacular mosaics in its dome. Lorenzo's project reflects a larger trend of renewed interest in mosaics during the Renaissance, one that provides a context for appreciating the value attached to micromosaic icons in particular.

Our knowledge of portable mosaic icons is based upon only about fifty surviving examples dating from the 11th through the 14th centuries, making them

18 The seminal work engaging this parallel to theorize both icons and painting is Herbert L. Kessler and Gerhard Wolf, eds., *The Holy Face and the Paradox of Representation: Papers from a Colloquium at the Bibliotheca Hertziana, Rome and the Villa Spelman, Florence, 1996* (Bologna, 1998).

19 Nagel and Wood insist that Giotto's mosaic version of the Mandylion should not be read as an invention, but rather as an historic image undergoing repair. Yet the fact that the top of the head is incomplete suggests that Giotto is in fact creating a new version.

a rare class of artifacts.[20] Even in Byzantium, they were a luxury good reflecting a considerable investment of materials and skilled craftsmanship.[21] Micromosaics, an even rarer genre that briefly flowered during the late Palaeologan Empire, were likely produced primarily in workshops in Constantinople.[22] These miniature panels required the insertion of tesserae of half-millimeter square in size or smaller. While most of the tesserae were enamel, some of the colors would have been rendered by semiprecious stones, such as lapis lazuli and malachite, along with silver and gold. The combination of precious materials and facture would have contributed to the value of micromosaic icons and their jewel-like effect, amplifying the sacredness of the subjects they depicted.[23] While we have little evidence of how micromosaic icons were used in practice, we can imagine the appeal of a portable icon whose precious, reflective materials recalled the monumental mosaic programs of Byzantine churches: a small piece of the sacred which brought the heavenly to hand to be closely engaged, appreciated and venerated. When surrounded by ornamented revetments, jeweled or precious frames, their preciousness and sacrality was further amplified. In the case of a micromosaic icon of St. Demetrios, an ampulla of his holy oil is included in the frame surrounding his image.[24]

The owner of the Demetrios icon, Basilios Bessarion, fleeing émigré from Byzantium, archbishop of Nicaea and later Roman humanist and cardinal, donated seven mosaic icons to Saint Peter's in Rome (1462 and 1467), among them micromosaics, which are described as *ex opera mosayco minuto*.[25] Rome is the site of a second wave of mosaic revival that took place in the first decades of the 16th century, coinciding with the period of Sebastiano's innovation of painting on stone. The artist's unfinished altarpiece, *The Nativity of the Virgin* (1530-34)—[**Plate 2**] centerpiece of the dazzling funerary chapel of Agostino

20 On portable mosaics, see Carlo Bertelli, "Portative mosaics," in *Mosaics*, ed. Bertelli, pp. 225-56; Otto Demus, "Two Palaeologan Mosaic Icons in the Dumbarton Oaks Collection," *Dumbarton Oaks Papers* 14 (1960), 87, 89-119.

21 Effenberger, "Images of Personal Devotion," pp. 209-14.

22 Effenberger, "Images of Personal Devotion," p. 209ff.

23 See discussion in Demus, "Palaeologan," p. 91. Important work on the synaesthetic effect of icons, activated by light and sound, is found in Bissera Pentcheva's *The Sensual Icon: Space, Ritual, and the Senses in Byzantium* (University Park, 2013); see also Glen Peers, *Sacred Shock: Framing Visual Experience in Byzantium* (University Park, 2004).

24 For catalog entry, see for example *Byzantium: Faith and Power (1261-1557)*, pp. 232-33. The interaction of icons, relics and frames is discussed below.

25 Effenburger, "Images of Personal Devotion," 210-11; and F. Lollini, "Bessarione e Perotti diffusori della cultura figurative bizantina," *Studi umanistici piceni* 11 (1991), 127-42.

Chigi at the church of Santa Maria del Popolo in Rome—is set within a rich complex of works in stone that resonate with his own invention.[26] Crowning the chapel is a Pantheon-like dome adorned with a spectacular mosaic program, designed by Raphael and executed by Luigi de Pace in 1516. That mosaics are featured in such an eminent chapel and commissioned of an artist of Raphael's stature is significant.

While rare, an important precedent for this revived use of mosaics is found at Santa Croce in Gerusalemme in Rome, the pilgrim church famous for its relics of the True Cross, purported to have been translated from the Holy Land by the Empress Helena, mother of Constantine I. Among the church's treasured objects is the Byzantine micromosaic icon of Christ [**Plate 27**] that became a legendary object of cult, as it was claimed to be the first *imago pietatis*: an authentic portrait of Christ witnessed by Pope Gregory during his celebration of Mass, when Christ appeared to him in bodily form. As I will argue, much like the Chigi chapel, the complex interplay of materials in the icon's probable setting at Santa Croce actively contribute to its authority and meaning.

From the late 1490s through the 1530s, Santa Croce is the site of an extensive project of renovation involving mosaics that sought to restore the ancient and prestigious origins of the church.[27] Only recently have scholars noted the importance of the mosaic program within a wider sphere of a cultural renewal in Rome based upon retrospective models.[28] The church renovations, inaugurated by the Spanish cardinal Pedro Gonzalez de Mendoza (1478-1495) and

26 On the altarpiece, see Costanza Barbieri, "La Pala della Concezione e Natività della Vergine di Sebastiano per la cappella Chigi e un disegno inedito," *Konsthistorisktidskrift* 81 (2012), 245-53. An analysis of the interplay of materials in the chapel has yet to be written. Nadia Baadj's essay for this volume engages these themes in her study of a Northern exemplar.

27 The most comprehensive study of the mosaic program is Cynthia Payne, "'In the fullness of time': the vault mosaic in the Cappella Sant'Elena, Santa Croce in Gerusalemme, Rome" (PhD, diss. University of Georgia, 2003). For the renovations, see also Payne, "Lux Mundi: the Vault Mosaic in the Cappella S. Elena, S. Croce in Gerusalemme, Rome," *Athanor* 17 (1999), 35-43. See also Marilena Luzietta, "La cappella di Sant'Elena in Santa Croce in Gerusalemme: il programma iconografico," in *Atti del XX colloquio dell'Associazione Italiana per lo Studio e la Conservazione del Mosaico*, ed, Claudia Angelelli and Andra Parbeni (Tivoli, 2015), pp. 587-93.

28 More recently, Jack Freiburg has engaged the chapel as part of a larger program of renewal in *Bramante's Tempietto, the Roman Renaissance, and the Spanish Crown* (Cambridge, 2014), pp. 9-36; both projects are connected by the patronage of Spanish Cardinals, a fact that bears relevance for Sebastiano's stone altarpieces. Nagel and Wood, *Anachronic Renaissance*, pp. 334-45, also discuss Santa Croce in the context of renovations in the Renaissance that sought to foreground their antique origins.

carried out by his successor, Cardinal Bernardino Lopez de Carvajal (1456-1523), include Antoniazzo Romano's dazzling and technically-complex fresco cycle, which depicts the legend of the Empress Helena's discovery of the True Cross in continuous narrative frieze over a curved surface of monumental scale. But equally ambitious was the mosaic program in Helena's ancient chapel, located underground beneath the main apse—what was known as "*Hierusalem*." Architecturally, the chapel is similar in design to Near Eastern models, and more specifically a martyrium erected by Constantine in Jerusalem to house a fragment of the True Cross, therefore typologically replicating that prestigious prototype.[29]

As its toponym suggests, the church of Santa Croce in Gerusalemme in Rome is a conscious imitation of Jerusalem for the pilgrim faithful. It was first constructed as a basilica on the site of the Sessorian palace owned by Helena.[30] According to the *Liber Pontificalis*, Helena's *cubiculum* or private chamber was converted into a chapel for prayer and featured relics of the True Cross she carried with her to Rome from the Holy Land, along with earth from Mt. Calvary, the historic site of Christ's crucifixion.[31] The basilica became the traditional site for the Liturgy of the *inventio crucis* and legendary for these precious relics.[32] In addition, the chapel became the site of the celebration of Mass by the Pope during Holy Week.

In 1507-1508, the chapel's mosaic vault, with images dating from the time of Emperor Valentinian III (425-455), was restored by an intricate mosaic design that has been attributed to Baldassare Peruzzi (1481-1536) [**Fig. 8.2**] and argu-

29 Richard Krautheimer, *Corpus Basilicarum Christianarum Romae. The Early Christian Basilicas of Rome* (*IV-IX Cent.*), vol. 1 (Vatican City, 1937), p. 194; on the substitutional process of transfer through architectural structure, see Krautheimer's seminal essay, "Introduction to an Iconography of Medieval Architecture," *Journal of the Warburg and Courtauld Institutes* 5 (1942), 1-33.

30 On the church, see Krautheimer, *Corpus Basilicarum Christianarum Romae*, 1:165-97; Anna Cavallaro, *Santa Croce in Gerusalemme* (Rome, 2009). Notably, the earliest reference to Helena as founder of Santa Croce is Flavio Biondo's *Roma Instaurata* 1444-1448. A crucial source which draws upon a now lost manuscript of 1475 is Raimondo Besozzi's *La storia della Basilica di S. Croce in Gerusalemme, Roma* (Rome, 1750) and Onofrio Panvinio, *Le sette chiese romane* (Rome, 1570).

31 Anastasio, *Liber Pontificalis* (Vita S. Gregoriii, II), (Paris, 1884). On the chapel, see also Claudio Varagnoli, *S. Croce in Gerusalemme: la Basilica restaurata e l'architettura del Settecento romano* (Rome, 1995); *Gerusalemme a Roma: la basilica di Santa Croce e le reliquie della Passione*, ed. Roberto Cassanelli and Emilia Stolfi (Milan, 2012).

32 On this history, see Sible de Blaauw, "Jerusalem in Rome and the *Cult of the Cross*," in *Pratum romanum, Richard Krautheimer zum 100. Geburtstag*, ed. Renate Colella et. al. (Wiesbaden, 1977), pp. 55-73.

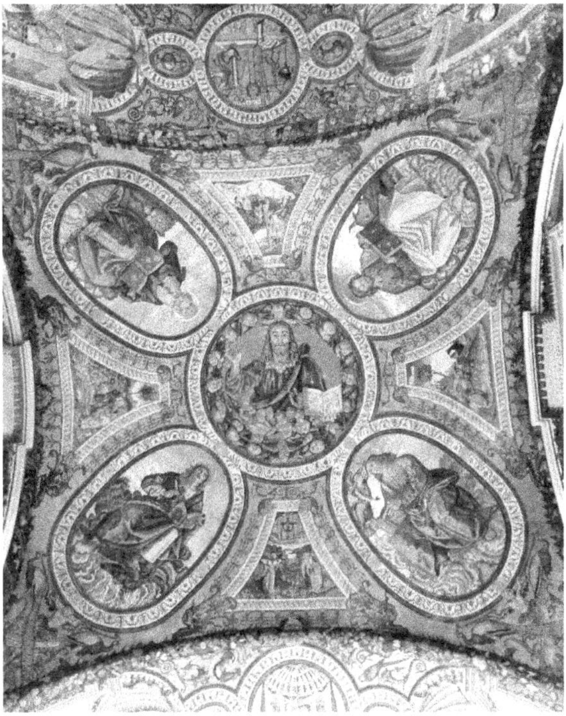

FIGURE 8.2
*Baldassare Peruzzi (attributed), mosaic vault, ca. 1500. Helena (Jerusalem) chapel, Rome, Santa Croce in Gerusalemme* (PHOTO: AUTHOR).

ably an attempt to recall Byzantine precedents. As Jack Freiburg has recently discussed, the narrative of the mosaic program united in a continuous history the chapel's origin and its present patronage under the Spanish cardinals, who sought to elevate the church's standing as a site of pilgrimage and cultic veneration in anticipation of the Holy Years of 1500 and 1525.[33] In Francesco Albertini's 1510 pilgrim guide to Rome, the chapel's mosaics are described as "most beautiful," and continued to be the subject of praise for over a century.[34] As one of the few major works of mosaic in Rome during the Renaissance, they clearly constitute a watershed.[35] But within the complex of Santa Croce itself, they also form part of a larger frame of material associations and substitutions that bear meaning for the micromosaic icon in particular.

In the unusual sloping stairwell that descends into the Helena chapel, a majolica inscription produced in the 1520s emphasizes the ancient origin of the site and reasserts the claim that "the chapel itself and the whole basilica

33 Freiburg, *Bramante's Tempietto, the Roman Renaissance, and the Spanish Crown*, pp. 9-36.

34 Franscisco Albertini, *Opusculum de mirabilibus novae et veteris urbis Romae* (1510), ed. August Schmarsow (Heilbronn, 1886), p. 7.

35 Cavallaro, *Santa Croce*, pp. 33-41.

and all Rome deserved to be called 'a second Jerusalem.'"[36] Within the subterranean chapel, pilgrims venerating fragments of the True Cross and other relics could walk across earth from Calvary that claimed to contain Christ's precious blood. Painted topographical vistas of the Holy Land on the chapel walls worked in concert with the locative earth-relic to enact a form of 'spatio-temporal transport' to Jerusalem, which at that time was inaccessible as a pilgrim destination.[37] By these material and visual substitutes, the 'heavenly Jerusalem' of the church above was mirrored in the 'earthly Jerusalem' of the chapel below. What I would emphasize here is the retrospective and literal grounding of the identity and authority of the church by Helena's small, underground chapel. Within this chapel of substitutive experience, rich with relics from the life of Christ, the micromosaic icon was likely first venerated. The specificity of its siting, and its experiential, material dimensions, played a further role in the image's reception and spiritual activation.[38]

## Mosaic, Mimesis, and Transubstantiation

The icon that became a legendary object of cult at Santa Croce in Gerusalemme is a rare miniature mosaic of Christ, thought to have originated circa 1300 from the luxury workshops of Constantinople [**Plate 27**]. Measuring less than 6 × 8

36 Illaria Toesca, "A Majolica Inscription in Santa Croce in Gerusalemme," in Douglas Fraser, Howard Hibbard, and Milton J. Lewine, eds., *Essays in the history of art presented to Rudolf Wittkower* (New York, 1969), pp. 101-05.

37 As Nagel has argued, rather than a mere "visual replica," "the constellation/installation initiates a process of activation. The potential for time-and-space travel is actualized in the experience of users who assemble the elements imaginatively and thus, for a time, inhabit a space … linked to Jerusalem through real conjunctures—the relics and the earth." Alexander Nagel, *Medieval Modern: Art Out of Time* (New York, 2012), pp. 100-01; on the chapel as "spatio-temporal transport to Jerusalem," see p. 114.

38 To date, there is no clear evidence concerning where the icon and reliquary were housed or venerated during this period, although most scholars assume its location in the Helena chapel. Besozzi, in *La storia della basilica*, 149, describes an "*apparuit in specie pastoris sub effigie pietatis beato Gregorio celebranti super altare Jerusalem in ecclesia s. Crucis.*" See Carlo Bertelli, "The *Image of Pity*," in *Essays in the History of Art Presented to R. Wittkower*, pp. 46ff, for discussion. Besozzi's description could refer to either the main altar of the church or the one in the chapel. In a recent study, Corinna T. Gallori proposes its storage in a relic chapel and display on the high altar during the celebration of *Laetare* Sunday; see "The Late Trecento in Santa Croce in Gerusalemme: Napoleone and Nicola Orsini, the Carthusians, and the Triptych of Saint Gregory," *Mitteilungen des Kunsthistorichen Instituts in Florenz* 58 no. 2 (2016), 181-86.

inches, the icon features a three-quarter-length image of the crucified Christ with the Greek inscription "BASILEUS TES DOXES" ("King of Glory") in gold still faintly legible above his head, and the bars of a cross visible behind his body. The announcement of the icon's status as Byzantine by this Greek inscription, and its specific figuration of Christ, marks it among a few rare instances of its kind. At some point after it came into the possession of the church, the icon was framed in an enormous and unusual triptych reliquary[39] [**Fig. 8.3**]. Approximately 98 × 63 cm. in its open form, the triptych structure is comprised of a central panel with two wings faced by a grid-like pattern created by strips of metal punctuated by small rosettes. The grid delineates over 200 miniature reliquary boxes, each containing a relic encased in paper and silk wrappings.[40] These include relics from the Holy Land that resonate particularly with their location in Santa Croce in Gerusalemme. The later addition of an aedicule and base gives the frame a distinctly cruciform dimension that also reflects its placement at the church of the Holy Cross. An inscription on the reliquary's curved pediment, "*Fuit S. Gregori Magni Papae*," likely an 18th-century addition, makes explicit the icon's identity as the vision of Gregory. The icon became famous as an object of cult: as the true image or *vera immagine* of Christ seen by Pope Gregory during his celebration of Mass. In a widely disseminated print of the 1490s by Israhel van Meckenem (1445-1503), the icon is celebrated as the first *imago pietatis* (*prime imaginis pie/tatis* ).[41]

39 The most comprehensive study of the icon and its frame is still Bertelli, "The *Image of Pity*," pp. 40-55. Gallori focuses on the patronage of Nicola Orsini, whom she proposes as contributor to the reliquary case; see "The Late Trecento in Santa Croce in Gerusalemme," pp. 156-87. The icon and frame have been featured in numerous recent exhibitions with catalogue entries: *Treasures of Heaven: Saints, Relics, and Devotion in Medieval Europe*, ed. Martina Bagnoli, et al. (New Haven and London, 2011); *Byzantium 330-1453*, ed. Robin Cormack and Maria Vassiliki (London, 2008); and Evans, ed., *Byzantium: Faith and Power.*

40 The contents of the reliquary are catalogued in Onofrio Panvinio, *Le sette chiese principali di Roma* (Rome, 1570), p. 281ff.

41 An inscription below the image of the icon claims: "*Hec ymago contrafacta est ad istar et similitudenem illus prime imagines pie/tatis custodie in ecclesia S. Crucis in Urbe romana, quam fecerat de pingi sanctissimi Gregorius papa Magnus, post habitam ac sibi ostensam desuper visionem.*" On van Meckenem's invention as possibly the first European example to claim to be 'contrafacta' (or an authentic reproduction of a prototype), see Peter Parshall, "*Imago Contrafacta*: Images and Facts in the Northern Renaissance," *Art History* 16, no. 4 (1993), 554-79; for the image more generally, David Landau and Peter Parshall, eds., *The Renaissance Print* (New Haven and London, 1994), p. 58; Gerhard Wolf, "Imago Pietatis-Israhel van Meckenems Konterfei eines Abbildes einer Erscheinung," in *Glaube, Hoffnung, Liebe, Tod*, ed. Christoph Geissmar-Brandi and Eleonora Louis, 2nd rev. ed. (Klagenfurt, 1995), pp, 274-76.

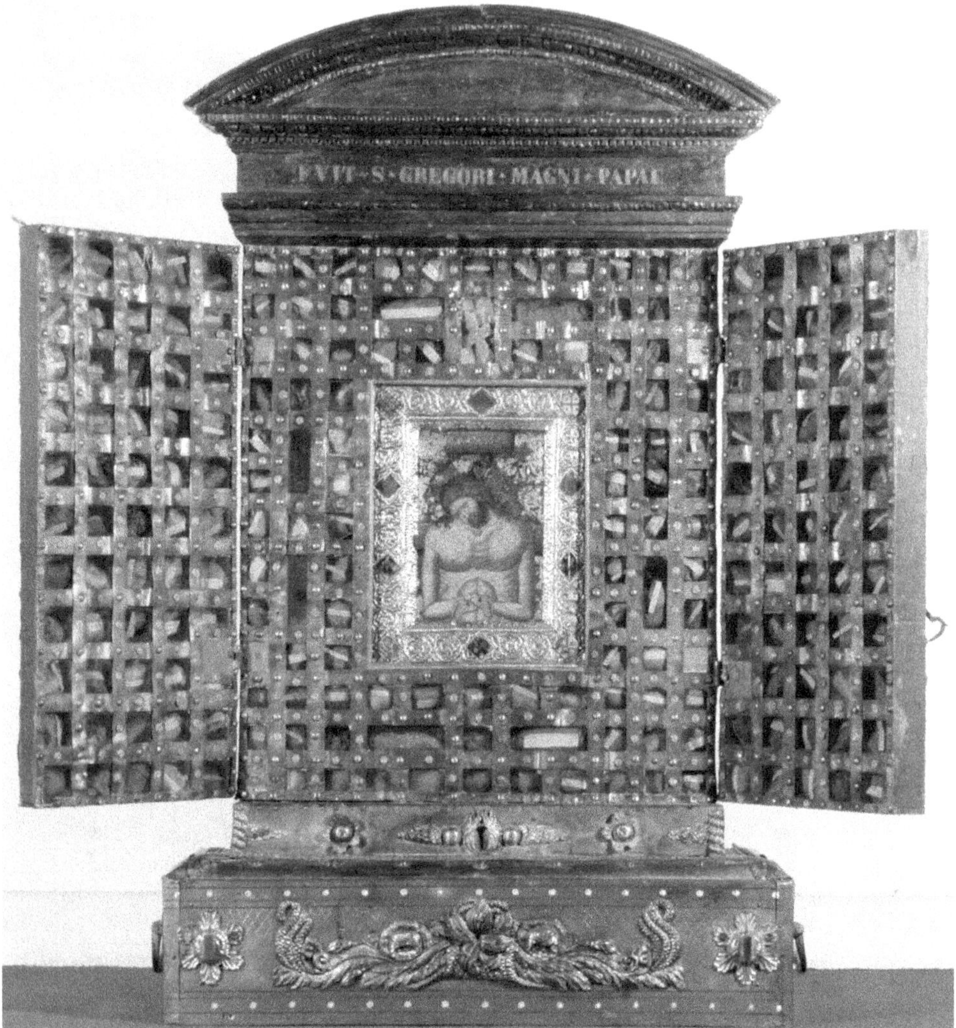


FIGURE 8.3 *Mosaic icon of Christ* Akra tapeinosis *and triptych reliquary frame, late 14th century, wood, metal, paper identification and silk wrappings for the relics, 98.7 × 97.1 cm (wings open). Santa Croce in Gerusalemme, Rome, Museo della Basilica* (PHOTO: ©FONDO EDIFICI DI CULTO).

Because pilgrim accounts specify the icon's location as the altar "*Hierusalem*," a toponym for both the chapel and the church as a whole, we cannot be entirely certain of the micromosaic's location throughout its history. However, it is highly likely that the icon and its reliquary, as a treasured relic with ancient authority and of holy origin, would have originally been venerated along with other precious relics displayed within the Helena chapel.[42] Within the intentionally archaizing program of the mosaic chapel, the micromosaic icon would have taken on further spiritual authority as an authentic artifact from the life of Gregory.[43] Eventually, the adjacent underground chapel came to be dedicated to Pope Gregory himself.

The complex experiential dimension of the Jerusalem chapel is key, I would argue, in understanding the function of this chapel dedicated to Gregory [**Fig. 8.4**], which to date has not been subject to substantive interpretation.[44] This *antecappella*, constructed around 1520, was in all probability originally conceived to enhance the flow of pilgrims to the more famous Helena chapel to which it connects. A sketch from the 1520s by Antonio da Sangallo reveals that the Gregory chapel was part of larger set of interventions in the church. Christoph Frommel sees in Sangallo's treatment an informed sensitivity to the site's ancient structure and program to restore the ancient prestige of the church, particularly as it bears relation to an Etruscan temple.[45] It is also men-

42 See n. 39 above.

43 Cynthia Hahn's observations on relic collections are relevant here: The "desire for Christ's body is precisely the element that structures and even impels the [relic] collection (perhaps explaining the prominence in the Late Middle Ages of the transformation of the Corpus Christi, the host into a relic). The assemblage of relics or *bricolage*, made up of imperfect and fragmentary parts, must both strive for and metaphorically indicate a more meaningful yet ultimately unattainable whole ... the forever absent and unattainable body of Christ. Each of the relics in the series represents this body, but it does so only imperfectly and incompletely." Hahn, "The Meaning of Early Medieval Treasuries," in *Reliquiare im Mittelalter*, ed. Bruno Reudenbach and Gia Toussant (Berlin, 2005), p. 11.

44 On the chapel, see Mario Sensei, "Dall'*Imago Pietatis* alle Cappelle Gregoriane, Immagini, Racconti e Devozioni per la 'Visione' e la Cristomimesi," *Collectanea Franciscana* 70 (2000), 93-104. For a description of the chapel with images, see Cavallaro, *Santa Croce*, pp. 55-63.

45 Christoph Frommel, "Projetto e archeologia in due disegni di Antonio da Sangallo il Giovane per Santa Croce in Gerusalemme," in *Roma, della cultura dell'Antico nei secoli XV e XVI Da Martino V al Sacco di Roma 1417-1525*, ed. Silvia Danesi Squarzina, (Milan, 1989), pp. 382-89. 178 U899A recto (reproduced on 390) gives a hypothetical reconstruction of the structure underlying Santa Croce. The so-called anticappella was added in 1520 by order of Carvajal, according to a lost inscription; the occasion for the sketch; see Frommel, p. 177, and Cavallaro, *Santa Croce*, pp. 42-47.

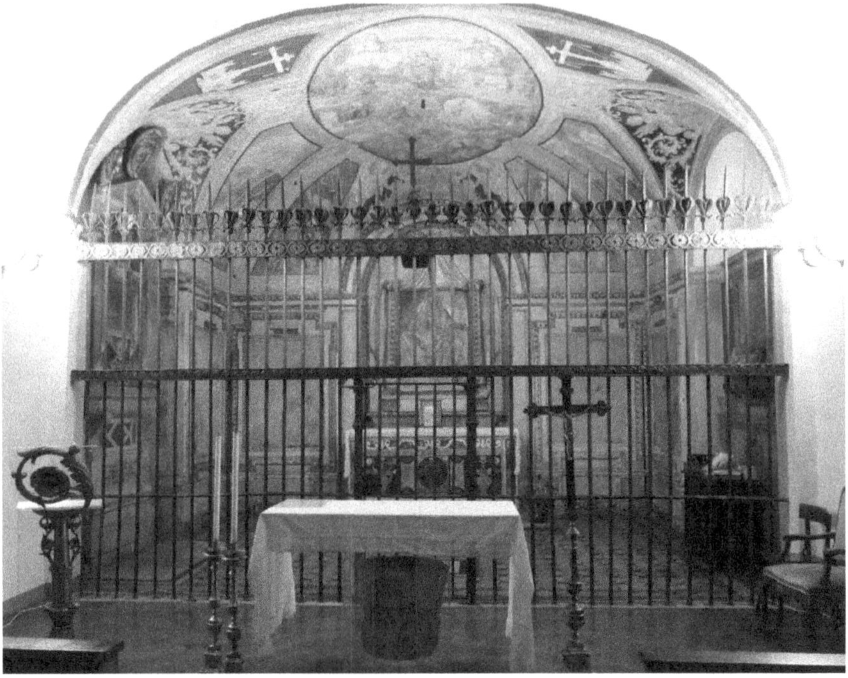

FIGURE 8.4 *Gregory chapel (view towards altar, present day). Rome, Santa Croce in Gerusalemme* (PHOTO: AUTHOR).

tioned in the 1575 testimony of an abbot of the church, Besozzi based upon a now lost manuscript of 1475.[46] With its Roman vaulting, the Gregory chapel is deeply connected to a wider archaeological initiative to literally unearth the paleo-Christian origins of the Roman church during the Renaissance. We may place it in the same moment as the exploration of the catacombs, for instance, by the humanist Pomponio Leto (d. 1498) and his followers, and with structures such as the crypt of Bramante's Tempietto, the presumed site of St. Peter's crucifixion. Freiburg has demonstrated the significance of this retrospective moment at the Roman church of San Pietro in Montorio for Santa Croce, drawing a parallel between the creation of the Tempietto-stone and the discovery of

46 The *cappella* is also mentioned in Sergio Ortolani, *Schede del Catalogo delle Opere d'arte contenute in Santa Croce in Gerusalemme*, (Ministero della P. I., Direzione Generale della B. A., 1922), n. 6: "Si sa dal Besozzi che su questo altare di S. Gregorio—Cappella construtta ex-nova nel 1520 dal Card. Caravaja (v. armi sulla volta) e da lui messa in communicazione con l'altra di S. Elena per mezzo di un vasto corridoio, dopo d' aver ambedue collegato al transetto con due cordonate e porte (v. armi) v'era una tela assai detrita della Pieta…."

the *titulus*.[47] The Spanish patronage common to both churches is yet another point of contact with Sebastiano's Roman career.

Geographically, a more immediate parallel to the Gregory chapel may be found close by: in the Sancta Sanctorum of the rival pilgrim church of San Giovanni in Laterano, which housed one of the most famous cult images in Rome. With new indulgences granted by Pope Sixtus V for the veneration of such *vera immagini* throughout the churches of Rome, it is likely that Santa Croce also sought to produce a miraculous image that would bring pilgrims to its doors. Given the identification of the *imago pietatis* of the vision of Gregory with the Santa Croce micromosaic icon, the freestanding reliquary came to be identified as the altar of Gregory himself. At some point in its history, it is likely that the ensemble of icon and frame was installed in the chapel that came to bear its name, upon an altar dedicated to Gregory behind a metal grate that also displayed relics. The vaulting of this area was later frescoed with images to further visualize and substantiate Gregory's legendary Eucharistic vision.

As the legend of Pope Gregory's mystical vision of Christ was propagated in devotional literature and print culture, the micromosaic icon was re-conceptualized within the frame of this narrative. Since 1475, the icon has been ensconced within its large reliquary frame, which is described in the church's history of Besozzi in 1575 as belonging to Gregory himself.[48] In addition to its grand scale and triptych form, the materials of the Santa Croce reliquary play a significant role in the activation of the icon. In jeweled reliquaries, such as those which housed famed relics of the True Cross at Santa Croce in Gerusalemme, the preciousness and 'eternal' durability of stone amplifies the perception of similar qualities of the less visible fragment contained within.[49] Stone and wood fragments are transformed in the process, becoming "vessels through which the divine could speak to humans."[50] As Thelma Thomas has

---

47 See Jack Freiburg, "Bramante's Tempietto and the Spanish Crown," *Memoirs of the American Academy in Rome* 50 (2005), 162-205 and esp. p. 166.

48 Besozzi, *La storia della basilica*, p. 149, describes an "*apparuit in specie pastoris sub effigie pietatis beato Gregorio celebranti super altare Jerusalem in ecclesia s. Crucis*." See Bertelli, "The *Image of Pity*," p. 46ff, for discussion.

49 For True Cross reliquaries, see Holger Klein, ed., *Byzanz, der Westen und das 'wahre' Kreuz: die Geschichte einer Reliquie und ihrer künstlerischen Fassung in Byzanz und im Abendland* (Wiesbaden, 2004); the Trier reliquary Klein reproduces provides a parallel worth pursuing.

50 Brigitte Buettner, "From Bones to Stones–Reflections on Jeweled Reliquaries," in *Reliquiare im Mittelalter*, p. 47. For related discussion, see Charles Freeman, *Holy Bones, Holy Dust: How Relics Shaped the History of Medieval Europe* (New Haven, 2011). For the signifying dimension of reliquaries more generally, see Cynthia Hahn, "What do Reliquaries do for Relics?" in *Strange Beauty: Issues In the Making and Meaning of Reliquaries, 400-circa 1204*

noted, reliquaries and even images of relics partook of the perceived potency of what they represented, evoking a "charged space … [or] signifying field to contain, represent, and otherwise manipulate sacred energies."[51]

In the ensemble of the Santa Croce icon and its frame, mosaic 'stones' and the bones of relics work together to dynamic, sacralizing effect. Here we may draw upon what Seeta Chaganti has usefully theorized as a 'poetics of enshrinement': the interplay of container and contained which provides "aesthetic and even epistemological structure … [and] the more complex effect whereby contained and containing are interchangeable, and the borders between them are indeterminate …."[52] In this way, the Eucharistic body of Christ depicted by the Santa Croce icon becomes more than depiction within the Communion of Saints in the surrounding relics.[53] Over time legend grew that Gregory had made the icon himself from the bones of saints contained in the reliquary frame, a story seamlessly uniting icon and its later supervening structure, transforming the micromosaic icon into a holy object of truly miraculous facture.[54]

(University Park, 2012); and Patricia Cox Miller, "'The Little Blue Flower is Red': Relics and the Poetizing of the Body," *Journal of Early Christian Studies* 8 (1997), 213-36.

51 Thelma K. Thomas, "Understanding Objects," in *Reading Medieval Images: The Art Historian and the Object,* ed. Elizabeth Sears and Thelma K. Thomas (Ann Arbor, 2002), pp. 9-15, here extending the argument of Patricia Cox Miller's excellent work.

52 Seeta Chaganti, *The Medieval Poetics of the Reliquary* (New York, 2008), pp. 19, 15. For the "poietic" dimension of Renaissance frames, see discussion by Rayna Kalas, *Frame, Glass, Verse: The Technology of Poetic Invention in the English Renaissance* (Ithaca and London, 2007), esp. pp. 19, 17; Kalas, p. 9, uses framing to chart a transformation in modern thought: from *poiesis*/poetic to technology (Heidegger, The Question Concerning Technology: 2 modes of revealing: poiesis, a bringing forth, and "enframing," (Ge-stell) of technology, preface xiii-xiv) or the 'picturing' that divorces interior from exterior rather than performs an active bringing into presence of matter in the world." See also her observation, p. 36, that the pre-modern "engaged frame does not isolate the unique and precious object as much as it establishes adjacencies between things."

53 For the related idea that relic treasuries represent a vision of the collection of saints that is the Heavenly Jerusalem, see discussion by Cynthia Hahn, "The Meaning of Early Medieval Treasuries," pp. 1-20, esp. p. 19.

54 Bertelli, "Attualità e nostalgia: ancora sull'immagine venerata in Santa Croce in Gerusalemme," in *Certose e Certosini in Europa: atti del convegno alla Certosa di San Lorenzo, Padula, 22, 23, 24 settembre 1988,* ed. Vega di Martini and Angelina Montefusco (Naples, 1990), p. 69. An interesting parallel is the micromosaic of the Virgin *Eleousa* currently in the Patriarchal Seminary of Santa Maria Salute in Venice. A parchment attached to the back of the icon attests that it was made centuries earlier and venerated as a relic by the Emperor Constantinople Emmanuel. Edmund C. Ryder, "Micromosaic Icons of the Late Byzantine Period" (PhD diss., New York University, Institute of Fine Arts), pp. 91-95.

As Cynthia Hahn has thoughtfully explored, reliquaries condition and circumscribe the approach of the believer to the holy. In particular they propose a complex interaction of the senses and the imagination, originating a discourse on the holy that encompasses space, time, and performance. Hahn further notes that all reliquaries in some sense seek to reconstruct a form of sacred space that originates in a typologically conceived relationship to the Ark of the Covenant and the sanctuary of the Temple in Jerusalem.[55] The point is especially relevant to the reinvention of the icon in its reliquary frame at Santa Croce in Gerusalemme, a place that sought to recreate the pilgrim's experience of the lost Holy Land through relics, images and strategies of mimetic substitution.[56]

Along with the narrative frame of the Gregory legend that conditioned the icon's viewing, the work of the reliquary in these varied dimensions would have provided a privileged moment of sacred revelation. Within the highly charged, signifying field of its unique framing, the devotional encounter with the iconic image of Christ would have simulated that of the saint himself: the flesh-like appearance of the body, as rendered by its micromosaic facture, would have not only symbolized, but actively staged, the Eucharistic vision of the transubstantiated Christ.[57] The viewer would have the potential to 'see' Christ as he was seen in the mystic moment of Gregory's vision.[58]

---

55 Hahn, *Strange Beauty: Issues In the Making and Meaning of Reliquaries*, p. 9.

56 On topomimesis or the imitation of topography through substitutive means, see for example Michele Bacci, "Performed Topographies and Topomimetic Piety: Imaginative Spaces in Medieval Italy, in Alexei Lidov, ed., *Spatial Icons: Performativity in Byzantium and Medieval Russia* (Moscow, 2011), pp. 101-18.

57 It is interesting to note in this regard that in German, the central part of an altarpiece is called the "*corpus*;" as cited in Kalas, *Frame, Glass, Verse*, p. 4. For the symbolic associations of steatite (a medium closely related to micromosaic), drawing upon poetic sources, see Ioli Kalavrezou, *Byzantine Icons in Steatite* (Vienna, 1985).

58 The question of whether the faithful viewer was meant to 'see' the miracle of transubstantiation during the so-called Gregorymass is a complicated one, as theological doctrine on the matter was complex. See the important discussion by Carol Walker Bynum in "Seeing and Seeing Beyond: The Mass of St. Gregory in the Fifteenth-Century," in *The Mind's Eye: Art and Theological Argument in the Middle Ages*, ed. Jeffrey Hamburger and Anne Marie Bouchè (Princeton, 2006), pp. 208-40. Aden Kumler also discusses the issue of the depictability of transubstantiation, and responds to Bynum's position, in *Translating Truth: Ambitious Images and Religious Knowledge in Late Medieval France and England* (New Haven, 2011), esp. pp. 158-59.

## Micromosaic Materiality and 'Reignited' Painting

In considering the role of technique and materials in 'painting the eternal,' specifically the eternal body of Christ, we might make two further observations. In characterizing mosaics' special capacity to resist the extinction or consumption (*si spegne*) over time to which ordinary painting was vulnerable, Vasari described their ability to be 'reignited' (*s'accende*), or constantly renewed:

> *E certo è che il musaico è a più durabile pittura che sia, imperò che l'altra col tempo si spegne e questa nello stare fatta di continuo s'accende, et inoltre la pittura manca e si consuma per se medesima, ove il musaico per la sua lunghissima vita si può quasi chiamare eterno.*[59]

Clearly Vasari had in mind the unique capacity of mosaic to be renewed in the sense of *restored*, piece-by-piece, in a way that ordinary painting could not. But the use of the word *accendere*—'to reignite'—is a highly unusual one in the context of art. From Boccaccio to Petrarch to Dante, and in common parlance as well, the term is most often associated with light (*lumi*) and spirit (*animi*). What might be meant then by 'reignited painting'? Here we might return to the example of Giotto who, through mosaic, made painting live or animate (*animi*) again, as well as to the luminous (*lumi*) effect specific to mosaic.[60] The language of consumption and renewal is also Eucharistic, as Christ's body is both consumed as Host and yet mysteriously endures, as an eternally renewed sacrifice.

With regard to micromosaic, the technique of embedding tesserae in mastic and wax produces a remarkably flesh-like appearance [**Plate 27**], in this case, one that seems to enact the very miracle it depicts. The flesh-like effect of micromosaic is notable in other examples from the same period, such as the Forty Martyrs icon currently housed at Dumbarton Oaks.[61] The St. Demetrios micromosaic was also highly valued specifically for its corporeal illusionism, as described by the Italian humanist Niccolò Perotti in a donation document of

59 Introduction to the *Lives*, ch. 29: Del musaico de'vetri, et a quello che si conosce il buono e lodato, 1:148: "E certo è che il musaico è a più durabile pittura che sia, imperò che l'altra col tempo si spegne e questa nello stare fatta di continuo s'accende, et inoltre la pittura manca e si consuma per se medesima, ove il musaico per la sua lunghissima vita si può quasi chiamare eterno."

60 Thunø notes the animating and amplifying effect of mosaics in the context of church decoration and the Eucharistic rite, as well as the dynamic relation to relics at the altar; see "Incorporation Becoming a Living Stone," *The Apse Mosaic in Early Medieval Rome*, pp. 119-71.

61 See example in *Byzantium: Faith and Power (1261-1557)*, pp. 224-25.

1472: "la sua effigie in finissimo musaico, ed in miniatura per le parti carnee visibili della persona."[62] Like the miracle of transubstantiation witnessed by Gregory, when Host became physical body, micromosaic effects a transformation of materials that produces a corporeal body of Christ for the believer. Here I would point to a mimetic relation between the icon's material and the material of the object it *presents*, not only represents.[63] In addition to this matching of aesthetic to spiritual purpose, we might also understand micromosaic—like the innovation of oil painting on stone—as a 'technology of enchantment' in the sense that Alfred Gell has theorized.[64] That is to say, the extraordinary quality of its facture—much like painting on stone—itself plays a significant role in constituting the mystery of the sacred.

## Bibliography

Alberti, Leon Battista. *Della pittura/De picture*. In *Opere volgari*, vol. 3. Edited by Cecil Grayson. Bari, 1973.

Albertini, Franscesco. *Opusculum de mirabilibus novae et veteris urbis Romae* (1510). Edited by August Schmarsow. Heilbronn, 1886.

Anastasius. "Vita S. Gregorii." In *Liber Pontificalis*. Edited by Louis Duchesne. Vol. 2. Paris, 1886-92.

Anderson, Christy, Anne Dunlop and Pamela H. Smith, eds. *The Matter of Art, Materials, Practices, Cultural Logics, c. 1250-1750*. Manchester, 2015.

Bacci, Michele. "Performed Topographies and Topomimetic Piety: Imaginative Spaces in Medieval Italy." In *Spatial Icons: Performativity in Byzantium and Medieval Russia*, edited by Alexei Lidov, pp. 101-18. Moscow, 2011.

Bagnoli, Martina, Holger Klein, C. Griffith Mann, and James Robinson, eds. *Treasures of Heaven: Saints, Relics, and Devotion in Medieval Europe*. New Haven and London, 2010.

Barbieri, Costanza. "La Pala della Concezione e Natività della Vergine di Sebastiano per la cappella Chigi e un disegno inedito." *Konsthistorisktidskrift* 81 (2012), 245-53

---

62 Quoted in Lollini, "Bessarione e Perotti diffusori della cultura figurative bizantina," p. 10. On wax as a material specifically associated with flesh and skin, see Roberta Panzanelli, "Compelling Presence: Wax Effigies in Renaissance Florence," in *Ephemeral Bodies: Wax Sculpture and the Human Figure* (Los Angeles, 2008), esp. pp. 30-31.

63 Such relations are taken up in *The Matter of Art, Materials, Practices, Cultural Logics, c. 1250-1750*, ed. Christy Anderson, Anne Dunlop and Patricia H. Smith, eds. (Manchester, 2015); see discussion in the Introduction, esp. p, 31.

64 Citing Gell, Ann-Sophie Lehmann offers a parallel discussion of oil paint as a technology of enchantment and form of transubstantiation in "The matter of the medium: some tools for an art-theoretical interpretation of materials," in *The Matter of Art*, pp. 30-33.

Barucca, Gabriele. "I Reliquiari da Niccolo Perotti a Sassoferrato." *Studi Umanistici Piceni* 12 (1992), 9-46.

Barocchi, Paola, ed. *Trattati d'arte del cinquecento fra manierismo e controriforma*. 3 vols. Bari, 1960-62.

Belting, Hans. *The Image and Its Public in the Middle Ages: Form and Function of Early Paintings of the Passion*. New Rochelle, 1990.

Bertelli, Carlo."Attualità e nostalgia: ancora sull'immagine venerate in Santa Croce in Gerusalemme." *Certose e certosini in Europa* 1 (1990), 65-73.

Bertelli, Carlo. "The *Image of Pity* in Santa Croce in Gerusalemme." In *Essays in the History of Art Presented to R. Wittkower*, edited by Douglas Fraser, Howard Hibberd, and Milton J. Lewine, pp. 40-55. New York, 1967.

Bertelli, Carlo, ed. *Il Mosaico*. Milan, 1997.

Besozzi, Raimondo. *La storia della Basilica di S. Croce in Gerusalemme*. Rome, 1750.

Blaauw, Sible de. "Jerusalem in Rome and the *Cult of the Cross*." In *Pratum romanum, Richard Krautheimer zum 100. Geburtstag*, edited by Renate Colella, Meredith Jane Gill, and Lawrence A. Jenkins, pp. 55-73. Wiesbaden, 1977.

Buettner, Brigitte. "From Bones to Stones—Reflections on Jeweled Reliquaries." In *Reliquiare im Mittelalter*, edited by Bruno Reudenbach and Gia Toussaint, pp. 43-60. Berlin, 2005.

Bynum, Caroline Walker. *Christian Materiality: An Essay on Religion in Late Medieval Europe*. New York, 2011.

Bynum, Caroline Walker. "Seeing and Seeing Beyond: The Mass of St. Gregory in the Fifteenth-Century." In *The Mind's Eye: Art and Theological Argument in the Middle Ages*, edited by Jeffrey Hamburger and Anne Marie Bouchè, pp. 208-240. Princeton, 2006.

*Byzantium: Faith and Power (1261-1557)*. Edited by Helen Evans. New York and New Haven:, 2004..

Calvillo, Elena. "Authoritative Copies and Divine Originals: Lucretian Metaphor, Painting on Stone, and the Problem of Originality in Michelangelo's Rome." *Renaissance Quarterly* 66, no. 2 (2013), 453-508.

Cassanelli, Roberto, and Emelia Stolfi, eds. *Gerusalemme a Roma: la basilica di Santa Croce e le reliquie della Passione*. Milan, 2012.

Cavallaro, Anna. *Antoniazzo Romano e gli Antoniazzeschi. Una generazione di pittori nella Roma del Quattrocento*. Udine, 1992.

Cavallaro, Anna. *Santa Croce in Gerusalemme*. Rome, 2009.

Chaganti, Seeta. *The Medieval Poetics of the Reliquary: Enshrinement, Inscription, Performance*. New York, 2008.

Cormack, Robin, and Maria Vassiliki, eds. *Byzantium 330-1453*. London, 2008.

Cutler, Anthony. "From Loot to Scholarship: Changing Modes in the Italian Response to Byzantine Artifacts, ca. 1200-1750." In "Symposium on Byzantium and the Italians, 13th-15th Centuries," special issue, *Dumbarton Oaks Papers* 49 (1995), 237-67.

Demus, Otto. *Byzantine Mosaic Decoration: Aspects of Monumental Art in Byzantium*. New Rochelle, 1976.

Demus, Otto. "Two Palaeologan Mosaic Icons in the Dumbarton Oaks Collection." *Dumbarton Oaks Papers* 14 (1960), 87-119.

Duits, Rembrandt. "Byzantine Icons in the Medici Collection." In *Byzantine Art and Renaissance Europe*, edited by Angeliki Lymberopoulou and Rembrandt Duits, pp. 157-88. Farnham and Burlington, 2013.

Durand, Jannic. "Precious-metal Icon Revetments." In *Byzantium: Faith and Power (1261-1557)*, edited by Helen Evans, pp. 243-51. New York and New Haven, 2004.

Effenberger, Arne. "Images of Personal Devotion: Miniature Mosaic and Steatite Icons." In *Byzantium: Faith and Power (1261-1557)*, edited by Helen Evans, pp. 209-44. New York and New Haven, 2004.

Falomir, Miguel. "Sebastiano and 'Spanish Taste'." In *Sebastiano del Piombo*, edited by Giuseppe Scandiani, Claudio Strinati, and Bernd Wolfgang Lindemann, pp. 67-71. Milan, 2008. Exhibition catalogue.

Filarete. *Trattato di Architettura*. Edited by Anna Maria Finoli and Lilana Grassi. Vol. 2. Milan, 1972.

Freeman, Charles. *Holy Bones, Holy Dust: How Relics Shaped the History of Medieval Europe*. New Haven, 2011.

Freiberg, Jack. "Bramante's Tempietto and the Spanish Crown," *Memoirs of the American Academy in Rome* 50 (2005), 162-205.

Freiberg, Jack. *Bramante's Tempietto, the Roman Renaissance, and the Spanish Crown*. Cambridge, 2014.

Frommel, Christoph. "Projetto e archeologia in due disegni di Antonio da Sangallo il Giovane per Santa Croce in Gerusalemme." In *Roma, centro ideale della cultura dell'Antico nei secoli XV e XVI Da Martino V al Sacco di Roma* 1417-1525, edited by Silvia Danesi Squarzina, pp. 382-389. Milan, 1989.

Frommel, Christoph, and Gerhard Wolf, eds. *L'immagine Di Cristo dall'Acheropita Alla Mano D'artista: Dal Tardo Medioevo All'età Barocca*. Vatican City, 2006.

Fusco, Laurie, and Gino Corti. *Lorenzo de'Medici, Collector of Antiquities*. Cambridge, 2006.

Gallori, Corinna T. "The Late Trecento in Santa Croce in Gerusalemme Napoleone and Nicola Orsini, the Carthusians, and the Triptych of Saint Gregory," *Mitteilungen des Kunsthistorichen Institutes in Florenz* 58, no. 2 (2016), 156-87.

Gill, Meredith. "Antoniazzo Romano and the Recovery of Jerusalem in Late Fifteenth-Century Rome." *Storia dell'Arte* 83 (1995): 28-47.

Grabar, Andrè. *Martyrium: recherches sur le culte des reliques et l'art chrétien antique: Album*. Paris: Collège de France, 1946.
Hahn, Cynthia J. "The Meaning of Early Medieval Treasuries." In *Reliquiare im Mittelalter*, edited by Bruno Reudenbach and Gia Toussaint, pp. 1-20. Berlin, 2005.
Hahn, Cynthia J. *Strange Beauty: Issues in the Making and Meaning of Reliquaries, 400-circa 1204*. University Park, 2012.
Hahn, Cynthia J. "What Do Reliquaries Do For Relics?" *Numen* 57 (2010), 284-316.
Hecht, Christian. *Katholische Bildertheologie im Zeitalter von Gegenreformation und Barock: Studien zu Traktaten von Johannes Molanus, Gabriele Paleotti und Anderen Autoren*. Berlin, 1997.
Hecht, Christian. "Von der Imago pietatis zur Gregorsmesse—Ikonographie der Eucharistie vom hohen Mittelalter bis zur Epoche des Humanismus." *Römisches Jahrbuch der Bibliotheca Hertziana* 36 (2005), 9-44.
Jacobs, Lynn F. *Opening Doors: the Early Netherlandish Triptych Reinterpreted.* University Park, 2011.
James, Liz. *Mosaics in the Medieval World from Late Antiquity to the Fifteenth Century*. Cambridge, 2017.
Kalas, Rayna. *Frame, Glass, Verse: The Technology of Poetic Invention in the English Renaissance*. Ithaca and London, 2007.
Kalavrezou, Ioli. *Byzantine Icons in Steatite*. Vienna, 1985.
Kalopissi-Verti, Sharon. "The Proskynetria of the Templon and Narthex: Form, Imagery, Spatial Connection and Reception." In *Thresholds of the Sacred: Architectural, Art Historical, Liturgical and Theological Perspectives on Religious Screens, East and West*, edited by Sharon Gerstel, pp. 107-34. Washington, DC, 2007.
Kessler, Herbert L., and Gerhard Wolf, eds. *The Holy Face and the Paradox of Representation: Papers from a Colloquium at the Bibliotheca Hertziana, Rome and the Villa Spelman, Florence, 1996*. Bologna, 1998.
Kim, Anna Marazuela. "Real Presence, Iconic Images, and Iconoclasm from Byzantium to the Reformation." PhD diss., University of Virginia, 2014.
Klein, Holger. "Eastern Objects and Western Desires: Relics and Reliquaries between Byzantium and the West." *Dumbarton Oaks Papers* 58 (2004), 283-314.
Klein, Holger, ed. *Byzanz, der Westen und das 'wahre' Kreuz: die Geschichte einer Reliquie und ihrer künstlerischen Fassung in Byzanz und im Abendland*. Wiesbaden, 2004.
Krautheimer, Richard. *Corpus Basilicarum Christianarum Romae. The Early Christian Basilicas of Rome (IV-IX Cent.)*. Vatican City, 1937.
Krautheimer, Richard. "Introduction to an Iconography of Medieval Architecture." *Journal of the Warburg and Courtauld Institutes* 5 (1942), 1-33.
Kumler, Aden. *Translating Truth: Ambitious Images and Religious Knowledge in Late Medieval France and England*. New Haven, 2011.

Landau, David and Peter Parshall, eds. *The Renaissance Print*. New Haven and London, 1994.

Lehman, Anne-Sophie. "The matter of the medium: some tools for an art-theoretical interpretation of materials." In *The Matter of Art Materials, Practices, Cultural Logics, c. 1250-1750*, edited by Christy Anderson, Anne Dunlop, and Pamela H. Smith, pp. 21-41. Manchester, 2015.

Lollini, F. "Bessarione e Perotti diffusori della cultura figurative bizantina." *Studi umanistici piceni* 11 (1991), 127-42.

Luzietta, M. "La cappella di Sant'Elena in Santa Croce in Gerusalemme: il programma iconografico." In *Atti del XX colloquio dell'Associazione Italiana per lo Studio e la Conservazione del Mosaico*, edited by Claudia Angelelli and Andrea Parbeni, pp 587-93. Tivoli, 2015.

Lymberopoulou, Angeliki, and Rembrandt Duits, eds. *Byzantine Art and Renaissance Europe*. Farnham and Burlington, 2013.

Miller, Patricia Cox. *The Corporeal Imagination: Signifying the Holy in Late Ancient Christianity*. Philadelphia, 2009.

Nagel, Alexander, and Christopher S. Wood. *Anachronic Renaissance*. New York, 2010.

Nagel, Alexander. *Medieval Modern: Art out of Time*. New York, 2012.

Nelson, Robert S. "The Italian Appreciation and Appropriation of Illuminated Byzantine Manuscripts, ca 1200-1450." In Robert S. Nelson, *Later Byzantine Painting: Art, Agency, and Appreciation*, pp. 209-235. Aldershot and Burlington, 2007. Originally published *Dumbarton Oaks Papers* 49 (1995).

Nordhagen, Per Jonas. "Byzantium and the West: with some remarks on the activity of Greek mosaic artists in Italy in the fourteenth century." In Per Jonas Nordhagen, *Studies in Byzantine and Early Medieval Painting*, pp. 447-84. London, 1990.

Ortolani, Sergio, *Schede del Catalogo delle Opere d'arte contenute in Santa Croce in Gerusalemme*. Ministero della P. I., Direzione Generale della B. A., 1922.

Panofsky, Erwin. "'*Imago Pietatis*': Ein Beitrag zur Typengeschichte des 'Schmerzensmanns' und der 'Maria Mediatrix'." In *Festschrift für Max J. Friedlander zum 60. Geburtstag*, pp. 261-308. Leipzig, 1927.

Panvinio, Onofrio. *Le sette chiese romane*, Rome, 1570.

Panzanelli, Roberta. "Compelling Presence: Wax Effigies in Renaissance Florence." In *Ephemeral Bodies: Wax Sculpture and the Human Figure*, edited by Roberta Panzanelli, pp. 13-39. Los Angeles, 2008.

Parshall, Peter. "*Imago Contrafacta*: Images and Facts in the Northern Renaissance." *Art History* 16, no. 4 (1993), 554-79.

Payne, Cynthia. "'In the fullness of time': the vault mosaic in the Cappella Sant'Elena, Santa Croce in Gerusalemme, Rome." PhD diss., University of Georgia, 2003.

Payne, Cynthia. "Lux Mundi: The Vault Mosaic in the Cappella S. Elena, S. Croce in Gerusalemme, Rome." *Athanor* 17 (1999), 35-43.

Peers, Glenn. *Sacred Shock: Framing Visual Experience in Byzantium.* University Park, 2005.

Pentcheva, Bissera. *The Sensual Icon: Space, Ritual, and the Senses in Byzantium.* University Park, 2013.

Reudenbach, Bruno, and Gia Toussaint. *Reliquiare im Mittelalter.* Berlin, 2005.

Ryder, Edmund C. "Micromosaic Icons of the Late Byzantine Period." PhD diss., New York University, Institute of Fine Arts, 2007.

Sciascio, Sofia Di. *Reliquie e reliquiari in Puglia fra IX e XV secolo.* Galatina, 2009.

Sensei, Mario. "Dall'*Imago Pietatis* alle Cappelle Gregoriane, Immagini, Racconti e Devozioni per la 'Visione' e la Cristomimesi." *Collectanea Franciscana* 70 (2000), 93-104.

Teuffel, Christa Gardner von. "Light on the Cross: Cardinal Pedro Gonzalez de Mendoza & Antoniazzo Romano in Sta. Croce in Gerusalemme, Rome." In *Coming About... A Festchrift for John Shearman*, edited by Lars Jones and Louisa Matthew, pp. 49-55. Cambridge, 2001.

Thomas, Thelma K. "Understanding Objects." In *Reading Medieval Images: The Art Historian and the Object,* edited by Elizabeth Sears and Thelma K. Thomas, pp. 9-15. Ann Arbor, 2002.

Thunø, Erik. *The Apse Mosaic in Early Medieval Rome: Time, Network and Repetition.* Cambridge, 2015.

Toesca, Illaria. "A Majolica Inscription in Santa Croce in Gerusalemme." In *Essays in the history of art presented to Rudolf Wittkower*, edited by Douglas Fraser, Howard Hibbard, and Milton J. Lewine, pp. 101-05. New York, 1969.

Varagnoli, C.S. *Croce in Gerusalemme: la Basilica restaurata e l'architettura del Settecento romano.* Rome, 1995.

Vasari, Giorgio. *Le vite de' più eccellenti pittori, scultori e architettori nelle redazioni del 1550 e 1568.* Edited by Rosana Bettarini and Paola Barocchi. 6 vols. Florence, 1966-87.

Wolf, Gerhard. "Imago Pietatis-Israhel van Meckenems Konterfei eines Abbildes einer Erscheinung." In *Glaube, Liebe, Hoffnung, Tod*, edited by Christoph Geissmar-Brandi and Eleonora Louis, pp. 274-76. 2nd rev. ed. Klagenfurt, 1995.

CHAPTER 9

# Invention, Ambition and Failure: Niccolò Tornioli (1606-51) and 'Il Segreto di Colorire il Marmo'

*Giulia Martina Weston*

A closer look into the little-explored career of the Sienese artist Niccolò Tornioli provides an original insight into the history, theory, and practice of painting on stone in the early modern period. By combining intellectual aspirations with material inventions, Tornioli carried out a prolonged period of investigation and experimentation with stone, which, as this essay argues, played a role in shaping the painter's artistic identity. By examining a *corpus* of Tornioli's letters and of Sienese manuscripts, his technique of painting on marble will be related to and understood in light of the Sienese tradition of marble inlays (*commesso*). This essay will also connect Tornioli's invention to the Florentine craft of *pietre dure* and paintings on precious and semiprecious stones, and the rise of Gian Lorenzo Bernini's *bel composto* in mid-century Rome. Finding its roots in Domenico Beccafumi's artistic legacy in Siena, as the 17th-century theorist Teofilo Gallaccini noted, Tornioli's lifelong practice of painting on stone will be examined according to its materiality and ideological content, and its two-fold potential for imperishable fame and professional failure.

Born in Siena in 1606 and settled in Rome in 1635, Niccolò Tornioli reached the peak of his career while a *pittore di casa* for the Spada family, as the leading works in his production attest: the *Astronomers* (Rome, Galleria Spada, *c.* 1645), the fresco decoration of the Sala Rossa in Santa Maria in Vallicella (1643-51), and the mosaic decoration of the Chapel of the Holy Sacrament in St. Peter's (1647-50).[1] These works, however, chronologically followed and

1 For Tornioli's catalogue raisonné see Marco Ciampolini, *I pittori senesi del Seicento*, 3 vols. (Siena, 2010), 3:869-895. The most relevant contributions for the artist's *corpus* are Federico Zeri, *La Galleria Spada in Roma* (Florence, 1954), pp. 134-39; Alessandro Bagnoli, *Mostra di opere d'arte restaurate nelle provincie di Siena e Grosseto II* (Genoa, 1981), pp. 212-14; Marzia Faietti, "Nuove acquisizioni conoscitive su alcuni disegni restaurati della Pinacoteca Nazionale di Bologna," *Bollettino d'Arte* 33-34 (1985), 217-32; Marco Ciampolini, "Niccolò Tornioli," in *Bernardino Mei e la Pittura Barocca a Siena*, ed. F. Bisogni and M. Ciampolini, exhibition catalogue, Palazzo Chigi Saracini, Siena, 1987, pp. 109-12; Marco Ciampolini, "Integrazioni al catalogo di Niccolò Tornioli," *Antichità Viva* 34 (1995), 26-34; Rita Randolfi, "Alcune precisazioni

 | DOI 10.1163/9789004361492_013

partially coincided with a series of experiments and failures that considerably complicate our understanding of Tornioli's Roman period. This essay concerns a previous, less studied phase of Tornioli's career in the Eternal City: his activity for Prince Cardinal Maurizio of Savoy (1593-1657). As stressed by Matthias Oberli, a fundamental key to reconstruct Maurizio of Savoy's patronage and self-fashioning is the rhetorical concept of *magnificentia*.[2] Paralleled by the erudite discourses developed in the *Accademia dei Desiosi* and Luigi Manzini's encomiastic work patronized by Maurizio of Savoy, Tornioli's visual tribute of Maurizio's *persona* and his court is epitomized by his preparatory drawings for the frontispiece and the illustrations of pyrotechnical machines included in Luigi Manzini's *Applausi Festivi* of 1637 and engraved by Luca Ciamberlano. Manzini's volume describes the spectacular feasts hosted by Maurizio to celebrate the election of Ferdinand III as Emperor of the Romans. These ephemeral spectacles, which occurred in Rome in February 1637, celebrated a strategic victory of the German Emperor during the Thirty Years War and a diplomatic success for Pope Urban VIII Barberini.[3] In Maurizio's program and Tornioli's inventions, the event offered an unprecedented opportunity to combine political discourses with the celebration of the Savoy court in Rome. Thus, from Manzini's volume, the patron's and the artists' fame emerge as interwoven in a mythological, superhistorical dimension.

Nevertheless, a letter written by Tornioli to Maurizio's agent Ferrante Carlo (Appendix 2) in 1637 records that, in addition—and in contrast—to this offi-

---

sull'attività romana di Niccolò Tornioli," *Studi di Storia dell'Arte* 7 (1996), 347-53; Francesca Curti, *Committenza, collezionismo e mercato dell'arte tra Roma e Bologna nel Seicento. La quadreria di Cristiana Duglioli Angelelli* (Rome, 2007); Salvador Salort-Pons, "La llegada de Niccolò Tornioli a Roma y el malogrado mecenazgo del abad Borromeo," *Antologia di Belle Arti, Studi Romani* 1 (2004), 9-13; Anna Elena Galli, "Federico IV Borromeo: scelte artistiche tra Milano e Siena," in *Atti delle Giornate di Studi sul Caravaggismo e il naturalismo nella Toscana del Seicento*, ed. P. Carofano (Pontedera, 2009), pp. 295-322; Simona Sperindei, "L'inventario di Niccolò Tornioli. Aggiornamenti e nuovi documenti," *Valori Tattili* 00 (2011), 115-23. For a monographic study of the artist see Giulia Martina Weston, *Niccolò Tornioli (1606-1651). Art and patronage in Seicento Rome* (Rome, 2016). For a recent discussion of Tornioli's *Astronomers* see Giulia Martina Weston, "After Galileo: The Image of Science in Niccolò Tornioli's *Astronomers*," *Art History* 36, no. 2 (2016), 302-17.

2 See Matthias Oberli, *"Magnificentia Principis." Das Mäzenatentum des Prinzen und Kardinals Maurizio von Savoyen (1593-1657)* (Weimar, 1999).

3 On this feast see Maurizio Fagiolo Dell'Arco and Silvia Carandini, *L'Effimero Barocco. Strutture della festa nella Roma del '600*, 2 vols. (Rome, 1977-78), 1:389-402. On the Piedmontese tradition of feasts and celebrations, see Mercedes Viale Ferrero, "Le feste e il teatro," in *Diana Trionfatrice. Arte di corte nel Piemonte del Seicento*, ed. M. Di Macco and G. Romano, exhibition catalogue, Parco del Valentino, Turin, 1989, pp. 74-81

cial commission, Tornioli was engaged in the most controversial and secret experiments on marble.[4] The importance of this document for our analysis is twofold. In the first place, it reveals that Tornioli's status as court artist was characterized by professional and financial difficulties. In fact, Tornioli complains about the lack of payments for the *Annunciation* commissioned by Maurizio.[5] Being denied adequate refunds for the cost of materials, colors and marble pieces, the painter explains that he owed a "Sig. Luca" 120 *scudi* (with an additional interest of 10 *scudi*) and that he did not know "what to do for a living" (*non sapevo come farmi a vivermi*). Further, the 'disgusting' conditions of his room forced him to leave the court palace in Monte Giordano to rent a room and were possibly responsible for several diseases (*tre malattie pericolose*; *di nuovo avevo avuto questa malattia*; *questa state ero ammalato*). Although the *contrapposto* between fame and famine was almost a *topos* in the biographies of the artists, Tornioli's case seems to epitomize the paradoxical status of the court painter, who struggled with fame and famine inside and outside of the prince's court.[6] Moreover, the letter attests that Tornioli was denied a reimbursement for the marble pieces (*pietre*) because he was accused of using them to please his own taste (*per mio gusto*) rather than Maurizio of Savoy's (*per ordine suo*). Another letter, written by the Sienese polymath and academician Teofilo Gallaccini (1564-1641) to Tornioli on 13 October 1640 (Appendix 1) sheds considerable light on this charge.[7]

---

4 Edited in Scipione Borghesi and Luciano Banchi, *Nuovi documenti per la storia dell'arte senese* (Siena, 1898), n. 350, 644-46.

5 The *Annunciation* was never delivered to Maurizio and was purchased by Bernardino Spada in 1643. On the identification of this work (Castel Viscardo, ss.ma Annunziata) see Bianca Tavassi La Greca, "Niccolò Tornioli e la famiglia Spada," in *Per la Storia dell'Arte in Italia e in Europa. Studi in onore di Luisa Mortari*, ed. M. Pasculli Ferrara (Rome, 2004), pp. 249-57.

6 See, for example, Giorgio Vasari's use of the terms *fama* and *fame*. On Vasari's idea that 'poverty perverts art' see Philip Sohm, "Introduction," in *Painting for Profit. The Economic Lives of Seventeenth-century Italian Painters*, ed. R.E. Spear and P. Sohm (New Haven and London, 2010), p. 4.

7 Edited in Guglielmo Della Valle, *Lettere Sanesi del padre maestro Guglielmo Della Valle minor conventuale socio delle Rr. Accademie delle Scienze e Agraria di Torino etc. Sopra le Belle Arti* (Rome, 1786), pp. 399-402; Giovanni Gaetano Bottari and Stefano Ticozzi, *Raccolta di Lettere sulla Pittura, Scultura e Architettura scritte da' più Celebri Personaggi dei secoli XV, XVI, e XVII*, 8 vols. (Milan, 1822), 1:421-31, n. 186; Ettore Romagnoli, *Biografia Cronologica de' Bellartisti Senesi dal secolo VII a tutto il XVIII* [ante 1835], anastatic copy (Florence, 1978), 10:551-66.

### Reinventing a Tradition: Painting on Stone between *Commesso* and *Tintura* in Renaissance Siena

A member of the Accademia dei Filomati in Siena and erudite author of architectural and scientific writings, Gallaccini played a leading role in strengthening the cultural and artistic bonds between Siena and Rome.[8] Responding to Tornioli's quest for 'the secret of coloring stone' ("il segreto di colorire il marmo"), Gallaccini's letter sketches the history of this typically Sienese technique, drawing a fundamental distinction between the execution of marble inlays (*commesso*) and the newly invented procedure of infusing color into stone (*tintura*). According to Gallaccini, the former was invented by Matteo di Giovanni, author of the *Massacre of the Innocents* (1481) on the marble pavement in the cathedral of Siena.[9] [**Fig. 9.1**] This technique was further refined by Beccafumi (1484-1551), especially praised by Gallaccini for the *Sacrifice of Abraham*, executed around 1547 in the pavement of the Duomo.[10] [**Fig. 9.2**] Regarded as the apex of the technique of marble inlays, this scene is especially praised for its chiaroscuro and dynamic qualities. As revealed by Maria Gibellino Krasceninnicowa's pioneering studies, Beccafumi enhanced the traditional execution of *commesso* by both increasing the number of marble varieties, introducing grey, green, yellow and red stones, and replacing the technique of *niellatura* with an improved *graffito* method which required marble

8 On Gallaccini's architectural and scientific works see Gabriele Morolli, ed., *Siena circa 1600: dimenticare Firenze. Teofilo Gallaccini e l'eclisse presunta di una cultura architettonica*, exhibition catalogue, Santa Maria della Scala and Magazzini della Corticella, Siena, 1999; Alina Payne and Giovanni Maria Fara, "Teofilo Gallaccini e la critica architettonica a Siena fra XVI e XVII secolo," in *Architetti a Siena. Testimonianze della Biblioteca Comunale tra XV e XVIII secolo*, ed. D. Danesi, M. Pagni, and A. Pezzo, exhibition catalogue, Biblioteca Comunale degli Intronati, Siena, 2009, pp. 141-89; and Alina Payne, *The Telescope and the Compass. Teofilo Gallaccini and the Dialogue between Architecture and Science in the Age of Galileo* (Florence, 2012). For an insight into Teofilo's library see Giovanni Maria Fara, "Teofilo Gallaccini in his library: reflections and critical catalogue," in *Teofilo Gallaccini. Selected writing and library*, ed. A. Payne (Florence, 2012), pp. 1-40. On Gallaccini's network of relationships, which included Cesare Ripa, Giovan Battista Marino, Giovan Battista Ferrari, Isidoro Ugurgieri Azzolini, Volumnio Bandinelli and Fabio Chigi, see Annalisa Pezzo, "Una rete erudita," in *Siena circa 1600*, pp. 57-76.

9 On Matteo di Giovanni's scene see Marilena Caciorgna and Roberto Guerrini, *Il pavimento del Duomo di Siena. L'arte della tarsia marmorea dal XIV al XIX secolo. Fonti e simbologia* (Milan, 2004), pp. 104-11.

10 For an updated account of Beccafumi's scenes for the Duomo pavement see Caciorgna-Guerrini, *Il pavimento del Duomo*, pp. 169-223.

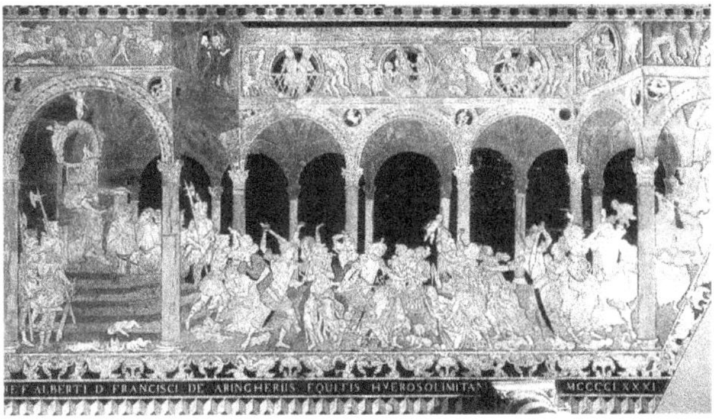

FIGURE 9.1
*Matteo di Giovanni,* Massacre of the Innocents, *1481, Siena, Duomo, pavement* (PHOTO: AUTHOR).

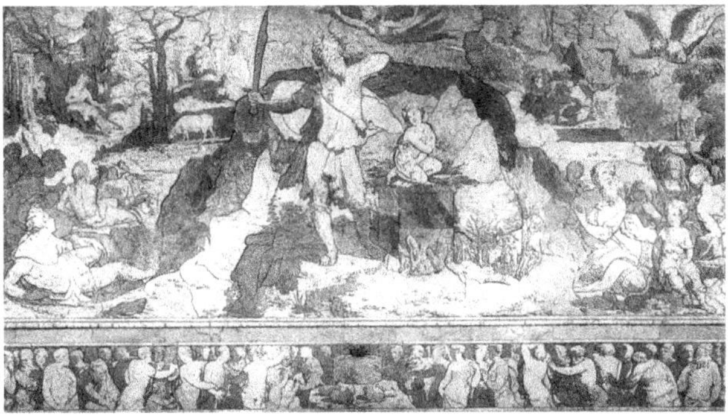

FIGURE 9.2
*Domenico Beccafumi,* Sacrifice of Abraham, *ca. 1547, Siena, Duomo, pavement* (PHOTO: AUTHOR).

inlays (*incrostature*).[11] These innovations increased Beccafumi's chromatic and luministic effects, leading to an unprecedented virtuosity in the technique.[12]

In a section dedicated to contemporary experiments in painting on stone, Gallaccini reports that both Giovan Battista Giustammiani and Ascanio Covatti laid claim to the invention of a procedure to infuse color into stone, or *tintura*. Remarkably, these two artists were active in the church of San Niccolò in Sasso in Siena, for which Tornioli executed a *Crucifixion* in 1631. The commission of this altarpiece, which was intended to be part of a significant renewal of the church, provided Tornioli with a chance to look closely at his rivals' output. Moreover, Gallaccini's links to Giustammiani were favored by their mutual acquaintance with Cassiano dal Pozzo. [**Plate 28**][13]

11 See Maria Gibellino Krasceninnicowa, *Il Beccafumi* (Florence, 1933), pp. 97-98.

12 See Donato Sanminiatelli, *Domenico Beccafumi* (Milan, 1967), p. 121, n. 76. For an insight on Beccafumi and the Sienese tradition see Gabriele Fattorini and Jennifer Sliwka, *Domenico Beccafumi and the Sienese tradition*, in *Renaissance Siena. Art for a City*, ed. L. Syson, exhibition catalogue, National Gallery, London, 2007, pp. 296-99.

13 A proof of Gallaccini's familiarity with the church is provided by the fact that he included Sassetta's signature on the (then in place) *Madonna and Child and two angels* (Siena, Pinacoteca Nazionale) in his lost manuscript *Iscrizioni Senesi*. On this hypothesis see John W.

Ultimately, Gallaccini doubts the 'secrets' of painting on stone, which he refers to as real 'modern yet groundless marvels' (*moderne meraviglie, come prive di fondamento reale*). He thus warns Tornioli about the ephemerality and unreliability of these 'secrets,' perhaps associating them with alchemical experiments. In this context, Gallaccini's use of the word *segreti* might be understood in relation to the books of the 'Professors of Secrets' which circulated in 16th-century Italy. As William Eamon has remarked, the authors of these books were amateurs, physicians and non-academics, who addressed everyday problems and provided middle-class readers with handbooks of remedies.[14] One might go so far as to suggest that Tornioli's invention of a procedure to paint 'in' stone, resulting from solitary and even self-referential experiments, could be equally related to the sphere of self-taught empiricism. That Ettore Romagnoli referred to the painter as the *professore del segreto* in 1832 suggests that Tornioli was long associated with his experimentation.[15]

### Struggling for Primacy: Michelangelo Vanni, Niccolò Tornioli, and the Veronica Legend

From Gallaccini's meticulous account, one becomes aware that the technique of painting on stone is connected to local identity and pride, even if in an anti-Florentine key. Gallaccini states that the encrusted pavement of the Sienese Duomo inspired the Florentine art of *pietre dure*. Paralleling the Sienese practice of painting 'in' stone, which involved Michelangelo Vanni and Niccolò Tornioli besides Giustammiani and Covatti, several Florentine or Florence-based artists were engaged in the production of paintings 'on' stone.[16] Antonio Tempesta, Alessandro Allori, Filippo Napoletano, Cigoli, Jacopo Ligozzi,

---

Pope Hennessy, *Sassetta* (London, 1939), p. 90, n. 21. This supposition has been rejected in Pietro Torriti, *La Pinacoteca Nazionale di Siena. I dipinti dal XV al XVIII secolo* (Genoa, 1978), 1:246, n. 325, but carefully reconsidered in Enzo Carli, ed., *La Chiesa di San Niccolò in Sasso a Siena* (Siena, 1995), 4.

14 On Gallaccini's drawings for Dal Pozzo's *Museum Chartaceum* see Patricia Collins, "Manoscritti autografi di Teofilo Gallaccini in Gran Bretagna," in *Siena circa 1600*, pp. 201-06. See William Eamon, "Science and Popular Culture in Sixteenth Century Italy: The 'Professors of Secrets' and Their Books," *The Sixteenth Century Journal* 16, no. 4 (1985), 471-85.

15 Romagnoli, *Biografia Cronologica*, 10:567.

16 For an insight into this production in Florence see Marco Chiarini, Annapaula Martelli Pampaloni, and Anna Maria Maetzke, eds., *Pittura su pietra*, exhibition catalogue, Palazzo Pitti, Florence, 1970; and Marco Chiarini and Cristina Acidini Luchinat, eds., *Bizzarrie di pietre dipinte dalle collezioni dei Medici*, exhibition catalogue, Galleria dell'Accademia and Museo dell'Opificio delle Pietre Dure, Florence, 2000.

Stefano della Bella, and Jacques Stella all produced works on stone that were promoted and collected by members of the Medici family.[17] These artists cleverly addressed the technical problems presented by the veins of marble and introduced a new and subtler level of competition between art and nature by imbuing the 'signs' of nature (such as veins or spots) with specific meaning (for instance, using curved signs to represent sea waves). As stressed by Paola Squellati Brizio, the topic of *natura sollecitata*, nature enhanced by art, finds its roots in a passage by Pliny the Elder, later included in ancient treatises on optical sciences and rediscovered by Leonardo da Vinci.[18]

The lively artistic scenario of *Seicento* Siena—particularly the fierce competition between Michelangelo Vanni and Niccolò Tornioli over the invention of a technique of painting 'in' stone—resulted in an extraordinary work by Tornioli. Sources report that Tornioli's innovation led to the creation of a marble work representing the *Veronica*, that is the miraculous appearance of Christ's face on Veronica's veil. Having infiltrated a marble block "of the width of a finger" with color, Tornioli cut it into two identical halves, thus obtaining two specular images from the color that had penetrated the stone. Although this work is now lost, it might be identified with one of the two "quadretti piccioli in marmo" listed in Tornioli's inventory of April 19, 1651: the one representing a *Testa del Salvatore*.[19] A consideration of the Veronica Legend may enhance our understanding of the genesis and technique of this work.

In his *Storia pittorica della Italia* of 1789, Luigi Lanzi revives Gallaccini's interest in the secret of painting in stone, stating that it was not in Beccafumi's times but, rather, in *Seicento* Siena that the technique was invented by Michelangelo Vanni.[20] Lanzi recalls that Tornioli was familiar with the practice, having executed the (two) *Veronica*(*s*) in 1640. Arguing that Tornioli learned this procedure in the workshop of the Vanni family, the author stresses that Michelangelo Vanni's primacy is documented by the self-celebratory inscription *novae hujus in petra pingendi artis inventor* on his father Francesco Vanni's tombstone in the church of S. Giorgio in Siena. [**Fig. 9.3**] Analogously, Ettore Romagnoli's *Biografia Cronologica de' Bellartisti Senesi*, reports that Tornioli

17 For a monographic overview on this topic see Hana Seifertova, *Malby na kameni: Umĕcký experiment v 16. Na začátku 17 století/ Painting on Stone: an Artistic Experiment in the 16th and Early 17th centuries* (Prague, 2007). On Antonio Tempesta's paintings on stone see Johanna Beate Lohff's essay in this volume.

18 See Paola Squellati Brizio, "Natura sollecitata," in *Bizzarrie di pietre dipinte*, pp. 32-38. On the rhetorics of painting on stone, *paragone*, and the dialectics between essence and representation see Christopher Nygren's essay in this volume.

19 Archivio di Stato di Roma-ASR, Notai del Tribunale AC, b. 3537, f. 499v; published in Sperindei, "L'inventario," p. 66.

20 Luigi Lanzi, *Notizie della scultura degli antichi e dei vari suoi stili* (Fiesole, 1824), 1:417.

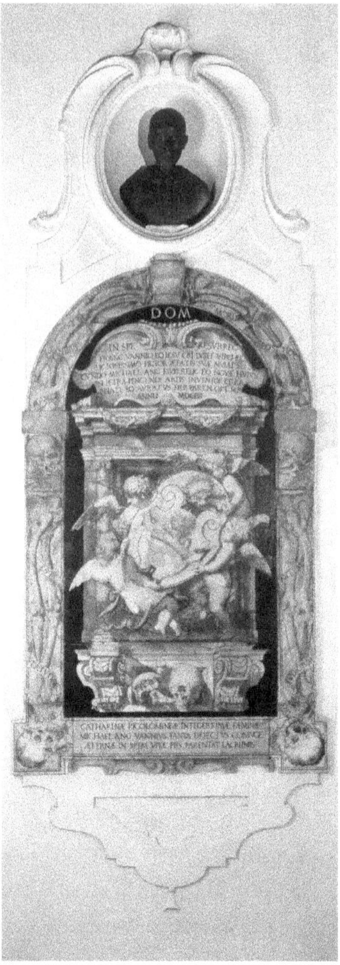

FIGURE 9.3
*Michelangelo Vanni,* Angels upholding the Vanni coat of arms, *1656, Siena, San Giorgio* (PHOTO: AUTHOR).

was "coloritore del marmo e professore del segreto di dipingere penetrabilmente quello," as demonstrated by the figure of *Veronica*, characterized by its vitality and expression (*vivezza ed espressione*).[21] Stefano Ticozzi's *Dizionario* of 1832 provides us with another version of the legend of Tornioli's *Veronica*, according to which the marble painting was jointly executed by Vanni and Tornioli.[22] Likewise, Giovan Battista Adriani mentions Tornioli as the co-

21 Romagnoli, *Biografia Cronologica*, 10:567.

22 Stefano Ticozzi, *Dizionario degli architetti, scultori, pittori* (Milan, 1832), 3:422: "Torniolo (Niccolò), operava in Siena sua patria nel 1610. Pare che imparasse da Michelangelo Vanni l'arte di colorire i marmi. Si dice, che avendo insieme al Vanni dipinta una Veronica, fece segare il marmo, e quella medesima figura restò nelle due superficie del diviso marmo."

author of a *Veronica* in 1856.[23] The most ancient reading of the Veronica legend is offered by Francesco Gori Gandellini's *Saggio d'istoria pittorica sanese* (*ante* 1784).[24] Assuming that Cardinal Savoy was the recipient of the work, Gori Gandellini believes that Tornioli did not only invent the technique, but also taught it to Michelangelo Vanni. Gori Gandellini persuasively argues that Tornioli's *Veronica* was executed in 1640 and Vanni's tombstone in 1656.

From a technical point of view, all these sources indicate that Tornioli's practice was identical to the one used by Michelangelo Vanni in his father's tomb in S. Giorgio in Siena. In both, colors, precisely *succhi d'erbe*, penetrate the marble surface throughout a mordant known as *aqua regia* and other acids.[25] Vanni achieved chiaroscuro and dimensional effects by means of controlling the penetration of these chromatic components. Guido Mazzoni has published a secret recipe mentioned by Ettore Romagnoli, which originally belonged to Vanni and was copied by the Sienese librarian Giuseppe Ciaccheri.[26] In addition, Guglielmo Della Valle reports that Ciaccheri had provided the French painter Pierre-Jean Mariette with a copy of this formula. However, the latter failed to reproduce the correct procedure in Paris because, in Della Valle's words, the secret is comparable to the *lapis philosophorum* (philosopher's stone), meant to drive one mad at his own expenses and labor.[27] The procedure of marble coloring was extremely complicated and consisted of a process based on the use of *succhi d'erbe*. *Succhi d'erbe* (literally "herbal juice") are mineral or vegetal pigments diluted in a mordant (usually alum), more frequently used to paint on silk. In Giovan Francesco Romanelli's *Battle of Mount Tabor* (Paris, Louvre, 1646), for example, *succhi d'erbe* are employed to imitate tapestry.

The subtle yet significant differences between the versions cited above the Veronica legend do not provide enough information to allow us to confirm Tornioli's role as the inventor of this technique. The fierce competition between the two Sienese painters, however, reveals the close connection

---

23 Giovanni Battista Adriani, *Memorie della vita e dei tempi di Monsignor Referendario Giansecondo Ferrero-Ponziglione. Memorie Storiche di Giovambatista Adriani* (Turin, 1856), p. 591.

24 Francesco Gori Gandellini, *Prospetto dei pittori, ante* 1784, pp. 183-84 (Biblioteca Comunale di Siena-BCS, L. V. 15).

25 See Cecilia Alessi in *Due restauri lapidei promossi dal Lions Club Siena. La lastra tombale di Francesco Vanni a San Giorgio e una iscrizione quattrocentesca nel Palazzo Pubblico di Siena* (Siena, 2009), pp. 7-10. On the tombstone see Ciampolini, *I Pittori Senesi*, 3:1017-18.

26 The document is preserved in the Biblioteca Comunale di Siena (BCS, L. V. 9, f. 49v) and was published in Guido Mazzoni, "Appunti su Michelangelo Vanni pittore e mercante d'arte," *La Diana* 2 (1996), 259.

27 Della Valle, *Lettere Sanesi*, 3:217.

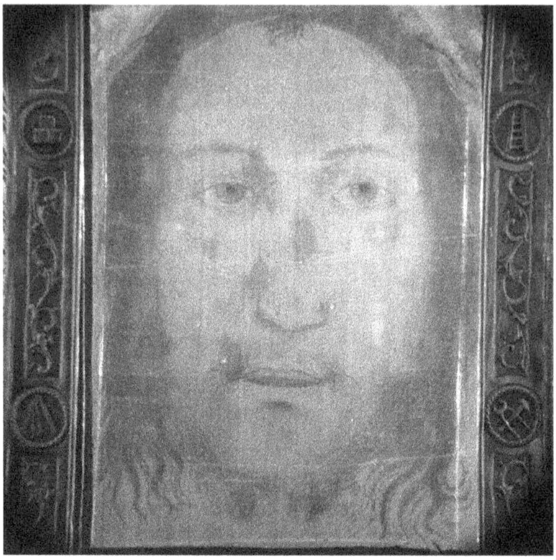

FIGURE 9.4
The Holy Face of Manoppello, *Manoppello, Basilica del Volto Santo* (PHOTO: AUTHOR).

between technical and intellectual achievements and artistic inventions in *Seicento* society. The subject of Tornioli's painting in stone invites further consideration of the crucial intertwining between medium and meaning. The iconographic fortune of the Veronica in the 17th century can be related to miraculous retrieval of the Veil of Manoppello in the Abbruzzi. [**Fig. 9.4**] Bearing the image of Christ's face, the Veil of Manoppello, which has been identified by some with the *Veronica* stolen from St. Peter's in Rome in the Sack of 1527, was venerated as an *acheropoietic* icon, an image not made by human hands.[28] Tornioli's technique, relying on color infusion and chemical reactions, was particularly suited to the miraculous appearance of Christ's face on Veronica's veil and could therefore be regarded as an *acheropoietic* work of art. Furthermore, the fact that Tornioli produced two images by splitting the stones recalls the Veronica's propensity to copy itself miraculously. This subject was also frequently combined with paintings on marble, as attested by two paintings on alabaster representing *Veronica* in Carlo Emanuele I of

28 For an iconographic analysis of the Veil of Manoppello, see Blandina Paschalis Schlömer, "La relique inconnue," in *L'identification scientifique de l'Homme du Linceul Jésus de Nazareth. Actes du symposium scientifique international, Rome 1993*, ed. A.A. Upinsky (Paris, 1995), pp. 73-79. For a historical account of the discovery of this relic see Heinrich Pfeiffer, "Il Velo di Manoppello, la Sindone e l'immagine autentica di Cristo nell'arte: saggio di identificazione e di distinzione," in *L'identification scientifique*, pp. 81-84. On the spiritual potency and authority of the Byzantine icon known as the Mandylion, and its role as iconographic prototype during the Renaissance, see Anna Marazuela Kim's essay in this volume. For an understanding of the links between material innovation and *paragone* in papal portraiture, see Elena Calvillo's essay in this volume.

Savoy's 1631 inventory, or by other paintings of the Holy Face recorded in the inventories of Maffeo Barberini, Taddeo Barberini, and Filippo I Colonna.[29] Possibly exploited by Tornioli as an alternative, secret means to achieve fame, the pioneering procedure of painting in stone might have been conceived by the artist as a virtuosic display of ingenuity to overcome the professional and financial difficulties he was facing in Maurizio of Savoy's court. Tornioli's case confirms the discrepancy found by Stephen Campbell between "the promise of distinction and fame offered by success at court" and "the less-than-glamorous rewards and working conditions."[30] Since fame was the court artist's most sought-after reward and Tornioli's socio-economic status by no means reflected his patron's splendor, it may be argued that experimentation on stone was aimed at achieving both intellectual and material goals. One last hope is therefore placed in the imperishability of painting in stone, which, as discussed, resulted from a series of competitions between cities and artists, and from a dialectical tension between nature and art.

### Technical Daring and Failure: The Mosaic Decoration of the Chapel of the Holy Sacrament in St. Peter's

Tornioli's experiments in stone intersected with the apex—and final act—of his Roman career. In 1647, shortly before the 1650 Jubilee, Tornioli was commissioned by Innocent X for the mosaic decoration of the Chapel of the Holy Sacrament in St. Peter's. This intervention was part of a greater project of renovation of the church, which included the mosaic decoration of the six chapels located in the side aisles.

Tornioli's most influential patron in Rome, Father Virgilio Spada (1596-1662), wrote an account (before 1650), which documents the original decorative scheme of the Chapel of the Holy Sacrament.[31] In addition to important his-

29 Carlo Emanuele I was Prince Cardinal Maurizio's father. On the presence of paintings on stone in these collections see Marco Bona Castellotti, "Dipinti su pietra nelle raccolte sabaude," in *Pietra dipinta. Tesori nascosti del '500 e del '600 da una collezione privata milanese*, ed. M. Bona Castellotti, exhibition catalogue, Palazzo Reale, Milan, 2000, pp. 199-200 and Laura Laureati, "Le pietre dipinte: oggetti o quadri? Alcuni esempi di collezionismo romano," in *Pietra dipinta*, pp. 209-24.

30 Stephen J. Campbell, "Introduction," in *Artists at Court. Image-making and identity*, ed. S.J. Campbell (Chicago, 2004), p. 18.

31 ASR, Fondo Spada Veralli-FSV, vol. 236, ff. 341-352. See Minna Heimbürger Ravalli, *Architettura, scultura e arti minori nel Barocco italiano. Ricerche nell'Archivio Spada* (Florence, 1977), pp. 201-04. The document published in Klaus Güthlein, "Quellen aus dem

torical information, this document records Spada's unclear feelings towards Tornioli, being in praise of the artist's *disegno*, but also an attack on his unreliability. It informs us that Innocent x commissioned Giovanni Antonio Galli and Niccolò Tornioli to carry out, respectively, the mosaic decoration of the Chapel of the Choir and the Chapel of the Holy Sacrament. Accordingly, both of the artists produced a set of highly finished, oil painted cartoons, 24 *palmi* high and 17 wide. The iconography of Tornioli's mosaic scenes centered on the sacraments, with the exclusion of the Eucharist, already symbolized by the altar of the chapel. According to Spada, the first and most prominent episode was to be the *Baptism of Christ* on the right of the chapel. In line with Tornioli's preparatory cartoon, the scene would have portrayed Innocent x genuflecting on a *prie dieu* covered in a sumptuous carpet, surrounded by the personifications of Justice and Charity.

New research conducted in the Archivio della Fabbrica di San Pietro provides the means to define further the artist's activity in the Chapel of the Holy Sacrament in the Vatican Basilica, in addition to Roberto Cannatà's isolated scholarly contribution.[32] Our knowledge of the commission is enriched by the technical data included in the records (such as the measurements and the material components of the cartoons), as well as by the artist's *suppliche*, consisting of specific requests made to the Congregation of St. Peter's. The payments to Tornioli, dating from 6 February 1647 to 28 April 1649, are quite detailed, indicating whether the sum was paid for a cartoon or a finished mosaic. On 2 April 1648, for instance, Tornioli received 51 *scudi* and 80 *baiocchi* for two mosaic scenes executed in the Chapel of the Holy Sacrament, representing the *Baptism of Christ* and the *Confirmation*, and another 50 *scudi* for the mosaics executed in the dome, described by Virgilio Spada as 'original and evocative.'[33] Consequently, on 7 August 1648, Tornioli was given 300 *scudi* for

Familienarchiv Spada zum römischen Barock. 1. Folge," *Römisches Jahrbuch für Kunstgeschichte* 18 (1979), 184-86 is a slightly different copy (ASR, FSV, vol. 186, ff. 29-39).

32 See Roberto Cannatà's entry, in *Laboratorio di Restauro 2* ed. D. Bernini, exhibition catalogue, Salone di Palazzo Barberini, Rome, 1988, pp. 176 and 179. It should be noticed, however, that the shelf marks indicated by Cannatà are no longer valid. Newly found archival records are published in Weston, *Niccolò Tornioli*.

33 Archivio Fabbrica di San Pietro-AFSP, Arm. 26, D, 276, f. 171. Virgilio reported that the execution of the mosaic in the dome was interrupted by the renovation of the floor (ASR, FSV, vol. 236, f. 349: "Fece anco il d.o Pittore il fregio della Cuppola di musaico con disegno nuovo, e vago; ma fù stimato più necessario applicare l'animo à finire il pavimento, per la gran polvere, e danno che porta il mattonato, potendosi poi con agio, e senza impedimento alcuno proseguire con commodità i musaici, conforme che hanno fatto i Predecessori Pontefici").

a second pair of cartoons, which would have represented the sacraments of *Extreme Unction* and *Penance*.[34] These episodes, however, were never executed in mosaic, nor does any record mention the scene with the kneeling pope surrounded by the personifications of Justice and Charity.

From a letter written by Virgilio Spada, we discover that Tornioli was banned from St. Peter's because of his lack of skill in the mosaic technique and, more specifically, because he had applied paint on the mosaic *tesserae*. Virgilio Spada wrote an account of what happened to counter accusations made by Gian Lorenzo Bernini against Tornioli. In it, Spada explains that he had provided Tornioli with the commission for the mosaic decoration because Pietro da Cortona was out of Rome. He also underlines that, despite the great strength of Tornioli's cartoons, the painter proved incompetent in the mosaic technique. Spada also reports that Tornioli blamed Bernini for having suggested the very practice of painting over the mosaics and concludes by saying that he had brutally dismissed the Sienese artist as 'an enemy.'[35]

Even though Spada's intention might have been to defend Tornioli, his words became a *damnatio memoriae* for Tornioli's critical fortune. The painter's name became ineluctably associated with this disastrous ending to an important commission for the Jubilee year, in the most important church in Christendom. This episode paved the way for both Lione Pascoli's and Giovanni Battista Passeri's erroneous supposition that the artist was discharged because his cartoons did not please the Congregation of St. Peter's and Innocent X.[36]

34 AFSP, Arm. 26, D, 276, f. 171.

35 ASR, FSV, vol. 186, f. 1112: "Cognosco bene che ne i musaichi del Tornioli danneggiai la Reverenda fabrica, ma in quel tempo non era in Roma il Cavalier Pietro da Cortona, et il Papa voleva che si facessero, ne io conoscevo pittore megliore, et i suoi cartoni dimostrano che nel disegno non m'ingannai, ma il non haver alcuna esperienza di musaici come egli prottestò prima, li fece fare de spropositi, oltre la temerità d'aver adoperato colori sopra essi musaichi il che se ben egli per calunnia l'attribuì al Cavalier Bernino nondimeno all'udir io ciò, me lo levai d'inanzi ne lo volsi mai più vedere, onde ivi mi divenne nemico." See Heimbürger Ravalli, *Architettura*, p. 205, and Güthlein, "Quellen aus dem Familienarchiv Spada," p. 226.

36 Lione Pascoli, *Vite de'Pittori, Scultori, ed Architetti Moderni scritte, e dedicate alla Maestà di Carlo Emanuel Re di Sardegna da Lione Pascoli* (Rome, 1736), 2:31-32: "E volendo che anche si desse principio alla cappella del Sagramento, ed all'altra a dirimpetto uffiziata da'canonici furono a Niccolò Tornioli Sanese, ed a Gianantonio Spadarino Romano commessi i cartoni. Ma non essendo riusciti di gusto, nè de'cardinali della congregazione, nè del pontefice quelli dell'ultimo, ne fu rivocata la commissione, e sospeso per allora il rimanente dell'opera"; Giovan Battista Passeri, *Vite de'Pittori Scultori ed Architetti che anno lavorato in Roma Morti dal 1641 fino al 1643 di Giambattista Passeri pittore e poeta* (Rome, 1772), 2:168: "Si diede principio nella Cappella del SS. Sagramento, ed in quella all'incontro

The representation of the scene with the kneeling pope surrounded by Justice and Charity, described in the smallest detail in the programmatic document in the Fondo Spada Veralli possibly written by Spada himself, was never executed. However, as has been noted, there are indications that at least some of Tornioli's work for this project was admired by his peers. A reflection of Tornioli's iconography is to be found in Bernini's *Study for a Monument to Innocent X*.[37] While Bernini's reference to Tornioli's original project proves that it might have been praiseworthy, Tornioli's contemporary *bozzetto* for the mosaic decoration of the Chapel of St. Charles Borromeo in Santa Maria in Vallicella testifies to the artist's pictorial use of mosaics and to the intended interaction between natural light and glazed *tesserae*.[38] **[Fig. 9.5]**

Following Tornioli's dismissal, new cartoons for the vault and the pendentives of Chapel of the Holy Sacrament were commissioned from Pietro da Cortona, while six lunette cartoons were painted by Raffaello Vanni between 1659 and 1663. Guido Ubaldo Abbatini, Matteo Piccioni, Fabio Cristofari, and Orazio Manenti executed the final mosaic decoration of the Chapel, in accordance with a revised iconographic scheme centered on the representation of the Holy Eucharist and its foreshadowing in the episodes of the Old Testament.[39]

---

della Pietà, nella quale officiano li Signori Canonici, e Chierici Beneficiati, e ne fecero li cartoni Nicolò Tornioli Senese, e Giovanni Antonio Spadarino Romano; ma per la poca soddisfazione che ne ricevé il Pontefice, e li Signori Cardinali della Fabbrica, se ne tenne sospesa l'operazione per qualche tempo."

37 See Jennifer Montagu, *Alessandro Algardi*, 2 vols. (New Haven and London, 1985), 2:433-34 and Aidan Weston-Lewis, ed., *Effigies and Ecstasies. Roman Baroque sculpture and design in the age of Bernini*, exhibition catalogue, National Gallery of Scotland, Edinburgh, 1998, pp. 152-54. Alessandro Algardi was commissioned to design and execute the tomb of Innocent X, but he died in 1654.

38 The unpublished archival record attesting the removal of the mosaic on 7 July 1654 is AFSP, Arm. 17, E, 28, f. 106. The sketch was sold at Sotheby's London, 12 July 1978, lot 120; for its attribution to Tornioli, see Ciampolini, "Integrazioni," p. 30.

39 On Pietro da Cortona's activity in the Chapel of the Holy Sacrament see Frank R. DiFederico, *The Mosaics of St. Peter's. Decorating the New Basilica* (University Park and London, 1983) pp. 61-62. The scenes in the pendentives are: *King Melchizedek Making an Offering of Bread; Elijah Restored by an Angel; A Priest Dispensing Ceremonial Bread*; and *Aaron Filling a Vase with Manna*. The lunettes present: *Uzzah Struck Dead While Attempting to Steady the Ark of the Covenant; An Angel About to Touch Isaiah's Lips with a Live Coal; High Priest Offering to God the Fruits of the First Grain Harvest; Return from Canaan with the Large Burch of Grapes; Jonathan Tasting the Honey*; and *The Idol Dagon Destroyed by the Presence of the Ark*. A pioneering study on the mosaic technique of St. Peter's is offered in Édouard Gerspach, *La mosaïque* (Paris, 1925), pp. 238-51. For an up-to-date reconstruction of the decorative phases of the Chapel see Assunta Di Sante, "Le immagini cristologiche

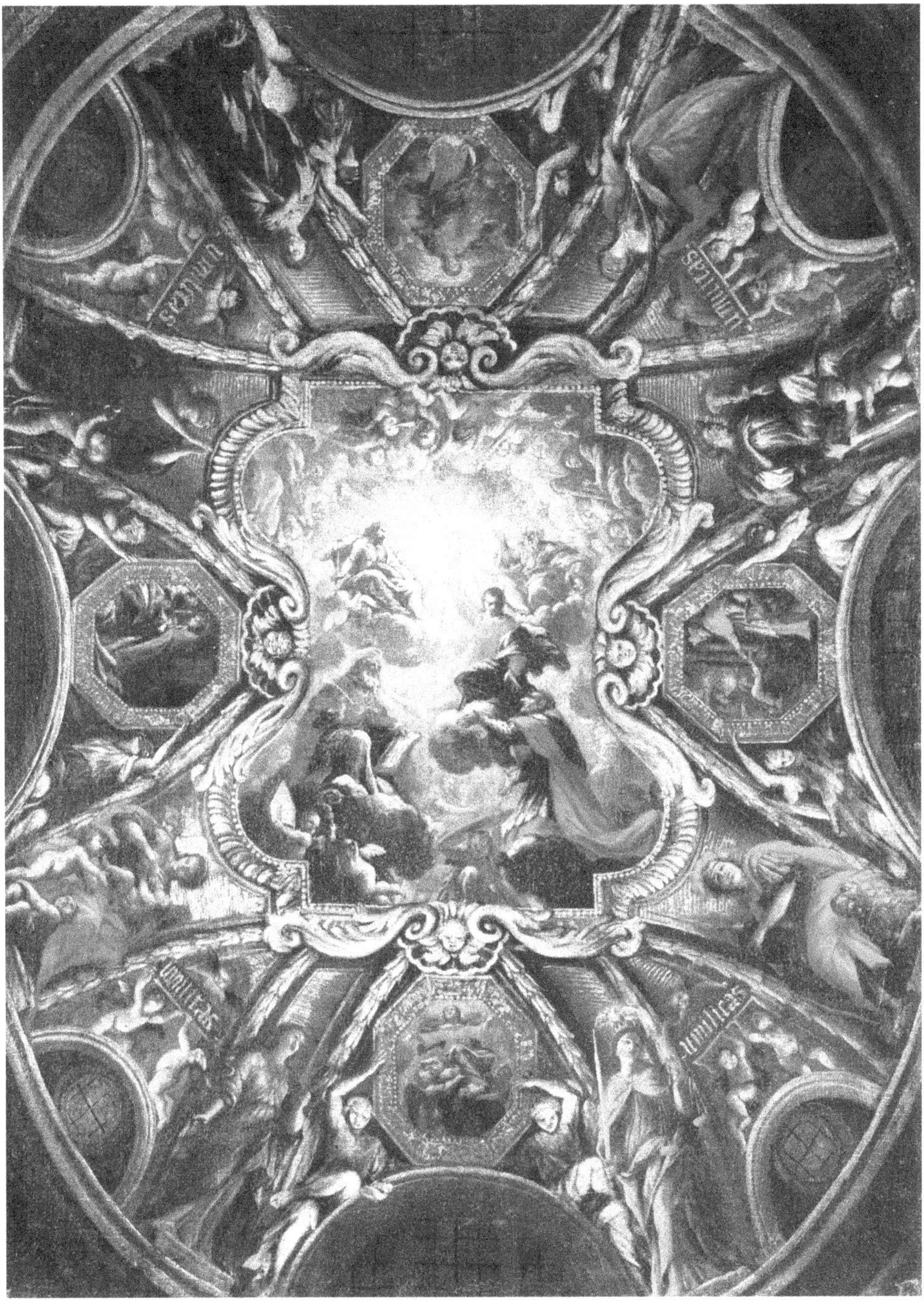

FIGURE 9.5 *Niccolò Tornioli,* Episodes of St. Charles Borromeo's life, *oil on canvas, formerly Sotheby's London* (PHOTO: SOTHEBY'S LONDON).

Tornioli's incompetence in mosaic technique calls for explanation. To some extent the practice of mosaic-making in mid-century Rome was characterized by a wealth of experimental procedures and difficulties concerning both materials and techniques. A *supplica* written before 3 February 1648 provides information about the relevant organizational and executional difficulties encountered by Tornioli in the execution of the mosaic decoration. Tornioli complains about the lack of the required conditions (*commodità necessarie*) to finish the work, reporting that his young collaborators (*suoi gioveni*) were inevitably wasting their time (*del continuo perdono il tempo*) because of the shortage of stretchers and mosaic *tesserae*.[40] Other records unveil additional details concerning Tornioli's preparatory cartoons, praised by Virgilio for the beauty of their *disegno*. A payment dating 28 March 1647, reporting that a certain 'Andrea' was given 28 *scudi* "per le tele fatte per il Signore Torniolli Pittore," attests that prototypes on canvases were later reproduced on the preparatory cartoons.[41] More notably, an account of the expenses made for Tornioli's cartoons for the dome of the Chapel records the materials used in the procedure, including canvases, stretchers, threads, priming (*mestica*), linseed oil, biacca, minium, *terra d'ombra di Venezia*, yellow and red pigments, and hot oil.[42] Additional information on the measurements and costs of Tornioli's mosaics with the *Baptism* and the *Confirmation* are provided by an estimate written on 29 February 1648 by Bernini and Pietro Paolo Drei. This records that Tornioli was paid 426 *scudi* and 80 *baiocchi* for a total mosaic surface of 766 *palmi*.[43] When reading these data in the broader context of the practice of

attraverso una lettura documentaria," in *San Pietro in Vaticano. I mosaici e lo spazio sacro*, ed. C. Thoenes et. al. (Milan, 2011), pp. 177-81, and Guido Cornini, "*Pittura per l'eternità*. Lo studio del mosaico e la decorazione a San Pietro da Gregorio XIII a Pio VII," in *La basilica di San Pietro. Fortuna e immagine*, ed. G. Morello (Rome, 2012), pp. 371-443.

40 AFSP, Arm. 11, E, 38, f. 670.

41 AFSP, Arm. 26, D, 277, f. 186.

42 AFSP, Arm. 2, B, 73, f. 911.

43 AFSP, Arm. 17, E, 24, f. 437. On the costs of mosaics in mid-century Rome see Richard Spear, "Rome: Setting the Stage," in *Painting for Profit*, pp. 61-62. These documents shed light on the figure of Scipione Santucci, a supplier of *smalti* for the mosaic *tesserae*, whose activity emerges as closely related to Tornioli's and Spadarino's work in St. Peter's. For the connection between Santucci and Tornioli see AFSP, Arm. 26, D, 277, f. 224, which attests that Santucci was paid 103 *scudi* and 52 *baiocchi* for the "smalti di far Musaico de diversi colori hauti per servitio del Musaico che fa il Signor Tornioli." An important document on Scipione Santucci's links to Tornioli and Spadarino is a *supplica* published in Maria Grazia D'Amelio and Carlo Stefano Salerno, "La tavolozza del mosaico a Roma e Venezia: classificazioni, catalogazioni e campionari dal primo tintometro (1655) di Pietro Paolo Drei della Fabbrica di San Pietro al campionario del 1889 di Angelo Orsoni di Murano," in *Atti del XVI*

mosaic-making in mid-17th-century Rome, we learn that the difficulties faced by Tornioli and his workshop were shared by other artists engaged with material innovation in the early modern period.

In 1607, for instance, Cavalier d'Arpino, responsible for the mosaic decoration of the main dome of St. Peter's, was brought to trial because of the accusation that he had failed to use first-rate materials.[44] As a consequence, the contracts between the artists and the Fabbrica started to include more detailed and severe clauses concerning the materials accepted or rejected in mosaic decorations, as in the prohibition of the use of colored terracotta as a substitute for glass *tesserae*.[45] As far as the preparation of the stucco is concerned, a technical milestone is represented by Girolamo Muziano's invention of an oil-based stucco used for the Gregoriana Chapel in St. Peter's, but it has nonetheless emerged that the recipes varied throughout the 17th century, possibly leading to uncertainties and mistakes. Further technical issues derived from the provenance of the glass tiles, produced in Venice, Umbria, Tuscany, and Rome.[46] Even the aesthetics of mosaics was facing a significant transformation, shifting towards a growing taste for chromatic innovations and technical experiments which would culminate, in the 18th century, in Alessio Mattioli's invention of the *pasta filata*.[47] More specifically, the mosaic-makers active in *Seicento* Rome began to imitate the visual effects of oil paintings. According to

---

*Colloquio dell'Associazione Italiana per lo studio e la conservazione del mosaico: Palermo, Piazza Armerina, 17-20 March 2010*, ed. C. Angelelli (Rome, 2011), pp. 77 (AFSP, Arm. 29, B, vol. 1, f. 435).

44 See Carmen Belmonte and Carlo Stefano Salerno, "Mosaic Glass made in Rome between the Sixteenth and Seventeenth centuries: Rome Glassmaker for the Fabbrica di San Pietro," *Journal of Cultural Heritage* 9 (2008): pp. e93-e96, and Alfredo Cirinei, "Conflitti artistici, rivalità e patronage nella Roma fra Cinque e Seicento. Il caso del processo criminale contro il Cavalier D'Arpino," in *La nobiltà romana in età moderna. Profili istituzionali e pratiche sociali*, ed. M.A. Visceglia (Rome, 2001), pp. 255-305.

45 Belmonte and Salerno, "Mosaic Glass," pp. e-94.

46 On this topic see D'Amelio and Salerno, "La tavolozza," pp. 55-70.

47 For an interesting comparison between the painter's and the mosaic-maker's palette, see D'Amelio and Salerno, "La tavolozza," pp. 71-84. On Mattioli's technical achievements in the mosaic technique and his activity in St. Peter's see Maria Grazia D'Amelio, "La tecnica degli smalti tagliati di Alessio Mattioli per lo Studio del Mosaico della Reverenda Fabbrica di San Pietro," in *I mosaici: cultura, tecnologia, conservazione. Atti del convegno di studi: Bressanone, 205 luglio 2002*, ed. G. Biscontin and G. Driussi (Venice, 2002), pp. 131-40, and Maria Grazia D'Amelio and Carlo Stefano Salerno, "Virtù e difetti degli smalti settecenteschi di Alessio Mattioli: la qualità delle tinte e la loro alterazione dalle prime critiche del chimico Alessandro Martelli alle indagini sui mosaici della Basilica Vaticana nel periodo dell'occupazione francese," in *Atti del primo convegno internazionale Ravenna Musiva.*

Alvar Gonzáles-Palacios, for instance, Giovan Battista Calandra's mosaic decoration of the Ubaldini Monument in Santa Maria Sopra Minerva (1642) should be read as a faithful transposition of an oil-painted model, possibly by Simon Vouet.[48] It might be proposed that Tornioli's mosaics fell into this category, as the highly finished, oil-painted cartoons were praised by Virgilio as accurate and beautifully designed, ready to be meticulously transposed into mosaic scenes. Moreover, this brief enquiry into the mosaic technique seems to suggest that Tornioli's ploy of painting on the *tesserae* was framed by a widespread technical uncertainty, further complicated by a multiplicity of materials and practices.

In other respects, the method adopted by Tornioli to paint on the mosaics, which, according to Virgilio Spada, the painter mendaciously attributed to Bernini, should be reconsidered in the light of the artist's lifelong interest in technical experimentation. Thus, Tornioli presumably attempted to replicate the newly invented practice of making mosaics imitate oil paint, by means of imbuing the tiles of the mosaic with color paint, possibly relying on the porosity of stone *tesserae*. This daring experiment reflects the artist's ambition to present himself as both *pictor* and *inventor*, capable of combining sophisticated iconographic schemes with extraordinary technical achievements. It is worth considering whether Tornioli's technical innovation intended to engage in a renewed *paragone* between painting and sculpture, and thus should be related to Bernini's *bel composto*. Both Bernini's and Tornioli's art might be understood in light of Baroque 'fascination with the *techne* of illusion.'[49] In contrast to Bernini's 'wholistic conception of the work of art,' Tornioli's invention could be read as a last, heroic episode of marvelous craftsmanship, rooted in Beccafumi's artistic legacy and in a taste for alchemical, pseudo-scientific *segreti*.[50]

Tornioli's encounter with Bernini at the end of his career and the disastrous experiments attempted in the mosaic decoration of the Chapel of the Holy Sacrament in St. Peter's compel us to reunite two complementary narratives in the history of *Seicento* art, both rooted in the search for fame through technical daring. The first, Bernini's, is a history of triumph, for *bel composto* fully

*Conservazione e restauro del mosaico antico e contemporaneo: Ravenna, 22-24 ottobre 2009*, ed. C. Fiori and M. Vandini (Bologna, 2010), pp. 521-32.

48 See Alvar Gonzáles-Palacios, "Giovan Battista Calandra, un mosaicista alla corte dei Barberini," *Ricerche di Storia dell'Arte* 1-2 (1976), 211-26. On Giovanni Battista Calandra see Arabella Cifani and Franco Monetti, *Giovanni Battista Calandra (1586-1644). Un artista piemontese nella Roma di Urbano VIII, di Maderno e di Bernini* (Turin, 2006).

49 See Genevievve Warwick, *Bernini. Art as Theatre* (New Haven and London, 2012), p. 13.

50 See Irving Lavin, *Bernini and the Unity of Visual Arts* (New York and London, 1980), 1:145.

exploited art's ability of being marvelous, deceptive, theatrical, and performative. The second, Tornioli's, is a history of failure, which, ironically, originated in a Sienese definition of success. Tornioli's rivalry with Michelangelo Vanni over the invention of a new method of painting 'in' stone goes beyond a concern to equal tradition. Technical virtuosity is linked to fame, for the imperishability of the artwork secures the longevity of its creator's reputation. (Art) history is written by the victors, and yet Tornioli's unfortunate case contributes to our understanding of the artist's identity as both *pictor* and 'inventor' in the age of the Baroque.

## Acknowledgements

I wish to thank Stefan Albl, Piers Baker-Bates, Elena Calvillo, Maria Grazia D'Amelio, Assunta Di Sante, James Hall, Philippa Jackson, Helen Langdon, Sheila McTighe, Scott Nethersole, Carlo Stefano Salerno, Anita Viola Sganzerla, Maria Cristina Terzaghi, and Simona Turriziani for sharing with me their knowledge on this fascinating topic.

## Appendix 1

### Teofilo Gallaccini's letter to Niccolò Tornioli [ed. Della Valle 1786, 399-402]

In somma per quanto compreso dal suo avviso, bisognerà, ch'io m'accordi con ciò, che disse Marziale poeta Spagnuolo, ciascuno esser ingegnoso sopra l'opere altrui. Il che è un voler tenere a sindacato, quanto da altri si fa; come se fosse loro per privilegio singolare conceduto il sovrastar col giudicio indiscreto, e senza senno alla censura dell'opere altrui. Questi nuovi nomi, o pure Sanesi, o Sanesi arromanescati non hanno altro gusto, che il sottrarre e degradare al valore altrui. Si confermano in così malnato, ed odioso costume dalle loro immaginate chimere, e dalle false oppinioni senza ragione alcuna, secondando il male affetto loro. Signor Niccolò mio, non se ne curi punto, e ne faccia quel conto, che dell'abbajar de'cani fa la Luna. In quanto all'invenzion del Francesino nello intenerire i marmi, le rispondo, siccome ho potuto ritrarre da Messer Bernardino Cortonese scarpellino in Pantaneto, esser vera, ed averla anche intesa da lui medesimo. Ciò faceva per non aver a servirsi dell'opera degli scarpelli d'altri artefici, potendo per se stesso agevolmente con stiletti di ferro fare i contorni, i lineamenti, e l'ombre delle figure. Non però tingeva il marmo per formare l'ombre e le mezze tinte; ma vi commetteva una pietra di color bigio, o piuttosto simigliante al color della liscìa, come già fece Domenico Beccafumi nel pavimento del Duomo di questa città. Questa pietra è quella stessa, che si adopera per arrotare ad olio i coltelli, i rasoi, e gli scarpelli da legname. Egli ne ha in casa un frammento lungo presso a due braccia. Di questa medesima pietra erano già fatti i gradi appie la colonna dell Imperadore lungo'l prato a Camollìa, davanti'l portone della Madonna; che conosciuti buoni per arrotare ad olio, furono in breve tempo levati, e portati via, siccome anche le pietre d'alcuni murelli di banchi. Tutto lo spazzo del duomo è adornato di marmi, sì nelle istorie come ne' fregi, e nè compartimenti, ed in alquante figure d'opera di commesso. Solamente vi è un'opera di musaico fatta di pezzetti di varj colori poco più oltre dell'entrar della porta maggiore, dove dentro un gran cerchio sono figurate l'armi delle Città confederate con la R. P. Sanese, postavi in mezzo la lupa lattante, insegna di Siena come colonia de' Romani. Ad imitazion di questo lavoro di musaico marmoreo, in Fiorenza al tempo del Gran Duca Ferdinando primo è stato inventato il musaico di pietre fine, di gioje, di pietre preziose, e di gemme, d'agate, di lapislazzuli, di diaspri, di ametisti, e d'altre pietre preziose. Formavano i paesi, le figure, l'istorie, e l'armi delle città principali degli stati di sua altezza Serenissima per collocarle fra gli ornamenti della cappella delle altezze Serenissime di Toscana in San Lorenzo. Non pertanto vi è parte alcuna di marmo artificiosamente colorata. La lunghezza del tempo consumatore delle cose, e signore di ciò, che si contiene sotto'l Cielo, non avrebbe conceduto, che si conservasse lungamente la maggior parte delle figure già con gli scalpelli intagliate a linee, a profili, a profondità d'incastri fatte con trapani; e così ancora sarebbero svaniti i colori. Il primo inventore degli ombreggiamenti nelle commettiture de'marmi per dar loro il rilievo, e la concavità fu Matteo di Giovanni pittore eccellentissimo Sanese, e'l primo ad introdur la buona maniera de' componimenti delle istorie, e del dipingere, il quale fiorì nel 1472. Questi osservava fra le più antiche figure del pavimento della nostra cattedrale, una figura di Davide giovanetto in atto di scaricar la frombole ed uccider il gigante Golia, vi conobbe una vena del marmo di color d'acquerella, che formava l'incavamento d'una piega della vesta fra le gambe, la quale faceva apparir come di rilievo il ginocchio, e la gamba sinistra. Così anche nella figura di Salomone, le pieghe della vesta, che avanzavano il manto, vide fatte di marmo quasi bigio. Quindi prese occasione di pensare ed aggiungerle a' commessi delle figure, e delle istorie, e particolarmente in quella degli Innocenti, ch'egli fece nel pavimento del Duomo, rappresentandovi la concavità del luogo con l'oscurezza del marmo bigio. Questi insegnò al Beccafumi il modo di fare istaccar le sue figure intagliate ne' bianchi marmi, e dar loro maggior forza con l'ombra, e con le mezze tinte, commettendo ne' proprj luoghi altre pietre oscure, e bigie fatte dalla natura. Non si trovò giammai alcuno artefice antico, che tingesse le pietre, ma solamente il legname, come chiaramente si vede nelle tarsie delle residenze nella chiesa de' Monaci di Monteoliveto fuor dalla porta Tufi, ed in duomo nella cappella di S. Giovanni di mano del Barilli. Così anche si tingeva diversamente l'osso, e l'avorio. Gli alchimisti sofistici tingono ad oro, ed argento in superficie sola. Ciò con ragione dicono tirare in pelle, mostrandosi l'oro, e l'argento in superficie. Si biancheggia, e s'indora il rame, questo con la giallamina, quello col bianchimento. Così fanno oggi alcuni emoli di V. S. e altri saccenti, che coloriscono il marmo solamente di sopra; ed io ne ho fatta la prova. Quando maestro Ascanio scarpellino Cortonese, che faceva troppo il valentuomo nella scultura, un giorno meco favellando dell'opere d'intaglio, e di scultura, esaltando l'opere de' suoi paesani, proponendoli io i bassi rilievi, i lavori di quadro, i corniciamenti della scala del pergamo di marmo, e que' dell'altar maggiore del Duomo, opera dei Marsini, e così l'ornamento dell'altar della Madonna a Fontegiusta, mostrò farne poca stima, costume naturale di chi soverchiamente la pretende. Venuto poi al proposito del pavimento, opera tanto celebrata, ed ammirata dal Mondo, fece l'istesso; e soggiunse, che le pietre bigie poste in vece d'ombre, e di mezza tinta, non fussero altre specie di pietre oscure per natura, e commesse nel marmo; ma che sieno una tintura fatta sopra il marmo. Volle ciò persuadermi col mezzo d'una sua sperienza spropositata, che quando si commettono le pietre, si scaldano molto bene; e postovi lo stucco e la pietra ben riscaldata, lo stucco liquefatto dal calore e sparso intorno il pezzo commesso, macchia il rimanente del marmo in guisa, che riceve la tintura. Chi non vede, che tal macchia non può in alcun modo penetrare, ma solamente spandersi nella superficie? Quindi egli trasse il segreto di colorir il marmo tanto mirabile, che diede al Sig. Cavalier Francesco Vanni. Ma se egli veramente aveva ritrovato questo segreto, perchè non lo messe in uso, quando fece la sepoltura nella Chiesa di Santa Maria della Scala al tempo del Signor Cavalier Agostino Chigi rettore dello Spedal grande? E pure vi ebbe a fare il ritratto del Beato Sorore istitutor di quello, ed alcune teste di morti, ed altre cose cavate dal disegno del Signor Cavalier Ventura Salimbeni. Dove oltre i tratteggiamenti fatti alla Mecherinesca, poteva aggiunger l'ombra con la sua nuova invenzione di tingere il marmo. Non sarebbe questa forse stata esquisita opportunità di fare acquisto di credito maggiore? E'l Francesino, quando fece nella stessa maniera la testa d'un Serafino nell'ornamento della sepoltura delle monache dello spedal di Madonna Agnesa detto altramente di San Niccolò in Sasso; perchè non si servì del suo segreto dello intenerire il marmo, per dare un saggio della sua nuova invenzione? Io per me credo, che il Cortonese non l'abbia fatto non essendo sicuro, che fosse per difendersi lungo tempo dal continovo consumamento cagionato dal camminarvi frequentissimamente sopra, conoscendo la pittura non penetrare oltre la superficie. Il Francesino non si valse della invenzione forse, perciocchè non ne aveva fatta la prova. Onde mi risolvo a concludere, che tali segreti non penetrassero più oltra della superficie. Se Maestro Ascanio tornasse in vita, il condurrei nella cattedrale di Siena, e li direi, che prendesse i migliori occhiali, che si trovassero, e con essi guardasse bene il detto pavimento, che vi troverebbe chiarissimo segno delle commettiture delle pietre. Io insieme col Sig. Tommaso scultore suo affezionatissimo, esaminato molto bene il pavimento del Duomo, mi son confermato nella mia opinione. Vi ho veduti espressamente i segni delle commettiture delle pietre postevi per l'ombre, o per le mezze tinture: cagione, che l'ombre vi appariscono terminate fra'l marmo e la pietra bigia. Se i marmi fossero stati colorati, come pretendeva il Cortonese, non vi apparirebbe terminazione alcuna. Chi avesse tinto il marmo vicino alla parte chiara, l'avrebbe colorito più dolce, e sfumato, come si usa ne' disegni per dar l'acquarelle. Vi apparisce il contrario. Però dunque concludo, la tintura de' marmi di questi così belli ingegni esser una gran mellonaggine e tanto più grossa, quanto più vi stanno ostinati, non si accorgendo essere in pelle in pelle. Mentre praticavo con la b. m. del Signor Vanni, di cui sono stato discepolo del disegno, e conversavo con lui: poi applicandomi agli studj della filosofia, e della medicina co' suoi famigliari, non udii già mai far alcuna menzione di tal segreto. Nè l'opinion dello scarpellino da Cortona mi potette mai persuadere a crederlo. Accade l'istesso della commettitura dell'ombre, che de' lumi. Il Beccafumi nel quadro dell'Abramo si ingegnò formare un perfetto chiaroscuro con l'ombre tratteggiate con le mezze tinte; e co' lumi; e il tutto eseguì con opera di commesso. Incassò nelle parti dell'ombra le pietre più oscure di marmo nero, o formando i tratti più vivi, e più ricacciati con lo stucco nero; ed appresso commettendovi le pietre bigie per la mezzatinta. Nelle parti del chiaro scuro, e del lume, scelti minuti pezzi di marmo più bianco, ve gli incastrava per formare i lumi, e per dar forza, e rilievo alle figure. Onde prendo a dire, che Mecarino, se avesse usato il tingere il marmo per fare l'ombre, avrebbe anche saputo schiarirlo per farlo candido, e per dare i lumi nelle parti più eminenti, e più esposte al lume. Egli nondimeno non pose in opera nè l'uno, nè l'altro, come si vede per esperienza. Nè si poteva così bene ed esquisitamente incassare un pezzo di marmo dell'altro, che non vi appariscono i commettimenti. Non voglio tacere, che egli avanti, che facesse l'opera degli spazzi del Duomo, fece per modello un quadro di tarsia di legname, dove con eccellenza formò l'istoria della Conversion di San Paolo con contorni, e tratteggiamenti neri, con ombre e mezzetinte di tavole commesse di colore oscuro, come si vede nel duomo. Questo quadro fu veduto da me nelle stanze del Vanni. Ora, siccome ho inteso, si ritrova in casa del Sig. Conte Fabio Delci. Questa è la somma, di quanto ho saputo dire in risposta della sua gratissima. V. S. non faccia altro risentimento intorno a queste menzogne, per non dar campo ad essere stimate. Così questi nuovi segreti, e queste moderne meraviglie, come prive di fondamento reale, svaniranno; e quello di V. S. rimarrà in piedi in codesta città, dove gli umani ingegni son graditi, ed esaltati, e dove il maggior acquisto del valore, e del nome prendono il vantaggio del premio, e della ricompensa conforme al merito: e la fama loro si dilata verso l'altre parti del Mondo, siccome la virtù del Capo, a cui corrisponde Roma, si diffonde per le membra, alle quali si confanno l'altre Città; e le provincie straniere. Io non ho altro da dirle sopra ciò. Se è conforme all'aspettazion sua, ne prenderò gusto, altramenti mi scusi, ed accetti il buon animo. Le invìo l'inclusa iscrizione fatta sopra la morte del Duca di Guisa accaduta in Cuna nel suo ritorno dal bagno, come avvenne a Pandolfo Petrucci tiranno di Siena, il quale tornando dal bagno morì nello Spedal di San Quirico. Quì facendo fine; che veramente ne sarebbe tempo dopo tanta lunghezza, le bacio le mani, pregandole da Dio il conservamento della sua felicità.

*Di Siena il dì 13 d'Ottobre 1640.*
D. Teofilo Gallacini

## Appendix 2

### Tornioli's letter to Ferrante Carlo [ed. Borghesi-Banchi 1898, n. 350, 644-646]

Dopo le molte vennero da me quei due ministri [del Duca di Savoia], il Vicario e il Controlor, ch'era quasi un'ora di notte. Primieramente cominciarono ad interrogarmi se avevo lettere del Sig. Conte Muriano, dove gi risposi che avevo lettere di S. A. e del Sig. Abate Soldati, il quale d'ordine del Sig. Conte mi aveva scritto. Le lettere di S. A. S. e dell'Abate non gliele volsi mostrare allora, per aspettare in quel che volesser dare questi ragionamenti, dove cominciarono in questa maniera, volendo veder le pietre, ma prima avevo esagerato che il mio male era cagionato dalle fatighe fatte per S. A. e per aver compito dopo tre malattie pericolose l'opera di che S. A. per sue lettere mi faceva istanza: dove le veddero e mostrorno averne gusto. Subito il Controlor mi disse: veniamo un poco ai nostri conti. Gli risposi che i conti eran così fatti, che da Maggio passato in qua, io delle mie provvisioni non avevo mai avuto niente, avendo io sempre operato per S. A. e per suo comandamento. Adesso di nuovo avevo avuta questa malattia per ultimar queste pietre. Mi soggiunse il Controlor che mi avrebber dato centoventicinque scudi: che era per tutto Ottobre che così avevano pagati gli altri, e che de' denari non ce n'era. Li dissi come dovevo fare, chè di questi che mi davano ne dovevo render centoventi al Sig. Luca, e che io dovevo darli di merito per i frutti scudi dieci: e che non sapevo come farmi a vivere, avendo speso tutto quello che avevo per dar gusto a S. A. et avendo sempre operato et speso del mio, non solo nelle tele e colori, ma nelle pietre. Mi rispose il medesimo che non occorreva che avessi fatte tante spese, che dovevo finir pezzo per pezzo, come la Nunziata; dove li risposi che se la Nunziata non era finita che n'era stata cagione il non aver avuto l'azzurro oltramare da tante volte che ne avevo fatto istanza, e che avevo operato in far questi altri quadri, acciò S. A. vedesse come ancora i suoi ministri, che non andavo a spasso. Non seppero che si rispondere, se non alle pietre; che quando li dissi che avevo fatto le medesime per S. A. e per ordine suo mi risposero che le avevo fatte per mio gusto, et io li dissi moltissime volte che le avevo fatte per S. A. e per ordine suo. Non credo che non sarebbe se non bene che si mostrasse al Sig. Vicario la lettera di S. A. e del Sig. Abate in confidenza oggi, quando io abbia avuto i danari, chè si rimase che questa mattina dovessi mandarli a pigliare. Desidero che V. S. dica sopra questo il suo parere, e se devo esagerare fingendo confidenza grande con il Vicario e dirli i molti disgusti che avevo ricevuto, e delle stanze cattive: come ancora che tenevo quella stanza fuori del Palazzo, e che ne pagavo la pigione: che per non aver mai voluto darmi stanza a proposito mi conveniva per servizio di S. A. e delle sue opere pagar quella pigione, ed esagerare che bene ai forestieri sono state date comodità, eccetto che a me, avendo S. A. così benignamente dato l'ordine, ad avendo loro avuto ordine dal Sig. Conte Muriano di satisfarmi, ed i medesimi ordini non solo li aveva dati a Lui, ma al Controlor ancora dove non vedo cosa alcuna. Esagerai ancora che il fogliame non valeva molto più di quel che non mi dava S. A. in un anno oltre all'altre (sic) che mandavo di qua. Rimanemmo che io dovessi aver pazienza, che come fussero venuti altri denari che mi avrebbero sodisfatto. Li dissi che sapevo che ci era ancora tre mila scudi, non volendo dir altro, ch'ero informato d'ogni cosa. Ei mi risposero ch'ero male informato. Dissi che averei scritto a S. A. come dovevo fare a vivere, frattanto non potendo operare per altri ed avere speso così crudelmente, pensando esser almeno pagato per tutto l'anno. Eglino mi dissero che scrivessi e che avrebbero essi ancora scritto e significato a S. A. le spese che avevo fatte, e conforme all'ordine di S. A. avrebbero subito soccorso. Frattanto bisogna aver pazienza. Ora Sig. Ferrante mio, credo che questi due furfantoni siano d'accordo. Desidero che V. S. consideri se sarà bene che esageri delle cose loro, cioè del traffico che fanno del denaro, del vivere, dello strapazzo che mi fanno, della casa fuori, e se devo dirli che ne debba avvisare S. A. e farli sapere della mia scomodità delle stanze, che nè anco posso aver cantina per la osteria come si è fatta: come ancora nel tempo che questa state ero ammalato mi messero con altro ammalato forestiero. Vanno alla mia camera quasi per dispetto, e gridando il ragazzo e la serva che andava a pigliar l'acqua, e minacciatoli di darli ancora, perché li guastava il sonno. Però prego V. S. per vita sua, senza che si scomodi di scrivere o altro, di dar di penna così – a quel che non devo ragionar col Vicario, e quello che devo ragionare lasciarlo nella lettera come sta, che subito intenderò; poichè oggi dopo il pranzo voglio parlargli, fingendo sempre in confidenza, mentre, per fine, facendole umilissima reverenza scusandomi se sono stato così lungo, e ne incolpi la mia mala disgrazia, chè se io avessi fortuna di poter venire io, non le avrei scritto: e le bacio le mani. Mentre le scrivevo mandai dal Sig. Controlor acciò desse i denari al Sig. Francesco, e che si facesse dare i danari dell'azzurro: mi disse che avrebbe dato i centoventicinque scudi, e dell'azzurro non voleva ingolfarsi, che non voleva darli. È tornato il Sig. Francesco con i danari, sicchè V. S. sente; e le bacio le mani. Ho reso i cento venti scudi al Sig. Luca. Adesso respiro un poco, non vedendo più quella pittima cordiale.

## Bibliography

Adriani, Giovanni Battista. *Memorie della vita e dei tempi di Monsignor Referendario Giansecondo Ferrero-Ponziglione. Memorie Storiche di Giovambatista Adriani*. Turin, 1856.

Bagnoli, Alessandro. *Mostra di opere d'arte restaurate nelle provincie di Siena e Grosseto II*. Genoa, 1981.

Belmonte, Carmen, and Carlo Stefano Salerno. "Mosaic Glass made in Rome between the Sixteenth and Seventeenth centuries: Rome Glassmaker for the Fabbrica di San Pietro." *Journal of Cultural Heritage* 9 (2008).

Bernini, Dante, ed. *Laboratorio di Restauro 2*. Exhibition catalogue, Salone di Palazzo Barberini, Rome, 1988.

Bona Castellotti, Marco. "Dipinti su pietra nelle raccolte sabaude", in *Pietra dipinta. Tesori nascosti del '500 e del '600 da una collezione privata milanese*, edited by M. Bona Castellotti, pp. 199-200. Exhibition catalogue, Palazzo Reale, Milan, 2000.

Borghesi, Scipione, and Luciano Banchi. *Nuovi documenti per la storia dell'arte senese*. Siena, 1898.

Bottari, Giovanni Gaetano, and Stefano Ticozzi. *Raccolta di Lettere sulla Pittura, Scultura e Architettura scritte da'più Celebri Personaggi dei secoli XV, XVI, e XVII*. 8 vols. Milan, 1822.

Caciorgna, Marilena, and Roberto Guerrini. *Il pavimento del Duomo di Siena. L'arte della tarsia marmorea dal XIV al XIX secolo. Fonti e simbologia*. Milan, 2004.

Campbell, Stephen J. Introduction to *Artists at Court. Image-making and identity*, edited by Stephen J. Campbell. Chicago, 2004.

Carli, Enzo, ed. *La Chiesa di San Niccolò in Sasso a Siena*. Siena, 1995.

Chiarini, Marco, and Cristina Acidini Luchinat, eds. *Bizzarrie di pietre dipinte dalle collezioni dei Medici*. Exhibition catalogue, Galleria dell'Accademia and Museo dell'Opificio delle Pietre Dure, Florence, 2000.

Chiarini, Marco, Annapaula Martelli Pampaloni, and Anna Maria Maetzke, eds. *Pittura su pietra*. Exhibition catalogue, Palazzo Pitti, Florence, 1970.

Ciampolini, Marco. "Integrazioni al catalogo di Niccolò Tornioli." *Antichità Viva* 34 (1995), 26-34.

Ciampolini, Marco. "Niccolò Tornioli", in *Bernardino Mei e la Pittura Barocca a Siena*, edited by F. Bisogni and M. Ciampolini, pp. 109-12. Exhibition catalogue, Palazzo Chigi Saracini, Siena, 1987.

Ciampolini, Marco. *I pittori senesi del Seicento*. 3 vols. Siena, 2010.

Cifani, Arabella, and Franco Monetti. *Giovanni Battista Calandra (1586-1644). Un artista piemontese nella Roma di Urbano VIII, di Maderno e di Bernini*. Turin, 2006.

Cirinei, Alfredo. "Conflitti artistici, rivalità e patronage nella Roma fra Cinque e Seicento. Il caso del processo criminale contro il Cavalier D'Arpino." In *La nobiltà romana in età moderna. Profili istituzionali e pratiche sociali*, edited by M.A. Visceglia, pp. 255-305. Rome, 2001.

Collins, Patricia. "Manoscritti autografi di Teofilo Gallaccini in Gran Bretagna." In *Siena circa 1600: dimenticare Firenze. Teofilo Gallaccini e l'eclisse presunta di una cultura architettonica*, edited by G. Morolli, pp. 201-06. Exhibition catalogue, Santa Maria della Scala and Magazzini della Corticella, Siena, 1999.

Cornini, Guido. "*Pittura per l'eternità*. Lo studio del mosaico e la decorazione a San Pietro da Gregorio XIII a Pio VII." In *La basilica di San Pietro. Fortuna e immagine*, edited by G. Morello, pp. 371-443. Rome, 2012..

Curti, Francesca. *Committenza, collezionismo e mercato dell'arte tra Roma e Bologna nel Seicento. La quadreria di Cristiana Duglioli Angelelli*. Rome, 2007.

D'Amelio, Maria Grazia, and Carlo Stefano Salerno. "La tavolozza del mosaico a Roma e Venezia: classificazioni, catalogazioni e campionari dal primo tintometro (1655) di Pietro Paolo Drei della Fabbrica di San Pietro al campionario del 1889 di Angelo Orsoni di Murano." In *Atti del XVI Colloquio dell'Associazione Italiana per lo studio e la conservazione del mosaico: Palermo, Piazza Armerina, 17-20 March 2010*, edited by C. Angelelli. Rome, 2011.

D'Amelio, Maria Grazia, and Carlo Stefano Salerno. "Virtù e difetti degli smalti settecenteschi di Alessio Mattioli: la qualità delle tinte e la loro alterazione dalle prime critiche del chimico Alessandro Martelli alle indagini sui mosaici della Basilica Vaticana nel periodo dell'occupazione francese." In *Atti del primo convegno internazionale Ravenna Musiva. Conservazione e restauro del mosaico antico e contemporaneo: Ravenna, 22-24 ottobre 2009*, edited by C. Fiori and M. Vandini, pp. 521-32. Bologna, 2010.

D'Amelio, Maria Grazia. "La tecnica degli smalti tagliati di Alessio Mattioli per lo Studio del Mosaico della Reverenda Fabbrica di San Pietro." In *I mosaici: cultura, tecnologia, conservazione. Atti del convegno di studi: Bressanone, 205 luglio 2002*, edited by G. Biscontin and G. Driussi, pp. 131-40. Venice, 2002.

Della Valle, Guglielmo. *Lettere Sanesi del padre maestro Guglielmo Della Valle minor conventuale socio delle Rr. Accademie delle Scienze e Agraria di Torino etc. Sopra le Belle Arti*. Rome, 1786.

Di Sante, Assunta. "Le immagini cristologiche attraverso una lettura documentaria." In *San Pietro in Vaticano. I mosaici e lo spazio sacro*, edited by C. Thoenes and A. Comastri, pp. 177-81. Milan, 2011.

DiFederico, Frank R. *The Mosaics of St. Peter's. Decorating the New Basilica*. University Park and London, 1983.

*Due restauri lapidei promossi dal Lions Club Siena. La lastra tombale di Francesco Vanni a San Giorgio e una iscrizione quattrocentesca nel Palazzo Pubblico di Siena*. Siena, 2009.

Eamon, William. "Science and Popular Culture in Sixteenth Century Italy: The 'Professors of Secrets' and Their Books." *The Sixteenth Century Journal* 16, no. 4 (1985), 471-85.

Fagiolo Dell'Arco, Maurizio, and Carandini, Silvia. *L'Effimero Barocco. Strutture della festa nella Roma del '600*. 2 vols. Rome, 1977-78.

Faietti, Marzia. "Nuove acquisizioni conoscitive su alcuni disegni restaurati della Pinacoteca Nazionale di Bologna." *Bollettino d'Arte* 33-34 (1985), 217-32.

Fara, Giovanni Maria. "Teofilo Gallaccini in his library: reflections and critical catalogue." In *Teofilo Gallaccini. Selected writing and library*, edited by A. Payne, 00. 1-40. Florence, 2012.

Fattorini, Gabriele, and Jennifer Sliwka. "Domenico Beccafumi and the Sienese tradition," in *Renaissance Siena. Art for a City*, edited by L. Syson, exhibition catalogue, National Gallery, London, 2007, pp. 296-99.

Galli, Anna Elena. "Federico IV Borromeo: scelte artistiche tra Milano e Siena." iIn *Atti delle Giornate di Studi sul Caravaggismo e il naturalismo nella Toscana del Seicento*, edited by P. Carofano, pp. 295-322. Pontedera, 2009.

Gerspach, Edouard. *La mosaïque*. Paris, 1925.

Gibellino Krasceninnicowa, Maria. *Il Beccafumi*. Florence, 1933.

González-Palacios, Alvar. "Giovan Battista Calandra, un mosaicista alla corte dei Barberini." *Ricerche di Storia dell'Arte* 1-2 (1976), 211-26.

Gori Gandellini, Francesco. *Prospetto dei pittori, ante* 1784, pp. 183-84 (Biblioteca Comunale di Siena-BCS, L. V. 15).

Güthlein, Klaus. "Quellen aus dem Familienarchiv Spada zum römischen Barock. 1. Folge." *Römisches Jahrbuch für Kunstgeschichte* 18 (1979), 173-246.

Heimbürger Ravalli, Minna. *Architettura, scultura e arti minori nel Barocco italiano. Ricerche nell'Archivio Spada*. Florence, 1977.

Lanzi, Luigi. *Notizie della scultura degli antichi e dei vari suoi stili*. Fiesole, 1824.

Laureati, Laura "Le pietre dipinte: oggetti o quadri? Alcuni esempi di collezionismo romano." In *Pietra dipinta. Tesori nascosti del '500 e del '600 da una collezione privata milanese*, edited by M. Bona Castellotti, pp. 209-224. Exhibition catalogue, Palazzo Reale, Milan, 2000.

Lavin, Irving. *Bernini and the Unity of Visual Arts*. New York and London, 1980.

Mazzoni, Guido. "Appunti su Michelangelo Vanni pittore e mercante d'arte." *La Diana* 2 (1996), 255-68.

Montagu, Jennifer. *Alessandro Algardi*. 2 vols. New Haven and London, 1985.

Morolli, Gabriele, ed. *Siena circa 1600: dimenticare Firenze. Teofilo Gallaccini e l'eclisse presunta di una cultura architettonica*. Exhibition catalogue, Santa Maria della Scala and Magazzini della Corticella, Siena, 1999.

Oberli, Matthias. *"Magnificentia Principis." Das Mäzenatentum des Prinzen und Kardinals Maurizio von Savoyen (1593-1657)*. Weimar, 1999.

Paschalis Schlömer, Blandina. "La relique inconnue." In *L'identification scientifique de l'Homme du Linceul Jésus de Nazareth. Actes du symposium scientifique international, Rome 1993*, edited by A.A. Upinsky, pp. 73-79. Paris, 1995.
Pascoli, Lione. *Vite de'Pittori, Scultori, ed Architetti Moderni scritte, e dedicate alla Maestà di Carlo Emanuel Re di Sardegna da Lione Pascoli*. Rome, 1736.
Passeri, Giovan Battista. *Vite de'Pittori Scultori ed Architetti che anno lavorato in Roma Morti dal 1641 fino al 1643 di Giambattista Passeri pittore e poeta*. Rome, 1772.
Payne, Alina, and Giovanni Maria Fara. "Teofilo Gallaccini e la critica architettonica a Siena fra XVI e XVII secolo." In *Architetti a Siena. Testimonianze della Biblioteca Comunale tra XV e XVIII secolo*, edited by D. Danesi, M. Pagni and A. Pezzo, pp. 141-89. Exhibition catalogue, Biblioteca Comunale degli Intronati, Siena, 2009.
Payne, Alina. *The Telescope and the Compass. Teofilo Gallaccini and the Dialogue between Architecture and Science in the Age of Galileo*. Florence, 2012.
Pezzo, Annalisa. "Una rete erudita." In *Siena circa 1600: dimenticare Firenze. Teofilo Gallaccini e l'eclisse presunta di una cultura architettonica*, edited by G. Morolli, pp. 57-76. Exhibition catalogue, Santa Maria della Scala and Magazzini della Corticella, Siena, 1999.
Pfeiffer, Heinrich. "Il Velo di Manoppello, la Sindone e l'immagine autentica di Cristo nell'arte: saggio di identificazione e di distinzione." In *L'identification scientifique de l'Homme du Linceul, Jésu de Nazareth: actes du symposium scientifique international, Rome 1993*, edited by A.A. Upinsky, pp. 81-84. Paris, 1995.
Pope Hennessy, John W. *Sassetta*. London, 1939.
Randolfi, Rita. "Alcune precisazioni sull'attività romana di Niccolò Tornioli." *Studi di Storia dell'Arte* 7 (1996), 347-53.
Romagnoli, Ettore. *Biografia Cronologica de' Bellartisti Senesi dal secolo VII a tutto il XVIII* [ante 1835], anastatic copy. Florence, 1978.
Salort-Pons, Salvador. "La llegada de Niccolò Tornioli a Roma y el malogrado mecenazgo del abad Borromeo." *Antologia di Belle Arti, Studi Romani* 1 (2004), 9-13.
Sanminiatelli, Donato. *Domenico Beccafumi*. Milan, 1967.
Seifertova, Hana. *Malby na kameni: Uměcký experiment v 16. Na začátku 17 století/ Painting on Stone: an Artistic Experiment in the 16th and Early 17th centuries*. Prague, 2007.
Sohm, Philip. Introduction to *Painting for Profit. The Economic Lives of Seventeenth-century Italian Painters*, edited by R.E. Spear and P. Sohm. New Haven and London, 2010.
Spear, Richard E. "Rome: Setting the Stage", in *Painting for Profit, Painting for Profit. The Economic Lives of Seventeenth-century Italian Painters*, edited by R.E. Spear and P. Sohm. New Haven and London, 2010.
Sperindei, Simona. "L'inventario di Niccolò Tornioli. Aggiornamenti e nuovi documenti." *Valori Tattili* 00 (2011), 115-23.
Squellati Brizio, Paola, "Natura sollecitata." In *Bizzarrie di pietre dipinte dalle collezioni dei Medici*, edited by M. Chiarini and C. Acidini Luchinat, pp. 32-38. Exhibition cata-

logue, Galleria dell'Accademia and Museo dell'Opificio delle Pietre Dure, Florence, 2000.

Tavassi La Greca, Bianca. "Niccolò Tornioli e la famiglia Spada." In *Per la Storia dell'Arte in Italia e in Europa. Studi in onore di Luisa Mortari*, edited by M. Pasculli Ferrara, pp. 249-57. Rome, 2004.

Ticozzi, Stefano. *Dizionario degli architetti, scultori, pittori*. Milan, 1832.

Torriti, Pietro. *La Pinacoteca Nazionale di Siena. I dipinti dal XV al XVIII secolo*. Genoa, 1978.

Vasari, Giorgio. *Le Vite de' piu eccellenti Pittori, Scultori, e Architetti*. Edited by C.L. Ragghianti. Milan, 1942.

Viale Ferrero, Mercedes. "Le feste e il teatro." In *Diana Trionfatrice. Arte di corte nel Piemonte del Seicento*, edited by M. Di Macco and G. Romano, pp. 74-81. Exhibition catalogue, Parco del Valentino, Turin, 1989.

Warwick, Genevieve. *Bernini. Art as Theatre*. New Haven and London, 2012.

Weston-Lewis, Aidan, ed. *Effigies and Ecstasies. Roman Baroque sculpture and design in the age of Bernini*, pp. 152-54. Exhibition catalogue, National Gallery of Scotland, Edinburgh, 1998.

Weston, Giulia Martina. "After Galileo: The Image of Science in Niccolò Tornioli's *Astronomers*." *Art History* 36, no.2 (2016), 302-17.

Weston, Giulia Martina. *Niccolò Tornioli (1606-1651). Art and Patronage in Baroque Rome*. Rome, 2016.

Zeri, Federico. *La Galleria Spada in Roma*. Florence, 1954.

CHAPTER 10

# Salvator Rosa: A Variety of Surfaces

*Helen Langdon*

Since he was a boy, his biographer tells us, the Neapolitan artist, Salvator Rosa, blessed by an *indole spiritosa*, had been so attracted by drawing that there was no wall in his home, or other places, which he did not cover with charcoal drawings, "creating little figures and small landscapes."[1] This literary topos, with its overtones of exuberant creativity, would have pleased Rosa, and it conveys something of his inventiveness and pleasure in witty experimentation. Unlike most of the artists in this volume, Rosa was not a specialist in painting on stone or copper, but he was constantly alive to the stimuli of varied materials. He produced a number of works on rare and precious surfaces to satisfy the tastes of his patrons, particularly in Florence, where refined Medici courtiers delighted in such art. At the same time, never entirely at ease at court, he created a world beyond its confines; he enjoyed the lively debates of the Florentine literary academies, and the pleasures of villa life in the remote Tuscan countryside. Witty and exuberant, the painter cultivated the persona of the Juvenilian satirist, mocking courtly riches. In this context, and for his intellectual friends, Rosa created, as gifts, many highly unusual works, among them landscape drawings on parchment, exotic heads in colored crayons, and, most striking of all, a series of drawings of philosophers and other solitary figures, in black ink and white chalk, on thin wooden panels. This essay has two aims; it sets Rosa's works on stone and copper in the context of the traditions established by such artists as Antonio Tempesta, Filippo Napoletano, and Stefano della Bella; but it also suggests how Rosa moved away from these traditions, to create works far removed from courtly luxury, whose frail and humble materials implied a criticism of court and city, and evoke an earlier age.

## Between Florence and Rome

In 1639 the Neapolitan Rosa, fresh from around five years in Rome, arrived in Florence in the service of Giovan Carlo de' Medici; he returned to Rome in

1 Filippo Baldinucci, *Notizie dei Professori del Disegno da Cimabue in qua*, ed. F.Ranalli, 6 vols. (Florence, 1846), 4:207.

 | DOI 10.1163/9789004361492_014

1649, and remained there for the rest of his life. In these years, in both Florence and Rome, collectors shared a passion for works on every kind of rare stone, on black stone, or *pietra di paragone*, alabaster, slate, lapis lazuli, agate, and *pietra d'Arno* or *pietra paesina*. They enjoyed the tensions between art and nature, taking pleasure in how skilled artists could use the evocative natural patterning of different surfaces to conjure up wonders or *meraviglie*. Their delight was enhanced by the long history of such pleasures, described in both ancient and Renaissance literature. Philostratus, (2nd/3rd century AD) in his *Life of Apollonius of Tiana*, had presented Apollonius discussing with Damis how it is in the nature of man to imitate, and to re-create the fantasies that the swirling clouds seem to conjure up in the heavens, of centaurs and stag antelopes, of wolves and horses.[2] In the Renaissance Leonardo da Vinci famously advised the artist to

> look into the stains on walls, or the ashes of a fire, or clouds, or mud, or like things, in which, if you consider them well, you will find really marvellous ideas. The mind of the painter is stimulated to new discoveries ... various compositions of landscapes and monstrous things, such as devils and similar creations.[3]

These sources privilege the same themes; battles, fires, the monstrous and grotesque, which often formed the subject matter of paintings on stone.[4]

Colored marbles, curiosities, and rare and precious stones held a particular attraction for the Florentines, where the Opificio delle Pietre Dure had been established by Ferdinand I in 1588. The treatise of Agostino del Riccio, *Trattato delle Pietre* (Florence, 1597), written when Francesco I de' Medici was encouraging the alchemical and naturalistic studies that found expression in his celebrated *Studiolo*, gives a vivid picture of a city gleaming with different varieties of stone, whose monuments and noble chapels vied with one another in the richness of their displays. Perhaps, many years later, Rosa recalled one of the paintings from the studiolo, Giovanni Fedini's dramatic *Polycrates and the Ring* (Florence, Palazzo Vecchio), in his own rendering of this highly unusual subject; Polycrates' precious sardonyx was the most celebrated jewel of the

---

2 Philostratus, *The Life of Apollonius of Tyana*, vol. 1, trans. Frederick Cornwallis Conybeare (London and New York, 1912), II, chap.xxii, pp. 175-6.

3 Leonardo da Vinci, *Treatise on Painting*, trans. A. Philip McMahon, (Princeton, 1956), 2:51.

4 For these ancient and Renaissance sources see Paolo Squellati Brizio, "Natura Sollecitata," in *Bizzarrie di pietre dipinte dalle collezioni dei Medici*, eds. Marco Chiarini and Cristina Acidini Luchinat, (Cinisello Balsamo [Milan], 2000), pp. 32-38.

ancient world.[5] Del Riccio delighted in the play between art and nature, writing of the "fantasie e scherzo che fa la madre natura." Very many paintings on stone are recorded in Florentine inventories, where their descriptions, such as "due quadretti di Pietre di alberetti e scogli fatti dalla natura, dipintovi più figurine" and others "fatti dalla natura e aiutati con pennello"[6] echo Del Riccio's language.[7] Such works were often displayed, in Florence and in Rome, in *künstkammers*, those collections abundant in curiosities, where small paintings hung with precious and semi-precious stones of all kinds, corals, ebony cabinets, golden medals, columns of alabaster, shells, and minerals. Often rare and suggestive stones were not clearly distinguished from paintings on stone, and collectors enjoyed setting the brush of nature beside the imagination of man.

There are many connections between Florence and Rome, both, between 1580 and 1650, centers of *pietre dipinte*. It is possible that the idiosyncratic painter, Leonard Bramer, also links Rome and Rosa's Naples, where François de Nomé and Jacopo van Swanenbergh's small paintings on copper, often of macabre scenes, were popular. Bramer specialized in small paintings on slate, and, in Rome in the 1620s, where he was a founder member of the Bentveughels, he became so well known for his night scenes that he was named 'Leonardo della Notte.' Bramer did at least twenty works on slate whilst in Italy, and a 1634 inventory of the celebrated collection of Gaspar Roomer lists forty small paintings by him; it has been suggested that the Dutch artist had a studio in Naples.[8] In Rome, Alessandro Turchi, Marcantonio Bassetti, and Pasquale Ottino, specialists in paintings on slate and black stone, had all been active in the city, and established there a taste for dark night time scenes, more austere and dramatic than the curiosities of Florence, and intended as aids to devotion. An inventory of the collection of the Contelori family records a *Pietà sopra pietra di para-*

5 Rosa's *Polycrates and the Fisherman*, which shows an earlier moment from the story illustrated by Fedini, is in the Art Institute, Chicago.

6 Antonio Del Riccio, *Istoria delle Pietre*, 1597, ed. Paola Barocchi (Florence, 1979), 33. The quotations from inventories are given by Marco Chiarini, "Pittura su pietra," in *Bizzarrie di pietre dipinte*, 14. See also Anne-Laure Collomb, *Splendeurs de l'Italie: La peinture sur pierre à la Renaissance* (Rennes and Tours, 2012), p. 191.

7 For paintings on stone in Florence see Anne-Laure Collomb, "Jacques Stella et la peinture sur pierre," in *Jacques Stella (1596—1657)*, ed. Sylvain Laveissière and Gilles Chomer, exhibition catalogue, Lyon, Musée des Beaux Arts and Toulouse, Musée des Augustins, 2006-7 (Paris, 2006). Collomb mentions that there were 43 paintings on stone listed in the inventory of the Medici Villa Imperiale in 1624.

8 Jane Ten Brink Goldsmith et al., *Leonaert Bramer, 1596-1674: Ingenious Painter and Draughtsman in Rome and Delft*, ed. Jane ten Brink Goldsmith (Zwolle and Delft, 1994). On Bramer's relationship with Roomer see p. 53.

*gone di Michelangelo da Caravaggio,* and some of these artists use the polished blackness of their grounds to create the darkness of night and a Caravaggesque chiaroscuro.[9] No painting on stone by Caravaggio is known, but it is of interest that, in a trial of 1603, Caravaggio's scanty list of good artists included Sigismondo Laire, a German artist renowned for wondrous paintings on precious surfaces.[10] Jacques Stella, who stayed in Florence (*c.* 1619-22) before a longer period in Rome (1623-34), is interesting in this context. He was celebrated for his rich, ornamental use of exquisite surfaces, of lapis lazuli, alabaster, slate and marble, but he also did a group of severe devotional works on stone. His *Rest on the Flight into Egypt* [**Fig. 10.1**] uses the surface to create the darkness of night. The mood is unusually direct and strong, and the putto, and the naturalistic, reflective water on the left, seem a homage to such a work as Caravaggio's *St John at the Fountain* (priv. coll.).[11] Highly influential in the diffusion of a more decorative aesthetic was Antonio Tempesta, sought after for his paintings on lapis lazuli, marble, alabaster, or Florentine *pietra paesina.*[12] His style spread to Florence, and Cardinal Carlo de' Medici, the brother of Grand Duke Cosimo II, arranged for some of his works to be sent there.

But the strongest link between Naples, Rome and Florence was Filippo Napoletano, whose career and interests in so many ways anticipated those of Rosa. In Rome, Filippo moved in the circles of Cassiano dal Pozzo and of the scientist Johann Faber; he had himself a celebrated cabinet of curiosities; he early won a reputation, Mancini tells us, for small paintings of fires, ships and animals; for battles and martyrdoms; and for "certe stravaganze di scheleti d'animali."[13] Filippo was associated with Cardinal Francesco Maria Del Monte, the inventory of whose collection lists twenty paintings on stone.[14] Del Monte

9 Renata Ago, *Il Gusto delle Cose: Una Storia degli Oggetti nella Roma del Seicento* (Rome, 2006), p. 140.

10 Helen Langdon, *Caravaggio: a Life* (New York, 1998), p. 269

11 For Stella's paintings in Rome see Laveissière and Chomer, eds., *Jacques Stella (1596-1657)*, pp. 79-109; for this work, p. 121.

12 On Rome as a center of 'pietra dipinta' see Laura Laureati, "Le pietre dipinte: oggetti o quadri? Alcuni esempi di collezionismo romano," in *Pietra Dipinta; Tesori Nascosti del '500 e del '600 da una collezione privata Milanese*, ed. Marco Bona Castellotti (Milan, 2000), pp. 209-24. For a series of extracts from inventories of Roman collections, see pp. 214-24. See also Collomb, *Splendeurs de l'Italie*, pp. 193-215. For Tempesta, pp. 197-203.

13 Giulio Mancini, *Considerazioni sulla pittura (Rome, 1616-24)*, ed. Adriana Marucchi and Luigi Salerno, 2 vols. (Rome, 1956-7), 1:255.

14 Laureati, "Le pietre dipinte," p. 211. Laureati identifies Filippo as the painter of battle scenes on copper from this collection, and also of "*sette paesini, sei dei quali sono in pietra di rosezzano, ch'e macchiata in modo di città, e fabbriche con historiette di variate figurini di*

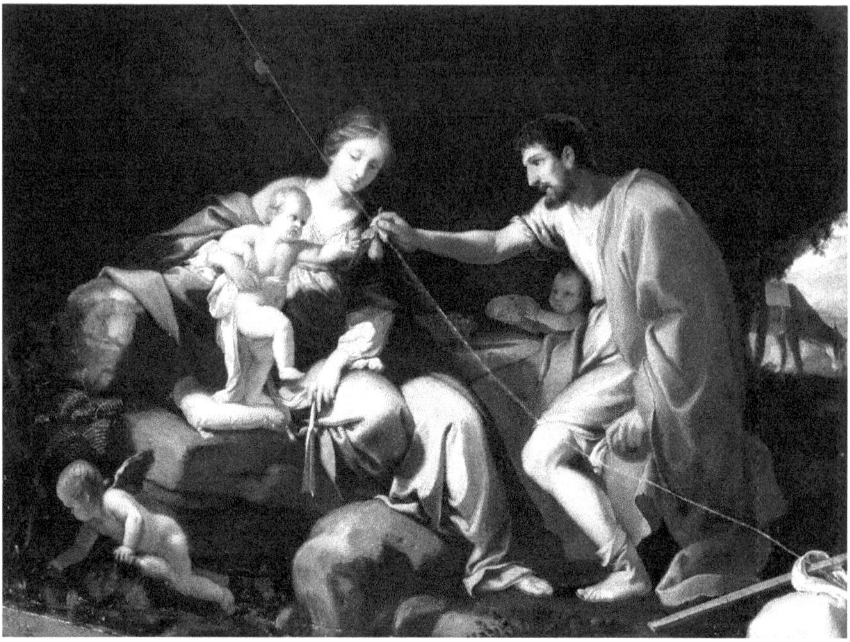

FIGURE 10.1 *Jacques Stella,* Rest on the Flight into Egypt, *oil on slate, ca. 1639,* GLASGOW, HUNTERIAN MUSEUM.

arranged for Filippo's move to Florence in 1617, where he entered the service of Cosimo II de' Medici, whose pleasure in rare surfaces and in northern art is evident in his acquisition, two years after Filippo's arrival, of Adam Elsheimer's small house altar (Frankfurt, Städel Museum).[15]

In Florence Filippo delighted the Grand duke with his paintings on copper and slate, landscapes and scenes of fires, which derive from the art of north European landscapists, perhaps particularly from the small Hellish scenes of Jan Breughel. The two enjoyed a close relationship, and Baldinucci tells us that Cosimo, confined to his bed towards the end of the life, took pleasure in watching Filippo paint landscapes with small and 'most beautiful figures.'[16] A small *City Burning,* [**Fig. 10.2**] which uses the darkness of the slate to establish the night time scene, may well be one such work. It suggests the paintings of Jan

*guerra.*" For a discussion of Del Monte's interest in paintings on copper see Francesca Cappelletti, "The Enticement of the North: Landscape, Myth and Gleaming Metal Supports," in *The Genius of Rome, 1592-1623*, ed. Beverly Louise Brown (London, 2001), p. 176.

15 For Filippo's move to Florence, and the taste of Cosimo II de' Medici, see Marco Chiarini, *Teodoro Filippo di Liagno detto Filippo Napoletano* (Florence, 2007), pp. 52-53.

16 Baldinucci, *Notizie dei Professori del Disegno*, 4:207.

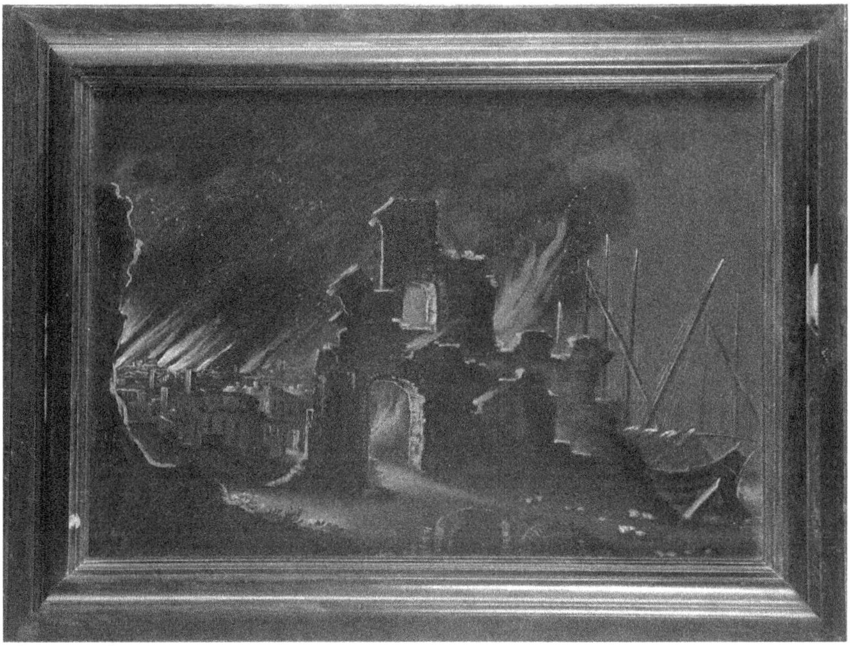

FIGURE 10.2 *Filippo Napoletano,* City Burning, *oil on slate, Florence, Palazzo Pitti, Pitti* (PHOTO: FIRENZE GABINETTO DEI DISEGNI E DELLE STAMPE DEGLI UFFIZI).

Breughel, and yet has a new reality; it shows not a fanciful Hell, but a real city engulfed by flame or war. The painting is distinguished by its exceptional freedom of touch and intensity of vision.[17]

A number of Filippo's works are on a type of limestone which was especially prized in Florence. These stones, Del Riccio tells us, were found in the Arno, "and today it is customary to cut into them, and then to polish them, so that there may be seen there the fantasies and caprices of Mother Nature, now small trees, at times woods ..."[18] Two types of stone are most common; *pietra paesina,* or 'ruin marble,' where the natural veining of iron oxides conjures up desolate ruin landscapes, with shattered cliff sides, castles, and mountain ridges crowned with trees, and *lineato d'Arno,* distinguished by flowing or undulating bands of varied shades, easily used for sea or sky. Such stones were often displayed untouched in collector's cabinets, as objects of wonder or contemplation, or masterpieces of nature, but in the 17th century it became

17 Chiarini,, "Pittura su Pietra," p. 70, no. 13, and Chiarini, *Teodoro Filippo di Liagno*, p. 285, cat. no. 59.

18 Del Riccio, *Istorie delle Pietre*, p. 33. "... che oggidi s'usano segare, cosi poi lustrare, che in essi si veggono varie fantasie e scherzi chef a la madre natura, or alberetti, talvolta selve..."

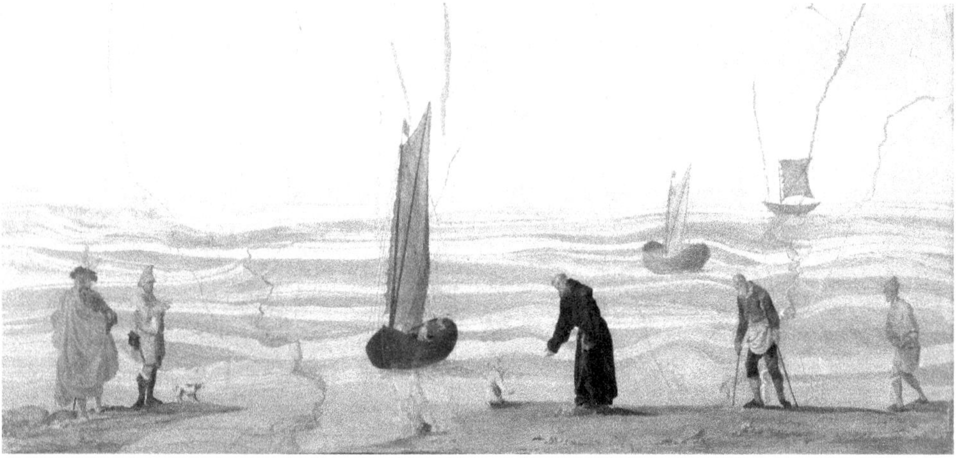

FIGURE 10.3 *Filippo Napoletano,* Vision of St Augustine, FLORENCE, MUSEO DELL'OPIFICIO DELLE PIETRE DURE. (PHOTO: SU CONCESSIONE DEL MINISTERO DEI BENI E DELLE ATTIVITA CULTURALI E DEL TURISMO-OPIFICIO DELLE PIETRE DURE DI FIRENZE)

common for the artist to see them as a spur to invention, and to set his own gifts against those of *natura artifex*.[19] Filippo's *Temptation of St Anthony* [**Plate 29**], and his *Vision of St Augustine,* [**Fig. 10.3**] suggest the range of effects offered by the two types of stone.[20] In the first the *pietra paesina* evokes a rocky and desolate landscape, and sets up a tension between the materiality of the surface, with its play of veined and fractured stone, which create an illusion of three dimensions, with the more painterly depth 'added by the brush.'[21] This stone was particularly suited to Hellish scenes, such as Dante's descent to hell, or, as here, to the lives of the early Christian hermits, in stony and demon infested surrounding.[22] The *Vision of St Augustine,* [**Fig. 10.3**] on *lineato d'Arno,* is very different. Here the viewer beholds the immensity of the sea, conjured

19 The landscape in stone was a common feature of natural history collections. It was illustrated and documented in Ulisse Aldrovandi's *Musaeum metallicum in libros IIII distributum bartholomaeum ambrosinus in patrio bono* (Bologna,1648) and in Athanasius Kircher's *Mundus Subterraneus in XII libros ...* (Amsterdam,1665), book 8. See *The Paper Museum of Cassiano dal Pozzo*, British Museum, Department of Prints and Drawings, (Milan, 1993), p. 215.

20 Chiarini, "Pittura su Pietra," p. 70, no.13, and Chiarini, *Teodoro Filippo di Liagno*, p. 285, cat. no. 59.

21 This painting is one of a group of seven small paintings on *alberese*, painted for Cosimo II; in the Medici archives they are described as "di scogli e lontananze fatti in d.a pietra dalla natura, e aiutati con il pennello." Chiarini, *Teodoro Filippo di Liagno*, p. 276. no. 47

22 Francesco Ligozzi's *Dante and Virgil in Hell* (Florence, Museo dell'Opificio delle Pietre Dure) is very close to Filippo's *Temptation of St Anthony.*

up from the wavy lines of the stone, the foam of breaking waves evoked by the bands of light color. A watcher on the shore, seen from the back, looks out to sea, and the diagonal line of boats leads his and our eye to the horizon. The work creates a tension between the intense naturalism of the vividly detailed and oddly unconnected figures—one elegantly cloaked, with feathered headwear, another old and on crutches—and the fantasy of the whole, as stone becomes sea, and the child of St Augustine's vision, a messenger from God, evokes the mysteries of religion, yet vaster than the sea. Filippo uses the natural quality of the stone to evoke the sense of looking into something small, and yet magically seeing an infinite and wide vista. These small paintings, or objects, which play on the tiny and the vast, on ideas of 'natural wonders and artistic wondrousness,' and on the phenomenon of metamorphosis, are *meraviglie,* of the kind admired in cabinets of curiosities.[23] They are suited, too, to scenes from Ariosto's *Orlando Furioso,* itself so rich in romance and marvel, and in magical transformations; Filippo's *Roger liberating Angelica* [**Fig. 10.4**] now uses the undulating lines of the stone as cloud, embellishing his surface with a tiny Roger, airborne on a winged Pegasus, and a wondrous sea monster, and cleverly setting a vividly naturalistic Angelica against the stone's evocation of rocky cliff. The work is a tour de force of enhancing the inherent nature of the rare and prized stone by a refined display of artistic skill.[24]

## At the Florentine Court

These then were Rosa's forerunners, and the traditions, both Roman and Florentine, which he inherited as court painter to the Medici. Both the Medici and their entourage enjoyed paintings on stone; Giovan Carlo's enthusiasm for works on Florentine stone, on copper and on black marble, is suggested by an inventory of 1647.[25] His taste was widespread in the Medici orbit. Giovanni Niccolini, father of Filippo, an important Rosa patron, founded the family chapel in Santa Croce in Florence, highly celebrated for its marble decoration. His collection included many works on stone, among them works by Jacopo Ligozzi and Orazio Gentileschi. From Tempesta he commissioned several works on

---

23 On this see James V. Mirollo, "Aesthetics of the Marvellous," in *Wonders, marvels and monsters in early modern culture,* ed. Peter G. Platt (Newark and London, 1999), pp. 24-44.

24 Chiarini, *Teodoro Filippo di* Liagno, p. 280, cat. no. 54.

25 Paolo Barocchi and Giovanni Gaeta Bertala, eds., *Collezionismo Mediceo e Storia Artistica. 3, Il Cardinale Giovan Carlo, Mattias e Leopoldo, 1628-1667* (Florence, 2007), 1:431-34.

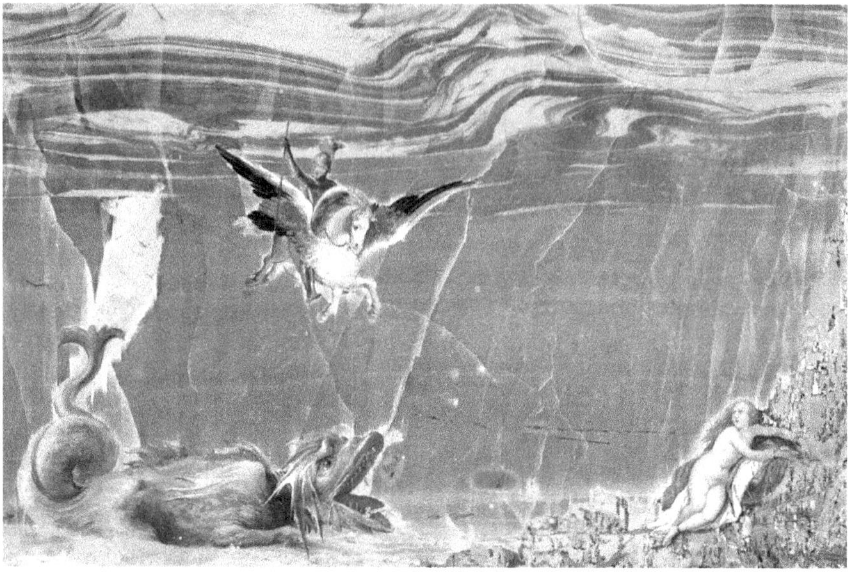

FIGURE 10.4 *Filippo Napoletano,* Roger liberating Angelica, FLORENCE, ISTITUTO DI STUDI ETRUSCHI. (PHOTO: SU CONCESSIONE DEL MINISTERO DEI BENI E DELLE ATTIVITA CULTURALI E DEL TURISMO – OPIFICIO DELLE PIETRE DURE DI FIRENZE E ISTITUTO DI STUDI ETRUSCHI ED ITALICI)

alabaster, some unusually large, as well as small coppers.[26] Stefano della Bella painted two series of works on *pietra paesina*, a commission associated with Cardinal Carlo de' Medici and with the Marchese Carlo Gerini, the latter a distinguished courtier and patron of Rosa.

Rosa moved in these Medici circles, and, for his patron, Giovan Carlo de' Medici, he painted an abundant variety of works, among them vast marines and landscapes, and, to hang in the small upper rooms by the terrace of his Casino in via della Scala, a wealth of informal works,—small oblong landscapes, landscapes in gouache, ceiling paintings, paintings of maps and cities.[27] He also restored some works for him, including one on black stone.[28] In the collection of the Marchese Gerini a mythological subject on slate (untraced) by Rosa is recorded, and it seems likely that he did more work of this kind than has survived.[29] This is also suggested by the recent discovery of a small

26 On Giovanni Niccolini and Tempesta see Riccardo Spinelli, "Documenti artistici dall'archivio Niccolini di Camugliano—1. Marmi antichi e 'moderne pitture' di Giovanni di Agnolo (Firenze, 1544-1611)," *Paragone, Arte* 61, no. 663 (2005), 80-103.

27 Barocchi and Bertala, eds., *Collezionismo Mediceo e Storia Artistica. 3,* 1:431-34.

28 Ibid., pp. 433-34.

29 Martina Ingendaay, *"Il migliori pennelli": I Marchesi Gerini mecenati e collezionisti nella Firenze barocca. Il palazzo e la galleria 1600-1825*, 2 vols. (Milan, 2013), 2:252.

oval painting on lapis lazuli, *Two Orientals* (priv. coll.), where two figures walk beside the seashore, one apparently a philosopher demonstrating a point to his companion, and behind them, the magical ultramarine blue of the sea, and the white edges of the clouds suggested by the limestone impurities in the stone.[30] The painting is close to Antonio Tempesta's two small ovals, The *Finding of Moses* and *Moses turning the waters of the Nile into Blood* (Milan, Giulini coll.), where Tempesta uses the lapis to create sea and sky, and an atmosphere of enchantment, in a similar way.[31] Rosa may too have known Cornelis Poelenbergh's small oval *Latona and the Lycian Peasants* (Florence, Opificio delle Pietre Dure) on lapis lazuli, in the Grand ducal collection.[32]

## Two Scenes of Witchcraft; Slate, Copper and the Bizarre

This Ovidian subject was a popular one; Filippo Napoletano painted at least three versions; it inspired Rosa's *Scene with Witches; Dawn* (Cleveland Museum of Art), one of four circular witchcraft scenes painted for the Marchese Filippo Niccolini, *maestro di camera* of Rosa's patron, Giovan Carlo de' Medici. Niccolini, the son of Tempesta's patron, Giovanni Niccolini, also owned a witchcraft scene on *pietra di paragone* by Filippo Napoletano, and, although Rosa's tondi are on canvas, rare surfaces seem often to have been favored for such subjects.[33]

Rosa became celebrated for scenes of witchcraft, which he painted in both Florence and Rome. Among them are two works, one on slate, one on copper, supports highly unusual in his art, and the eccentricity of the themes may have suggested their use. Both supports, flat and hard, encourage the rendering of detail with the finest of brushes, and, with their overtones of the stones used by witches and magi, and of the transformation of metals, themselves suggest themes of sorcery. This is heightened by the unusual shapes, oval and round, evoking the magic circles of witchcraft.

The oval *Scene of Witchcraft* [**Plate 30**] on slate suggests Rosa's early familiarity with cabinet paintings of macabre nocturnal scenes, by Francois de

30 This painting was recently published by Caterina Volpi; see Caterina Volpi, *Salvator Rosa (1615-1673) "pittore famoso,"* (Rome, 2014), p. 161, fig.132, cat. no. 156

31 For illustrations see Johanna Beate Lohff, *Malerei auf Stein: Antonio Tempestas Bilder auf Stein im Kontext der Kunst-und Naturtheorie seiner Zeit* (Munich, 2015), p. 218.

32 For the attribution of this work see Marco Chiarini, "Pittura su Pietra," *Antichita Viva* 2, (1970), fig.5

33 For the relevant parts of Niccolini's inventory see Sara Fabbri, "I Quattro Tondi con incantesimi di Salvator Rosa nella collezione del Marchese Filippo Niccolini: Il Volto Oscuro della Toscana del Seicento," *Notizie da Palazzo Albani* 38 (2009), 81-104.

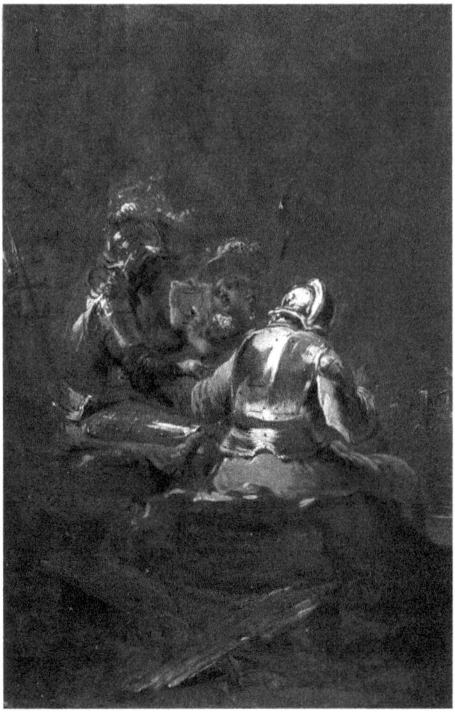

FIGURE 10.5
*Leonard Bramer,* Soldiers Resting, *oil on slate, 1626, The Hague, Museum Bredius* (PHOTO: MUSEUM BREDIUS).

Nomé and Jacopo van Swanenbergh.[34] But most important for his use of the support were the small paintings on slate by Leonard Bramer whose works, as we have seen, Rosa may have known in Naples and in Rome. Bramer's first signed and dated work, *Soldiers Resting,* [**Fig. 10.5**] is on slate and is so strikingly close to some of Rosa's early genre scenes that it seems likely Rosa knew this, or a similar, work.[35] Rosa's two small paintings, *A Witch*, and *A Soldier,* (Rome, Pinacoteca Capitolina)[36] may have been inspired by him and, in the Strozzi-Ridolfi collection in Florence were paintings by Rosa listed as "quadri in lavagna rappresentanti soldati che si scaldano, ed altre figure in riposo."[37] Volpi has suggested that the dark background of a newly discovered early work, *Soldiers playing jacks* (priv. coll.), may have been the young Rosa's attempt to

34 Volpi, *Salvator Rosa*, cat. no. 154.

35 Michael Kitson ed., *Salvator Rosa* (London, 1973), 70, first drew attention to Bramer as a possible influence on Rosa. This painting is discussed in ten Brink Goldsmith et al., *Leonaert Bramer, 1596-1674*, pp. 9-10.

36 Volpi, *Salvator Rosa*, cat. nos. 124 and 125. The dating of these works to *c.* 1645 is convincing.

37 Volpi, *Salvator Rosa*, p. 614.

imitate the paintings on stone of both Bramer and Filippo Napoletano.[38] Bramer had a highly idiosyncratic feeling for eerie scenes, where spectral figures are conjured up by the flares of torches or small fires, lit against the greyness of large areas of unpainted slate. Often his figures are sketched in with the lightest of touches, strung out across the foreground. Rosa's *Scene of Witchcraft*, with its expressive and caricatured figures, is close to the Dutch artist in the way that Rosa uses the support; it is painted with an unusually delicate brush. Carlo de'Rossi, who owned witchcraft scenes by Rosa, also had, in his museum of rare and bizarre objects, works by Bramer and other works on stone.[39]

In this painting the presence of a priest and Franciscan friar, unique in Rosa's witchcraft scenes, gives the picture a lively topicality. In Rome, in 1635, a group of priests had been condemned for plotting to murder the Pope with black magic in a trial that attracted immense publicity; there are many references, in Giacinto Gigli's *Diario di Roma*, to priests who were sorcerers and necromancers.[40] And yet the feeling of Rosa's painting, with its acrobatic and comic figures, so awkward and so clumsy, is amusing, close in mood to the mock-heroic poetry of his Florentine friend, Lorenzo Lippi. It unites humor and menace, weaving amongst the antic figures a wealth of repulsive detail, body parts, bones, a bloodstained sword, a sacrificial ram, a hanged corpse, a piling up of horror on horror. His imagery recalls that of Giambattista Marino, who, with his creation of Falsirena, in *L'Adone,* aimed to surpass and to exhaust the horrors of the entire tradition of literary witches, from Circe to Alcina and Armida. Marino presents an encyclopedia of black magic, a hyperbolic collection of the repulsive ingredients of spells; Falsirena, transformed into a dark bird of night, or astride a black ram, journeys through clouds and whirlwinds to glean body parts from a desolate battle field. Marino intended to terrify and to intrigue with his *novità,* and to play on the contrast between beauty and horror; Rosa shared this desire, and set the dark slate surface of his painting against the golds of a stupendous and costly frame, exploring an Aristotelian aesthetic of horrible beauty.[41] The slate support seems to have been favored by Rosa for

38 Volpi, *Salvator Rosa*, p. 389, cat. no. 24.

39 Volpi, *Salvator Rosa*, p. 328.

40 On this trial see Giacinto Gigli, *Diario di Roma*, ed. Manlio Barberito (Rome, 1994), p. xxxv, and Peter Rietbergen, *Power and Religion in Baroque Rome* (Leiden and Boston, 2006), pp. 356-61.

41 Volpi, *Salvator Rosa*, p. 474, cat. no. 154, illustrates the painting as it is now framed. The frame is 17th century, and it is possible, as Volpi implies, that this is the painting described in Carlo de' Rossi's inventory as a small oval scene of witchcraft with a gilt frame. Aristotle, *Poetics,* ed. L. Golden and O.B. Hardison (Englewood Cliffs, 1968), p. 7, discusses pleasure in the representation of animals and corpses.

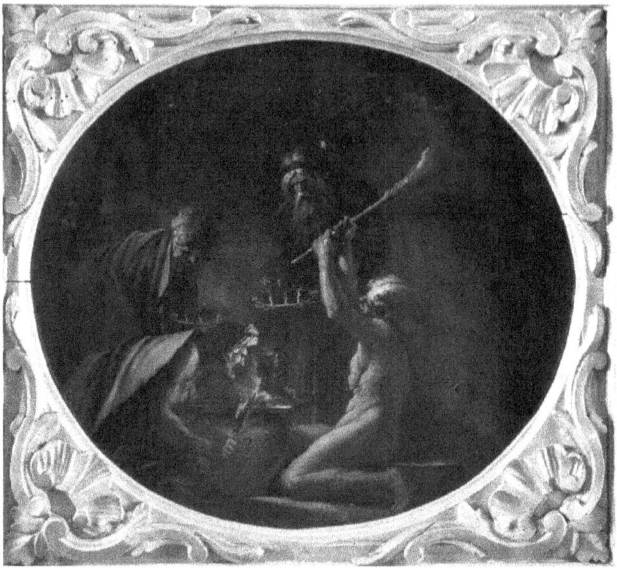

FIGURE 10.6
*Salvator Rosa,* Scene of Witchcraft, *oil on copper, ca. 1645-49, Rome, Private Collection* (PHOTO: PRIVATE COLLECTION).

these mysterious objects; his friend, Ascanio della Penna, had small witchcraft scenes by him, "sopra alcuni pezzi di lavagna," whose fantasy and caprice, wrote an 18th-century commentator, could not be surpassed.[42]

His round *Scene of Witchcraft* [**Fig. 10.6**] is painted on copper, a support rare in Rosa's work. Here too he was anticipated by Bramer, who, in his *Scene of Witchcraft* [**Fig. 10.7**], uses the glow of the copper to suggest flickering effects of light which enhance the mystery of his macabre figures set in ghostly ruins.[43] In Rosa's painting a hideous old witch squeezes a bloody body part onto the smoking surface of a stone altar, balanced on an urn decorated with kabbalistic symbols. Her companions, among them a magician or alchemist, form a circle around her, while lit candles, fiery branches and flaming broomstick, create, against the burnished and reflective warmth of the copper, a circle of

42 Volpi, *Salvator Rosa*, 617. On Rosa's gifts to Ascanio della Penna of drawings, and works on slate and *pietra paesina*, see Franco Paliaga, "Salvator Rosa, I soci dell'Accademia pisana degli Stravagenti e quella volterrana dei Sepolti: amicizie e relazioni artistiche," in *Salvator Rosa e il suo Tempo*, ed. Sybille Ebert-Schifferer, Helen Langdon, and Caterina Volpi (Rome, 2010), p. 80.

43 Olivier Le Bihan, *L'Or et L'Ombre; Catalogue critique et raisonné des peintures hollandaises du dix septième et du dix huitième siècles, conservées au Musée des Beaux-Arts de Bordeaux* (Bordeaux, 1990), pp. 66-69, suggests that this painting was done in Delft, *c.* 1630, which makes it impossible for Rosa to have seen it. But Bramer's work is steeped in reminiscences of Neapolitan painting, and Rosa may well have seen similar works, and, more generally, the copper support seems was particularly suited to witchcraft scenes.

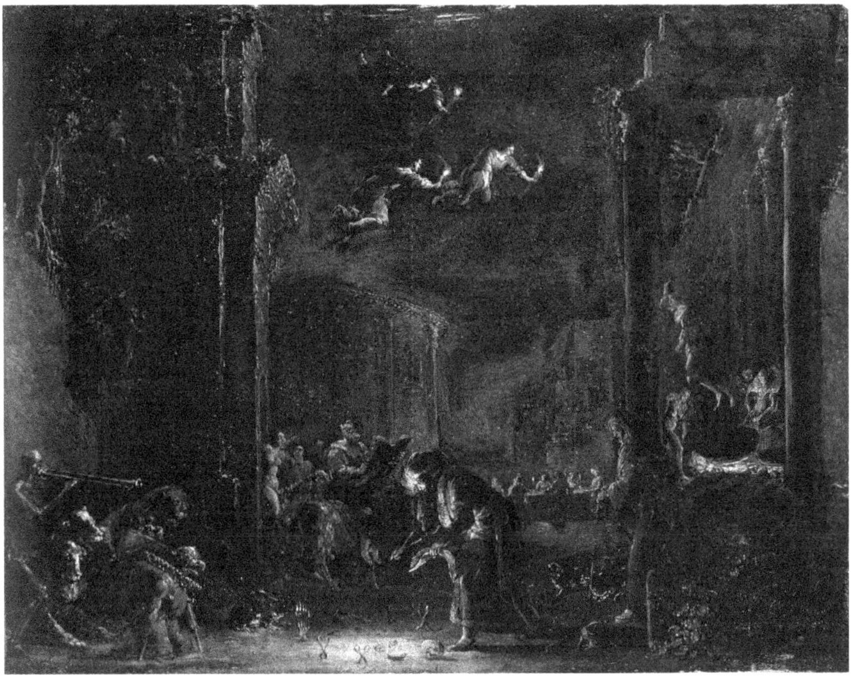

FIGURE 10.7 *Leonard Bramer,* Scene of Witchcraft, *oil on copper, ca. 1635, Bordeaux, Musée des Beaux-Arts* (PHOTO: MUSÉE DES BEAUX-ARTS DE BORDEAUX).

red light on the dark background, very different from the eerie greyness of the slate painting. Stripped, too, of its elaborate paraphernalia and comic mood, this work, with its larger figures, has a concentration new to Rosa's witchcraft scenes. It has been suggested that the copper support itself, a metal important in the practice of alchemy, was part of its meaning, and that the witches are transforming the altar stone into a magical object, suited for display in a *wünderkammer*.[44] This may have appealed to a Roman collector, where an interest in alchemy flourished. It is possible that Del Monte's interest in paintings on stone and copper was indebted to his interest in alchemy and to the distillery in his palace.[45] Athanasius Kircher, in the 1650s and 1660s, dedicated extensive writings to alchemy, and performed wonder-inducing experiments, and scary magic lantern displays, in his celebrated museum at the Collegio Romano, so rich in "oggetti meravigliosi," in stones, fossils, plants, skeletons,

44 Sybille Ebert-Schifferer, "Il teatro della vanitá: le iconografie di Salvator Rosa," in *Salvator Rosa tra mito e magia,* ed. Nicola Spinosa (Naples, 2008), p. 168, cat. no. 41. Ebert-Schifferer proposes a date after Rosa's return to Rome in 1649.

45 Cappelletti, "The Enticement of the North," p. 176.

and herbs. In *Oedipus Aegyptiacus* Kircher devoted a section to 'hieroglyphic alchemy,' writing of the delusional and absurd beliefs of a school of alchemists associated with the famed Emerald Tablet; later, in the *Mundus Subterraneus*, Kircher was to draw a clear distinction between true and false alchemists.[46] The latter, he writes, used every kind of material to make gold, and they were lured by demons, diviners, cabalisters, and superstitious magic of all kinds. Rosa's painting is hard to date, but in Rome he moved in Kircher's circle, and it seems likely that the painting, with its larger figures, and less satirical mood, was done in Rome rather than Florence. His courtly Florentine patrons had been succeeded by Carlo de' Rossi, whose collection included many curiosities, and Cardinal Flavio Chigi. Flavio Chigi, probably advised by Athanasius Kircher, had built up a celebrated *Museo della Curiosità;* he liked painters of rare and esoteric themes, such as Angelo Caroselli and Johann Heinrich Schönfeld. The two paintings together reveal how brilliantly Rosa moved between Naples, Florence, and Rome, picking up topical motifs and literary conceits, yet transforming all into some of the most curious paintings of the Seicento.

## The Philosophical Pleasures of Solitude

In his later years in Florence Rosa had increasingly withdrawn from the Rosa court, and found stimulus in the literary academies of Florence and in the country pleasures of villa life, above all at Barbaiano and Monterufoli near Volterra, villas which belonged to his friends, the Maffei family. Monterufoli lies in an area of geological richness, celebrated, Baldinucci tells us, for "la gran quantità de'Calcedoni e dell'altre pietre vaghe e durissimi."[47] By this date Rosa was well known as a satirist, relentless in his criticism of the luxury of the court; he was the center of a brilliant intellectual society, and a touching drawing, *Rosa in the countryside* (Karlsrühe, private collection), evokes this rural life, whose participants invoked the ancient virtues of solitude, of frugality and simplicity. Here Rosa, in monk's robe, pens a letter, while before him a young man imprints on a rock Statius' line, *Hic aevi mihi prima dies* (This day is my

46 On Kircher and alchemy see Anna Maria Partini, *Athanasius Kircher e l'alchimia* (Rome, 2004) and Daniel Stolzenberg, *Egyptian Oedipus: Athanasius Kircher and the Secrets of Antiquity* (Chicago, 2013). The most relevant section of the *Mundus Subterraneus* is Book 11, section 3, *De Alchymia Sophisticata.*

47 Baldinucci, *Notizie dei Professori del disegno,* p. 462. Francesco Rovai, poet and friend of Rosa, was particularly interested in crystals and precious stones; see Volpi, *Salvator Rosa,* p. 218.

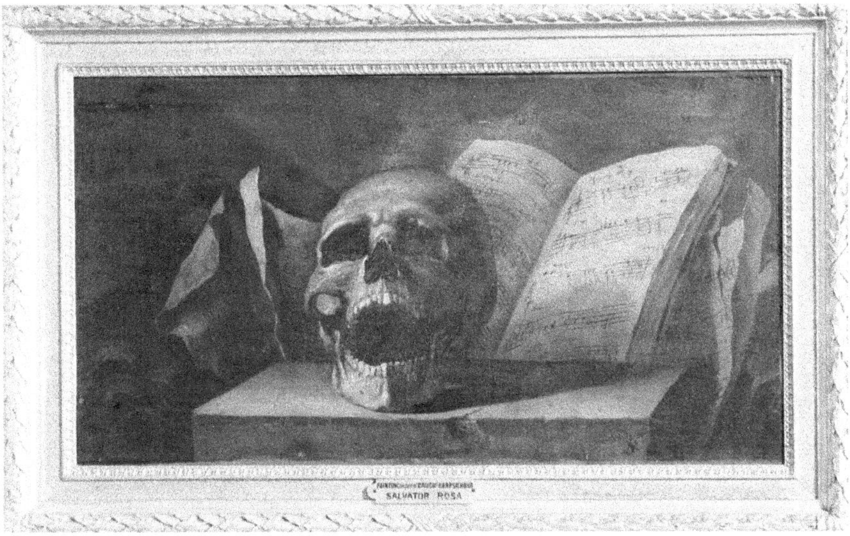


FIGURE 10.8 *Salvator Rosa,* A Singing Skull, *oil on panel, ca. 1649-50, Haddo House; National Trust* (PHOTO: NATIONAL TRUST FOR SCOTLAND, HADDO HOUSE).

first).[48] Statius was celebrating his acceptance by the court of Domitian; Rosa wittily inverts his meaning, and rejoices in his own freedom from the court, in a rocky landscape. An accompanying letter, almost certainly by Rosa's closest friend, the Pisan poet and intellectual Giovanni Battista Ricciardi, describes their pastoral life; they read philosophical works in the shade beside the river, and sometimes climb a mountain and look down on the follies of the world.[49] The two enjoyed an intensely sociable solitude, and for Ricciardi and other friends Rosa displayed the full range of his inventiveness, creating a dazzling array of informal works, which explore different media and supports.

Many of these were intended to amuse; he inscribed words on the rocks of the forest; at Barbaiano, he created, as though hanging from nails on the villa walls, small scenes and figures, all in "carbone lumeggiata con biacca."[50] A *Singing Skull* [**Fig. 10.8**], painted on the inside lid of a harpsichord, would have startled and amused the viewer as he raised the lid; its mouth gaping wide, the skull confronts the viewer with startling directness, a shocking

48 Statius, *Silvae,* 4.13.

49 For the full text of this letter, and for the identification of the writer as Ricciardi, see Francesco Paliaga, *Pittori, incisori e architetti pisani nel secolo di Galile,* (Pisa, 2009), 159. The drawing is reproduced in Walter Regel and Hartmut Köhler, *'....hoch gerühmt, fast vergessen, neu gesehen...'; Der Italienische Mahler und Poet Salvator Rosa* (Wurzburg, 2007), p. 37.

50 Baldinucci, *Notizie dei Professori del disegno*, p. 462.

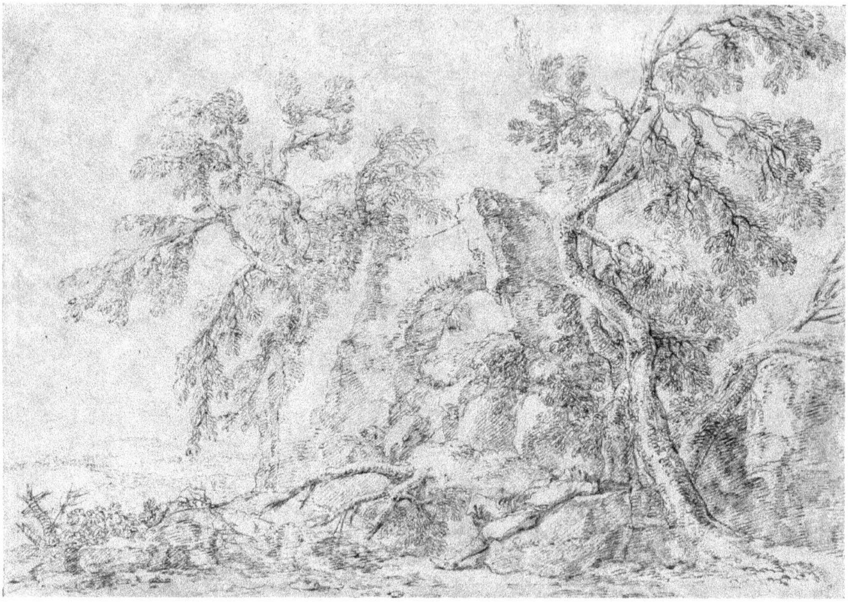

FIGURE 10.9 *Salvator Rosa,* Landscape with a Man sleeping under some Trees, *pen and brown ink on vellum, London British Museum* (PHOTO: © TRUSTEES OF THE THE BRITISH MUSEUM).

reminder that his musical pleasures would be all too brief.[51] The lid interior of the harpsichord had long been a tempting surface for painting, but Rosa subverts the tradition of showing soothing landscapes with mythical subjects, as in the splendid late 17th-century *Harpsichord* (New York, Metropolitan Museum) decorated for Lorenzo Onofrio Colonna by a follower of Gaspard Dughet.[52] He gave to his friends an unusual type of drawing, heads of exotic eastern figures, in colored crayons, perhaps, as Volpi has suggested, showing his response to such Florentine artists as Francesco Furini.[53] Two landscape drawings, presentation drawings on vellum, show men reading alone in an elaborately patterned framework of dead branches and splintered trunks (London, British Museum). One of these, the *Landscape with a man sleeping under some trees*, [**Fig. 10.9**] is signed S. Rosa on a rock directly below the figure,

---

51 Jonathan Scott, *Salvator Rosa: his Life and Times* (New Haven and London, 1995), p. 67, illus. 82.

52 J. Kenneth Moore, Jayson Kerr Dobney, and E. Bradley Strachen-Scherer, *Musical Instruments; Highlights of the Metropolitan Museum of Art* (New Haven, 2015), p. 70

53 Volpi, *Salvator Rosa,* p. 223.

so that we read the image as a memory of Rosa himself sleeping and dreaming in the rocky landscape of Monterufoli.[54]

Most unusual is a series of small drawings on thin wood panels, immediate and fresh, where Rosa uses the reddish rusty ground to unify his composition, creating chiaroscuro effects with black ink and white chalk. These are not panels prepared for painting, but the remnants of packing cases, their surfaces entirely untouched, and Rosa uses the rough textures of the wood to create unusual effects; they are hard to classify, as they fall between drawing and painting, and Rosa probably enjoyed exploring these boundaries.[55] He sent these drawings to his friends, and they share the informality and spontaneity of the poems and letters that they exchanged. Giovanno Battista Ricciardi, his closest friend, owned at least ten, alongside "due pietre di paragone da dipingervi sopra,"[56] and there were others in the collections of Carlo de' Rossi and the Maffei family.

These drawings were produced at different times. First comes a pair of small drawings, *A three-quarter length man standing in a landscape,* and *Landscape with seated figure, seen from the rear,*[57] [**Fig. 10.10**] where the figures and the forms of the landscape, with exposed tree roots and mossy, crackling branches, are reminiscent of Flemish and Germanic landscape and close to Rosa's tree drawings of the 1640s, such as the drawings on parchment mentioned above.[58] Freer and more painterly, *Two men in a rocky landscape,* [**Fig. 10.11**] and *A man seated, facing left, in a spacious landscape,* seem a little later, maybe from the

54 Regel and Köhler, "...hoch gerühmt," pp. 40-41.

55 For a discussion of the medium see Michael Mahoney, *The Drawings of Salvator Rosa,* 2 vols. (New York and London, 1977), Group 11, 12-14.

56 For the inventory of Ricciardi's collection see Paliaga, *Pittori, incisori, e architetti pisani,* pp. 229-30. Volpi, *Salvator Rosa,* 186, has suggested that at least ten entries may refer to Rosa's chiaroscuro drawings. Some are also mentioned in the collection of Carlo De Rossi, and others in the collection of his close friends the Maffei family. For De Rossi's inventory see Caterina Volpi, "Salvator Rosa e Carlo de Rossi," *Storia dell'Arte* 93-94 (1998), 356-73. See also Volpi, *Salvator Rosa,* pp. 191-92, for a general discussion of these drawings and their collectors.

57 These drawings are in the Gabinetto dei Disegni e delle Stampe at the Uffizi, but they are displayed at the Palazzo Pitti, Galleria Palatina.

58 Mahoney, *The Drawings of Salvator Rosa,* Group 68.6 and 68.5. For the tree drawings of the 1640s see Groups 23 and 27. Mahoney gives a later date to the Pitti drawings, as does Andreas Stolzenberg, ed. *Salvator Rosa: Genie der Zeichnung: Studien und Skizzen aus Leipzig und Haarlem* (Cologne, 1999), p. 148. For an earlier dating, the late 1640s or early 1650s, with which I am in agreement, see Volpi, *Salvator Rosa,* p. 191.

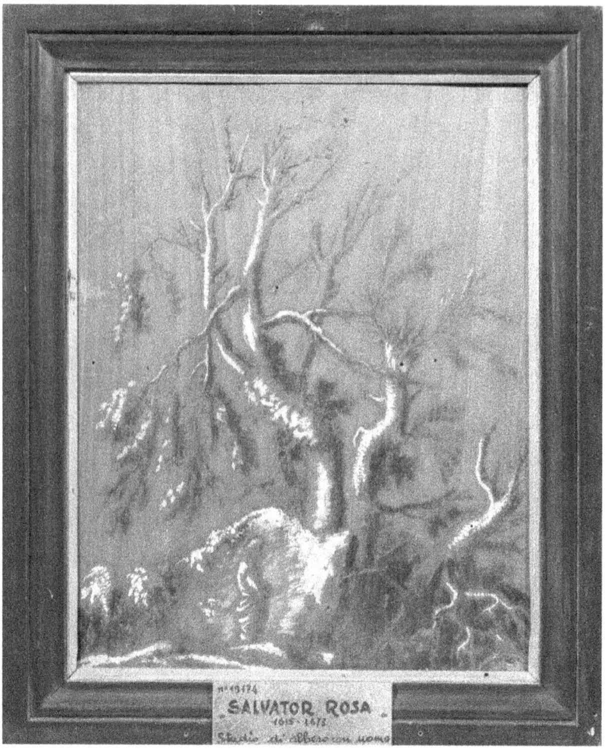


FIGURE 10.10
*Salvator Rosa,* A three quarter length man in a landscape; Landscape with seated figure, *Florence, Palazzo Pitti* (PHOTO: FIRENZE GABINETTO DEI DISEGNI E DELLE STAMPE DEGLI UFFIZI).

late 1640s [**Plate 31**].[59] These drawings suggest Rosa's days at Monterufoli, with figures immersed, conversing and meditating, in a rocky landscape, where the surfaces of the wood conjure up the movement of clouds and water. They are intensely poetic, and the untouched wood seems to evoke a closeness to nature, and to suggest a simple way of life; maybe the humble materials that he here used, showing solitary figures alone in rocky landscapes, themselves praise a simpler way of life. Far removed from the costliness of lapis lazuli or copper, these works startle with their unexpectedness; Rosa shows a response to texture and surface, to the lines of the wood, as Filippo Napoletano had done to the undulating lines of *pietra paesina*; and perhaps, for he moved in Galilean circles, he may have been aware of the Lincean interest in fossilized wood, with its wandering wave like patterns and the sense of wondrous *scherzi della natura.*[60]

59 Mahoney, *Drawings,* Group 80.14 and 80.15. Mahoney suggests a late date, close to *The Death of Empedocles,* between 1660-73.

60 For drawings of fossilized woods see David Freedberg, *The Eye of the Lynx* (Chicago, 2002), pp. 305-45.

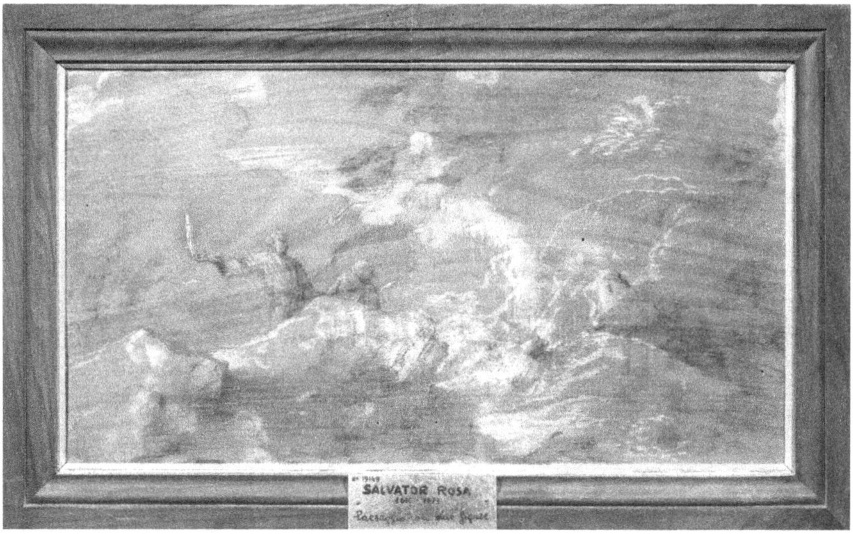


FIGURE 10.11 *Salvator Rosa,* Two Men in a Rocky Landscape, *Florence, Palazzo Pitti* (PHOTO: FIRENZE GABINETTO DEI DISEGNI E DELLE STAMPE DEGLI UFFIZI).

At the same time these drawings strongly recall the chiaroscuro drawings on intensely colored paper that had been made famous by Danube artists in the 16th century. It is not clear whether Rosa would have known chiaroscuro drawings and woodcuts by Albrecht Altdorfer, though both were prized by collectors, and the color of his unifying ground is reminiscent of Altdorfer. But he would surely have known chiaroscuro woodcuts and drawings by Italian artists, such as Ugo da Carpi, or Parmigianino, or perhaps some of Domenico Beccafumi's colored prints, or extraordinary drawings on colored papers.[61]

Francesco Cordini, Baldinucci tells us, owned five small paintings by Rosa, "parte di paesi, e parte di teste, fatte ad imitazione di maniere di pittori antichi";[62] Cordini, Rosa's friend in both Florence and Rome, was the owner of his most erudite allegory, *Moral Philosophy* (Caldaro, Palazzo Enzelberg). It is hard to identify these works, but this comment places Rosa firmly in the antiquarian circles around Cassiano dal Pozzo, whose members had been, since the 1640s, concerned to document every aspect of ancient pictorial art, and it suggests that Rosa shared with Gaspard Dughet and with Claude Lorrain an interest in recreating ancient landscape. His drawing, *Rosa in the countryside*, recalls Poussin's *Et in Arcadia Ego* (1637-9; Paris, Louvre). It has been suggested

61 On chiaroscuro prints and drawings by German and Austrian artists see David Landau and Peter Parshall, *The Renaissance Print*, 1470-1550 (New Haven and London, 1994), pp. 183-202. For Beccafumi, see pp. 274-83.

62 Baldinucci, *Notizie dei Professori del disegno*, p. 456.

that the kneeling shepherd in Poussin's work, his shadow so prominent against the stone, evokes the myth of the origins of painting in the tracing of the outline of a cast shadow.[63] Rosa too suggests a primordial age, and, as Poussin was to do in the early 1650s, he enjoyed discussions of ancient painting with his learned friend, Carlo Dati. Dati, a classical scholar with an immense range of literary and scientific interests, was to publish his *Vite de' Pittori Antichi* in 1667, but he had begun to work on this by 1647.[64]

Dati was particularly interested in the monochrome palettes of ancient painters, which he had read about in Pliny's *Natural History*.[65] For Pliny such austere colors had a moral virtue; he preferred their austerity to the gaudy colors of his contemporaries. Scholars debated the role of chiaroscuro in the earliest phase of Greek painting that Pliny describes, and in 1653 Poussin, encouraged by Cassiano dal Pozzo, wrote to Dati, supporting the view that Pliny's *monocromati* would have used chiaroscuro, and would have been like "i nostri disegni ombrati con acquerella et allumati con bianco per colorirli poi." In his *Vite* Dati picked out as the most splendid masters of monochrome Andrea del Sarto, famous for the monochrome frescoes in the *Chiostro dello Scalzo*, and Albert, that is Dürer.[66] Dati may well have meant the chiaroscuro drawings and prints done in Dürer's circle.[67]

In such a context it seems that Rosa's drawings on wood may have evoked, for his contemporaries, the chiaroscuro drawings of the ancients. His interest in *monocromati* is confirmed by a group of unusual paintings, executed in a narrow range of colors, which reveal Rosa's very evident interest in the discussions of Dati and Poussin. Most striking are *Terence and Plautus* (*La Menzogna*) (Florence, Palazzo Pitti), where a figure shows a mask to his companion, and

63 Elizabeth Cropper and Charles Dempsey, *Nicolas Poussin; Friendship and the Love of Painting*, (Princeton, 1996), pp. 201-03.

64 On Rosa and Carlo Dati see Caterina Volpi, "Salvator Rosa, nuovi documenti e riflessioni sul primo periodo romano e su quello fiorentino," *Storia dell'Arte* 119 (2008), 90, n. 19.

65 Pliny, *Natural History*, 33.117, 35.50. For a discussion of Pliny's writings on the four-color palette and on monochrome see John Gage, "A *Locus Classicus* of Colour Theory: the Fortunes of Apelles," *Journal of the Warburg and Courtauld Institute* 41 (1981), 1-26, and Sarah Blake McHam, *Pliny and the Artistic Culture of the Italian Renaissance* (New Haven and London, 2013), pp. 234-36. Both Dati and Rosa would have known Giorgio Vasari's rust colored monochrome depictions of Pliny's anecdotes from the lives of Greek painters in his house at Arezzo (see McHam, pp. 243-48).

66 Carlo Dati, *Vite de' Pittori antichi* (Milan, 1806), p. 67.

67 Ingo Herklotz, "Poussin et Pline l'Ancien: à propos des monocramata," in Olivier Bonfait, et al., eds., *Poussin et Rome; Actes* du colloque *a l'Académie de France à Rome et à la Biblioteca Hertziana, 1994*, (Paris, 1996), pp. 13-29. Dati and Poussin's correspondence is given in full on pp. 22-23.

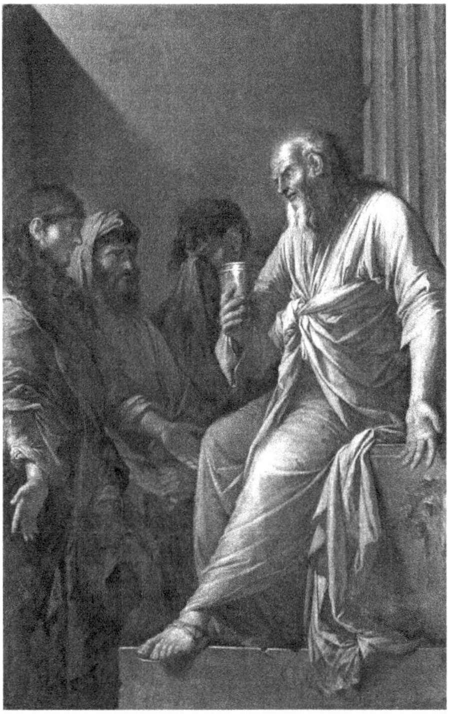

FIGURE 10.12
*Salvator Rosa,* Death of Socrates, *oil on canvas, ca. 1652-55, Private Collection* (PHOTO: PRIVATE COLLECTION).

*The Death of Socrates*.[68] [**Fig. 10.12**] Here the heads of the philosopher and his companions relate to a group of idealized heads in colored chalks, close to the painted heads which Rosa often gave to his friends; others, including an impressive example in the British Museum, are done on wooden panels, and explore the way in which the pen and ink were diffused by the rough surface of the wood.[69] These works create a Plinian sobriety and an aesthetic ideal of simplicity entirely consonant with his classical subjects, as the drawings on wood,

68 For the identification of the subject matter, and a summary of recent literature, see Volpi, *Salvator Rosa*, p. 508. Volpi dates the painting to *c.* 1650-55; the present writer argues for an earlier date, around 1648. The painting leads onto similar works certainly from the 1650s, *Joseph interpreting Dreams* (priv. coll.) and *Socrates drinking poison* (priv. coll; Volpi, *Salvator Rosa*, p. 494) but has a lightness and clarity that remain close to some Florentine works, such as the *Self Portrait as Pascariello.* (London, priv. coll.).

69 For a discussion of these drawings of heads see Mahoney, *Drawings*, Group 31. The British Museum drawing is Mahoney 30.12; Mahoney, p. 364, describes it as "a novel and impressive exploitation of the absorbent quality of a wooden panel for a pen drawing." Mahoney also catalogues two such drawings, 31.13 and 31.14, of which no photographs are available, and which the present writer has not seen. He gives evidence which suggests that these were originally part of Ricciardi's collection of Rosa's drawings.

some of landscapes, some of heads, evoked the simplicity of an earlier age. Such monochrome paintings recur in his art of the 1660s, reflecting contemporary scholarly interest; in 1663 Dati sent him a manuscript copy of his *Vite de' Pittori*, and an interest in ancient pictorial art received encouragement both from the writings of Bellori and the creative patronage of Cardinal Camillo Massimi.[70]

There are no surviving drawings on wood from the 1650s, but in the early 1660s, when Rosa became fascinated by ancient philosophers and magicians, he revived this interest, and again presented his friends with witty and informal works. He may have been stimulated by Castiglione, whose brush drawings in oil on unprepared paper from this decade show a similar attraction to the *novità* of unusual media and to exploring the boundaries between them. Rosa's *A figure kneeling under a tree* (coll. Incisa della Rocchetta), was done on a wooden container originally sent to his son, Augusto Rosa, which Rosa then sent on, decorated with this drawing, to Nicolò Simonelli, a member of the Chigi household, and a long standing friend and collector of curiosities.[71] A drawing in the Nationalmuseum in Stockholm, in ink and white wash on thin wooden panel, reproduces the etching, *The Abandoned Oedipus* with an exceptionally poetic, Venetian landscape.[72] The figures have very strong outlines, suggesting that this is a preparatory drawing, but it is more likely to have been done after the etching. Rosa usually sent his preparatory drawings for etchings to his friend, Giovanni Battista Ricciardi, but in this case he did not, explaining that, as the subject was mainly landscape, he had etched, with exceptional freedom, directly onto the plate; perhaps he later sent this more finished drawing as a compensatory gift.[73] His monochrome *Death of Empedocles* [**Plate 32**] takes its place in a long sequence of drawings associated with the large painting of this subject (Eastnor Castle coll.).[74] They include dark inchoate wash drawings, where the patterns of wash on the paper themselves seem to suggest the subject;[75] far removed from the naturalism of his Florentine drawings, Rosa at this date began with wash and used it to stimulate his imagination. The reduced monochrome version of the *Empedocles* is a drawing entirely about

70 On Massimi, Bellori and ancient painting see Lisa Beaven, *An Ardent Patron; Cardinal Camillo Massimo and his antiquarian artistic circle* (London, 2010).

71 Mahoney, *Drawings*, 80.16. Francesco Petrucci, "Salvator Rosa per casa Chigi e nuovi contributi su Francesco Rosa," in *Salvator Rosa e il suo Tempo*, ed. Ebert-Schifferer et al., p. 383.

72 Daniel Prytz and Johan Eriksson, "The Martelli Collection: Further Notes towards its History," *Art Bulletin of Nationalmuseum, Stockholm* 16 (2009), 77-86.

73 Salvator Rosa, *Lettere*, ed., Gian Giotto Borrelli (Bologna, 2003), p. 310.

74 Mahoney, *Drawings*, Group 80. For a catalogue of the eleven drawings for the Empedocles see Chris Fischer and Joachim Meyer, *Neapolitan Drawings* (Copenhagen, 2006), p. 91.

75 Mahoney, *Drawings*, 80.8 and 80.9.

texture and pattern, and, very late in Rosa's career, it is a brilliant summation of how, like contemporary painters on stone, he used the rough surface to stimulate invention, and it perhaps looks back to the sources of his art, to such a Filippo Napoletano as the *Roger liberating Angelica*, with its tiny figures set against the variegated textures of the surface. It is also reminiscent of Gaspard Dughet's small gouaches on panel, used as window shutters; Dughet's *Orpheus and Eurydice* includes a flaming Vesuvius, a highly unusual motif.[76] Rosa's drawing decorated the inside lid of a packing case, which Rosa had sent to his friend Cosimo Fabretti. Fabretti was the *segretario* of Ricciardi, for whom the work was almost certainly intended, and the sketch, in spirit close to the *Singing Skull,* would have startled them both as the lid was opened; Rosa perhaps intended to amuse his friends with the playful small scale rendering, on a novel support, of a subject then absorbing him.[77] Painters often made smaller versions of their works on precious surfaces such as lapis lazuli, but Rosa, to his philosopher companion, sends a version on a packing case.

The story of Empedocles' death held many meanings. He might be accused of trickery and pride, or defended as the forerunner of the heroic scientists of the 17th century, a fearless searcher for truth. And perhaps, for Rosa, he was an image of the burning creativity of such an artist as he felt himself to be, "tutto fuoco, tutto spirit, tutto bile,"[78] fearless in his crossing of boundaries.

## Bibliography

Ago, Renata. *Il Gusto delle Cose: Una Storia degli Oggetti nella Roma del Seicento*. Rome, 2006.

Aldrovandi, Ulisse. *Musaeum metallicum in libros IIII distributum bartholomaeum ambrosinus in patrio bono*. Bologna, 1648.

Alloisi, Sivigliano. *Arcadie e Vecchi Merletti, Paesaggi della Collezione Corsini*. Rome, 2002.

Aristotle. *Poetics.* Edited L. Golden and O.B. Hardison. Englewood Cliffs, 1968.

Baldinucci, Filippo, *Notizie dei Professori del disegno da Cimabue in qua*. Edited by F. Ranalli. Florence, 1847.

Barocchi, Paola, and Giovanna Gaeta Bertelà, eds. *Collezionismo Mediceo e Storia Artistica. 3, Il Cardinale Giovan Carlo, Mattias e Leopoldo, 1628-1667*. Florence, 2007.

76 For these small panels see Sivigliano Alloisi, *Arcadie e Vecchi Merletti, Paesaggi della Collezione Corsini* (Rome, 2002), pp. 52-59.

77 On Ricciardi's collection of works by Rosa see Franco Paliaga, "Salvator Rosa, I soci dell'Accademia Pisana," pp. 84-87.

78 Salvator Rosa, *Lettere,* p. 200.

Beaven, Lisa. *An Ardent Patron; Cardinal Camillo Massimo and his antiquarian artistic circle*. London, 2010.
Bonfait, Olivier, Christoph L. Frommel, and Michel Hochmann, eds. *Poussin et Rome; Actes du colloque à l'Academie de France à Rome et à la Biblioteca Hertziana, 16-18 novembre 1994*. Paris, 1996.
Brizio, Paola Squellati. "Natura Sollecitata." In *Bizzarrie di pietre dipinte: dalle collezioni dei Medici*, eds. Marco Chiarini and Cristina Acidini Luchinat, pp. 32-38. Cinisello Balsamo (Milan), 2000.
Brown, Beverly Louise, ed. *The Genius of Rome, 1592-1623*. London, 2001.
Cappelletti, Francesca. "The Enticement of the North: Landscape, Myth and Gleaming Metal Supports." In *The Genius of Rome, 1592-1623*, edited by Beverly Louise Brown. London, 2001.
Castellotti, Marco Bona, ed. *Pietra dipinta: Tesori Nascosti del '500 e del '600 da una collezione privata Milanese*. Milan, 2000.
Chiarini, Marco. "Pittura su pietra." In *Bizzarrie di pietre dipinte dalle collezioni dei Medici*. Edited by Marco Chiarini and Cristina Acidini Luchinat. Cinisello Balsamo (Milan), 2000.
Chiarini, Marco. *Teodoro Filippo di Liagno detto Filippo Napoletano*. Florence, 2007.
Chiarini, Marco, and Cristina Acidini Luchinat, eds. *Bizzarrie di Pietre Dipinte dalle collezioni dei Medici*. Cinisello Balsamo, 2000.
Collomb, Anne-Laure. "Jacques Stella et la peinture sur pierre." In *Jacques Stella (1596-1657)*. Edited by Sylvain Laveissière et Gilles Chomer. Exhibition catalogue, Lyon, Musée des Beaux Arts and Toulouse, Musée des Augustins, 2006-07. Paris, 2006.
Collomb, Anne-Laure. *Splendeurs de l'Italie: La Peinture sur pierre à la Renaissance*. Rennes and Tours, 2012.
Cropper, Elizabeth, and Charles Dempsey. *Nicolas Poussin; Friendship and the Love of Painting*. Princeton, 1996.
Dati, Carlo. *Vite de' Pittori antichi*. Milan, 1806.
Del Riccio, Agostino. *Istoria delle Pietre*, 1597. Edited by Paola Barocchi. Florence, 1979.
Ebert-Schifferer, Sybille, Helen Langdon, and Caterina Volpi, eds. *Salvator Rosa e il suo Tempo 1615-1673*. Rome, 2009.
Ebert-Schifferer, Sybille. "Il teatro della vanitá: le iconografie di Salvator Rosa." In *Salvator Rosa tra mito e magia,* edited Nicola Spinosa, pp. 66-82, 166-81. Naples, 2008.
Fabbri, Sara. "I Quattro Tondi con incantesimi di Salvator Rosa nella collezione del Marchese Filippo Niccolini: Il Volto Oscuro della Toscana del Seicento." *Notizie da Palazzo Albani* 38 (2009), 81-104.
Fisher, Chris, and Joachim Meyer. *Neapolitan Drawings*. Copenhagen, 2006.
Freedberg, David. *The Eye of the Lynx, Galileo, His Friends, and the Beginnings of Modern Natural History*. Chicago, 2002.

Gage, John, "A *Locus Classicus* of Colour Theory: the Fortunes of Apelles." *Journal of the Warburg and Courtauld Institute* 41 (1981), 1-26

Gigli, Giacinto. *Diario di Roma*. Edited by Manlio Barberito. Rome, 1994.

Herklotz, Ingo. "Poussin et Pline l'Ancien: à propos des monocramata." In *Poussin et Rome; Actes* du colloque *a l'Académie de France à Rome et à la Biblioteca Hertziana, 1994*, edited by Olivier Bonfait et al. Paris, 1996.

Ingendaay, Martina. *"Il migliori pennelli": I Marchesi Gerini mecanati e collezionisti nella Firenze barocca. Il palazzo e la galleria 1600-1825*. 2 vols. Milan, 2013.

Kircher, Athanasius. *Mundus Subterraneus in XII libros* [...]. Amsterdam, 1665.

Kitson, Michael, ed. *Salvator Rosa*. London, 1973.

Landau, David, and Peter Parshall. *The Renaissance Print, 1470-1550*. New Haven and London, 1994.

Langdon, Helen. *Caravaggio: a Life*. New York, 1998.

Laureati, Laura. "Le pietre dipinte: oggetti o quadri? Alcuni esempi di collezionismo romano." In *Pietra Dipinta; Tesori Nascosti del '500 e del '600 da una collezione privata Milanese*, ed. Marco Bona Castellotti, pp. 209-24. Milan, 2000.

Laveissière, Sylvain, and Gilles Chomer, eds. *Jacques Stella (1596—1657)*. Exhibition catalogue, Lyon, Musée des Beaux Arts and Toulouse, Musée des Augustins, 2006-07. Paris, 2006.

Leonardo da Vinci. *Treatise on Painting*. Translated by A. Philip McMahon. Princeton, 1956.

Le Bihan, Olivier. *L'Or et l'Ombre: catalogue critique et raisonne des peintures hollandaises du dix-septième et du dix huitième siècles, conservés au Musée des beaux-arts de Bordeaux*. Bordeaux, 1990.

Lohff, Johanna Beate. *Malerei auf Stein: Antonio Tempestas Bilder auf Stein im Kontext der Kunst-und Naturtheorie seiner Zeit*. Munich, 2015.

Mahoney, Michael. *The Drawings of Salvator Rosa*. 2 vols. New York and London, 1977.

Mancini, Giulio. *Considerazioni sulla pittura (Rome, 1616-24)*. 2 vols. Edited by Adriana Marucchi and Luigi Salerno. Rome, 1956-57.

Mirollo, James V. "Aesthetics of the Marvellous." In *Wonders, marvels and monsters in early modern culture*, ed. Peter G. Platt, pp. 24-44. Newark and London, 1999.

Moore, J. Kenneth, Jayson Kerr Dobney, and E. Bradley Strachen-Scherer. *Musical Instruments; Highlights of the Metropolitan Museum of Art*. New York and New Haven, 2015.

Paliaga, Franco. *Pittori, incisori e architetti pisani nel secolo di Galileo*. Pisa, 2009.

Paliaga, Franco. "Salvator Rosa, I soci dell'Accademia pisana degli Stravagenti e quella volterrana dei Sepolti: amicizie e relazioni artistiche." In *Salvator Rosa e il suo Tempo*, edited by Sybille Ebert-Schifferer, Helen Langdon, and Caterina Volpi. Rome, 2010.

*The Paper Museum of Cassiano dal Pozzo*, British Museum, Department of Prints and Drawings. Ivrea (Milan), 1993.

Partini, Anna Maria. *Athanasius Kircher e l'alchimia*. Rome, 2004.

Petrucci, Francesco. "Salvator Rosa per casa Chigi e nuovi contributi su Francesco Rosa." In *Salvator Rosa e il suo Tempo 1615-1673*, edited by Sybille Ebert-Schifferer, Helen Langdon, and Caterina Volpi. Rome, 2009.

Philostratus. *The Life of Apollonius of Tyana*. Vol. 1. Translated by Frederick Cornwallis Conybeare. London and New York, 1912.

Platt, Peter G., ed. *Wonders, Marvels and monsters in early modern culture*. Newark and London, 1999.

Prytz, Daniel, and Johan Eriksson. "The Martelli Collection: Further Notes towards its History." *Art Bulletin of Nationalmuseum, Stockholm* 16 (2009), 77-86.

Regel, Walter, and Hartmut Köhler. *"... hoch gerühmt, fast vergessen, neu gesehen...": Der Italienische Mahler und Poet Salvator Rosa.* Wurzburg, 2007.

Rietbergen, Peter. *Power and Religion in Baroque Rome: Barberini Cultural Policies*. Boston, 2006.

Salvator Rosa. *Lettere.* Edited by Gian Giotto Borrelli. Bologna, 2003.

Scott, Jonathan. *Salvator Rosa: his Life and Times*. New Haven and London, 1995.

Spinelli, Riccardo. "Documenti artistici dall'archivio Niccolini di Camugliano—1. Marmi antichi e 'moderne pitture' di Giovanni di Agnolo (Firenze, 1544-1611)." *Paragone, Arte* 61, no. 663 (2005), 80-103.

Spinosa, Nicola, ed. *Salvator Rosa tra mito e magia.* Naples, 2008.

Stolzenberg, Andreas, ed. *Salvator Rosa: Genie der Zeichnung: Studien und Skizzen aus Leipzig und Haarlem*. Cologne, 1999.

Stolzenberg, Daniel. *Egyptian Oedipus: Athanasius Kircher and the Secrets of Antiquity*. Chicago, 2013.

Ten Brink Goldsmith, Jane, ed. *Leonaert Bramer, 1596-1674: ingenious painter and draughtsman in Rome and Delft*. Zwolle and Delft, 1994.

Volpi, Caterina. *Salvator Rosa, (1615- 1673) "pittore Famoso.".* Rome, 2014.

Volpi, Caterina. "Salvator Rosa e Carlo de Rossi." *Storia dell'Arte* 93-94 (1998), 356-73.

Volpi, Caterina. "Salvator Rosa, nuovi documenti e riflessioni sul primo periodo romano e su quello fiorentino." *Storia dell'Arte* 119 (2008), 85-116.

Wichmann, Heinrich. *Leonaert Bramer: sein Leben und sein Kunst, Ein Beitrag Zur Geschichte der Holländischen Malerei zur Zeit Rembrandts*. Leipzig, 1923.

# Index

Illustrations are indicated in *italic*.
Colour plates are indicated in ***bold italic***.

www.ingramcontent.com/pod-product-compliance
Lightning Source LLC
LaVergne TN
LVHW010542300926
849162LV00002B/208